The Stability of Minerals

Mineralogical Society Series

Series editor
Professor Geoffrey D. Price

TITLES AVAILABLE

1. **Deformation Processes in Minerals, Ceramics and Rocks**
   Edited by D. J. Barber and P. G. Meredith
2. **High-temperature Metamorphism and Crustal Anatexis**
   Edited by J. R. Ashworth and M. Brown
3. **The Stability of Minerals**
   Edited by Geoffrey D. Price and Nancy L. Ross

The Stability of Minerals

Edited by

Geoffrey D. Price MA, PhD
Professor of Mineral Physics

and

Nancy L. Ross BS, MSc, PhD
Lecturer in Mineral Physics

Crystallography and Mineral Physics Unit
Department of Geological Sciences
University College London

Published in association with
The Mineralogical Society of Great Britain and Ireland

CHAPMAN & HALL
London · Glasgow · New York · Tokyo · Melbourne · Madras

Published by Chapman & Hall, 2–6 Boundary Row, London SE1 8HN

Chapman & Hall, 2–6 Boundary Row, London SE1 8HN, UK

Chapman & Hall, 29 West 35th Street, New York NY10001, USA

Chapman & Hall Japan, Thomson Publishing Japan, Hirakawacho Nemoto Building, 6F, 1-7-11 Hirakawa-cho, Chiyoda-ku, Tokyo 102, Japan

Chapman & Hall Australia, Thomas Nelson Australia, 102 Dodds Street, South Melbourne, Victoria 3205, Australia

Chapman & Hall India, R. Seshadri, 32 Second Main Road, CIT East, Madras 600 035, India

First edition 1992

© 1992 Geoffrey D. Price, Nancy L. Ross and the contributors

Typeset in 10/12p Times by Interprint Ltd, Malta
Printed and Bound in Great Britain by Hartnolls Limited, Bodmin, Cornwall.

ISBN 0 412 44150 0

Apart from any fair dealing for the purposes of research or private study, or criticism or review, as permitted under the UK Copyright Designs and Patents Act, 1988, this publication may not be reproduced, stored, or transmitted, in any form or by any means, without the prior permission in writing of the publishers, or in the case of reprographic reproduction only in accordance with the terms of the licences issued by the Copyright Licensing Agency in the UK, or in accordance with the terms of licences issued by the appropriate Reproduction Rights Organization outside the UK. Enquiries concerning reproduction outside the terms stated here should be sent to the publishers at the London address printed on this page.

The publisher makes no representation, express or implied, with regard to the accuracy of the information contained in this book and cannot accept any legal responsibility or liability for any errors or omissions that may be made.

A catalogue record for this book is available from the British Library

Library of Congress Cataloging-in-Publication data

The Stability of Minerals / edited by Geoffrey D. Price and Nancy L. Ross. — 1st ed.
    p. cm. — Mineralogical Society series)
  'Published in association with the Mineralogical Society of Great Britain and Ireland.'
  Papers presented at a meeting, held at University College London, England, in December 1989, under the auspices of the Mineralogical Society of Great Britain and Ireland.
  Includes bibliographical references and index.
  ISBN 0-412-44150-0
  1. Minerals—Congresses. 2. Molecular structure—Congresses.
I. Price, Geoffrey D. II. Ross, Nancy L. III. Mineralogical Society (Great Britain) IV. Series.
QE364.S76 1992
549'.131—dc20   92-20179
                                                              CIP

Contents

Contributors		ix
Preface		xi

1 The stability of minerals: an introduction   1
    Nancy L. Ross and Geoffrey D. Price

    1.1 Introduction   1
    1.2 Theories of crystal structures   2
    1.3 Geometrical constraints on crystal structures   3
    1.4 Electronic factors and mineral stability   10
    1.5 Thermodynamics and crystal stability   15
    1.6 Conclusion   21
    References   21

2 Bond topology, bond valence, and structure stability   25
    Frank C. Hawthorne

    2.1 Introduction   25
    2.2 Structures as graphs   26
    2.3 Topological aspects of molecular orbital theory   28
    2.4 Topological aspects of crystal chemistry   34
    2.5 Bond-valence theory   37
    2.6 Structural hierarchies   45
    2.7 (OH) and (H_2O) in oxysalt structures   59
    2.8 (H_2O) as a bond-valence transformer   66
    2.9 Binary structural representation   67
    2.10 Interstitial (H_2O) in minerals   69
    2.11 Bond-valence controls on interstitial cations   70
    2.12 Structural controls on mineral paragenesis   72
    2.13 Summary   75
    References   85

3 Electronic paradoxes in the structures of minerals   88
    Jeremy K. Burdett

    3.1 Introduction   88
    3.2 Orbitals of molecules   89

3.3	Energy bands of solids	94
3.4	Localized or delocalized bonds in silicates and metallic oxides?	99
3.5	Interatomic distances and overlap populations	105
3.6	Crystal-field and molecular-orbital stabilization energy	109
3.7	Pauling's third rule: the instability of edge-sharing tetrahedra	115
3.8	The Jahn–Teller effect	123
3.9	Conclusion	130
	Acknowledgements	130
	References	131

4 Lattice vibration and mineral stability 132
Nancy L. Ross

4.1	Introduction	132
4.2	Background	133
4.3	Vibrational density of states, $G(w)$	136
4.4	Approximations of $G(w)$	139
4.5	Comparison of vibrational models	146
4.6	Application of vibrational models to mineral stability	158
4.7	Lattice dynamics from atomistic simulations	161
4.8	Conclusion	168
	References	169

5 Thermodynamics of phase transitions in minerals: a macroscopic approach 172
Michael A. Carpenter

5.1	Introduction	172
5.2	The order parameter, Q	174
5.3	Landau free-energy expansions	179
5.4	Elaboration of single-order parameter expansions: coupling with strain and composition	185
5.5	Comparison of Landau and Bragg–Williams treatments of order/disorder transitions	193
5.6	Phase transitions involving only one order parameter	197
5.7	Systems with more than one phase transition: order-parameter coupling	200
5.8	Kinetics: the Ginzburg–Landau equation for time-dependent processes	207
5.9	Conclusion	211
	Acknowledgements	211
	References	211

CONTENTS

6	The stability of modulated structures *J. Desmond C. McConnell*	216
	6.1 Introduction	216
	6.2 The theory of phase transformations	219
	6.3 The theory of incommensurate (IC)-phase transformations	226
	6.4 The symmetry of modulated structures	228
	6.5 Gradient invariants	229
	6.6 Examples of modulated structures	233
	References	240
7	Thermochemistry of tetrahedrite–tennantite fahlores *Richard O. Sack*	243
	7.1 Introduction	243
	7.2 Systematics	244
	7.3 Thermodynamics	246
	7.4 Crystallochemical controls on metal zoning	259
	Acknowledgements	263
	References	264
8	Thermodynamic data for minerals: a critical assessment *Martin Engi*	267
	8.1 Introduction	267
	8.2 Thermodynamic measurements on chemical species, phases, and equilibria	274
	8.3 Data retrieval and data analysis	281
	8.4 Quality of optimized thermodynamic properties	289
	8.5 Equation of state for fluids and aqueous species	291
	8.6 Complications due to solid solution, ordering, and phase transitions in minerals	297
	8.7 A comparison of widely used data bases	299
	8.8 General assessment and outlook	320
	Acknowledgements	322
	References	323
9	The stability of clays *Bruce Velde*	329
	9.1 Introduction	329
	9.2 Time–temperature control of clay stability	330

9.3	Chemical effect on clay stability	342
9.4	Phases and internal chemical equilibrium	346
9.5	Conclusion	349
References		350

Index 352

Contributors

Jeremy K. Burdett
Professor
Department of Chemistry
James Frank Institute
University of Chicago
USA

Michael A. Carpenter
Department of Earth Sciences
University of Cambridge
UK

Martin Engi
Professor
Institute of Mineralogy and Petrology
University of Bern
Switzerland

Frank C. Hawthorne
Professor
Department of Geological Sciences
University of Manitoba
Canada

J. Desmond C. McConnell FRS
Professor
Department of Earth Sciences
University of Oxford
UK

Geoffrey D. Price
Professor
Department of Geological Sciences
University College London, and
Department of Geology
Birkbeck College
London
UK

CONTRIBUTORS

Nancy L. Ross
Department of Geological Sciences
University College London
UK

Richard O. Sack
Professor
Department of Earth and Atmospheric Sciences
Purdue University
Indiana
USA

B. Velde
Professor
Département de Géologie
URA 1316 du CNRS
École Normale Supérieure
Paris
France

Preface

The aim of this monograph is to provide an introduction to the problems of mineral stability. The chapters do not attempt to provide comprehensive reviews of the subject, but rather are tutorial in nature. It is hoped that anyone with a moderate background in mineralogy, chemistry or crystallography will be able to use this book as a general guide to the subject, and as an introduction to the current frontiers of research. Although focusing on mineralogical problems, this book should be of interest to material scientists, solid state chemists, and condensed matter physicists. It is hoped that this book will appeal to senior undergraduate students as well as those involved in research.

This monograph has its origins in a two-day meeting with the same title, held at University College London, England, in December 1989. The meeting was held under the auspices of the Mineralogical Society of Great Britain and Ireland, with contributions from its groups and from the British Crystallographic Association. The meeting, and hence this monograph, benefited greatly from the support of the Royal Society, Academic Press, Longman Scientific and Technical, Oxford University Press, Unwin Hyman, Wild-Leitz, Schlumberger, Bruker Spectrospin, the English China Clay Group Charitable Trust, ISA Instruments, and the Royal Bank of Scotland. We would also like to thank the authors and participants for contributing to a very successful meeting. Finally, we are indepted to our colleagues at UCL (Ross Angel, Monica Mendelssohn, Judith Milledge, Sally Price and Ian Wood) who helped in the organization of the meeting and in reviewing the manuscripts in this volume, and to Beth Price, whose birth on the first day of the meeting made 18 December 1989 an even more memorable occasion!

Geoffrey D. Price and Nancy L. Ross
University College London

CHAPTER ONE
The stability of minerals: an introduction
Nancy L. Ross and Geoffrey D. Price

1.1 Introduction

Historically, the elucidation of the fundamental factors that determine the stability of crystal structures has been one of the primary objectives of crystallographic and mineralogical research. The fact that this not insubstantial goal has yet to be fully attained was noted recently by the editor of *Nature* (Maddox, 1988), who went as far as to write:

> One of the continuing scandals in the physical sciences is that it remains in general impossible to predict the structure of even the simplest crystalline solids from a knowledge of their chemical composition.

Maddox further contended that:

> one would have thought that by now, it should be possible to equip a sufficiently large computer with a sufficiently large program, type in the formula of the chemical and obtain, as output, the atomic coordinates of the atoms in a unit cell. ... To reach this goal it should be possible ... to select from a listing of all possible structures of an arbitrary material those that can be considered plausible on crystallographic and other grounds.

It is perhaps not surprising that such a challenging article solicited some considerable response from, amongst others, Cohen (1989), Hawthorne (1990), and Catlow and Price (1990). These various authors pointed out the advances that have been made in the *ab initio* and semi-classical prediction of the relative stability of crystalline solids, but stressed the two major difficulties of any structure prediction strategy, namely:

1. establishing 'a listing of all possible structures';
2. differentiating between metastable structures, occupying local energy minima, and the thermodynamically stable phase, occupying the global energy minimum for that system.

Without wishing the belittle the major advances made in the computational

The Stability of Minerals. Edited by G. D. Price and N. L. Ross.
Published in 1992 by Chapman & Hall, London. ISBN 0 412 44150 0

analysis of solid-state phases, it is perhaps to miss the point of the crystallographers' search to suggest that a full computer analysis of the energetics of a system can provide the entire answer to the question of crystal stability. Indeed, it has been suggested by some, that although solving Schrödinger's equation could be used to establish the relative stabilities of potential crystal polymorphs, the many pages of numerical output may actually obscure rather than illuminate the reasons for the stabilization of one phase over the others. Historically, solid-state scientists have chosen to look for semi-quantitative but more tangible factors, such as 'atomic size' or 'bond type', to help them rationalize crystal structural architecture. In this chapter, and throughout this volume, we hope to show how the more classical concepts used to discuss the stability of minerals, can be put on a firm microscopic footing, and we will attempt to link quantum-mechanical ideas to concepts based upon atomic or polyhedral scale analyses, and thence show how these lattice-scale models relate to phenomenological or thermodynamic approaches.

1.2 Theories of crystal structures

Soon after the explosion in the number of known crystal structures caused by the discovery of X-ray diffraction, scientists began to attempt to answer the question 'why does this compound adopt this structure, rather than another structure exhibited by some isoelectronic phase?' One of the first, comprehensive approaches to this question was presented by Goldschmidt (1929), who analysed crystal structures in terms of:

1. the stoichiometry of the compound;
2. the relative sizes of the component atoms; and
3. their polarization.

The first two factors still form the basis for most analyses of crystal structure stability, but today we would substitute the terms 'the electronic structure' or 'the nature of bonding' for the term polarizability.

Further seminal contributions to this topic were made by Pauling (1960), who inferred a series of rules to describe the packing of cations and anions in predominantly ionic compounds in terms of their radius ratios, ionic charges, polyhedral linkage, and stoichiometry, and by Hume-Rothery and coworkers (1963) who attempted to rationalize the structure of metals and metallic compounds, again in terms of atomic size, electronegativities, and valence-electron concentration. These essentially qualitative analyses were synthesized by Laves (1955), who enumerated three general principles behind the stability of crystal structures, namely:

1. the principle of space, emphasizing the tendency of atoms in crystals to pack in such a way as efficiently to fill space;
2. the principle of symmetry, requiring atomic packing to tend towards high symmetry; and

3. the principle of connection, requiring that the number of connections between components be maximized.

Subsequent advances in our understanding of the stability of crystal structures have come from somewhat more quantitative considerations, such as those outlined by Mooser and Pearson (1959) and by Phillips and Van Vechten (1969; 1970). These workers used the ideas of *structural sorting diagrams*, to establish indices (such as average principal quantum number, the total number of valence electrons, the electronegativity difference, or average band gap, etc.), which when graphically plotted separate one from the other the different structural types of an isoelectronic family (e.g. the AB or AB_2 full-octet compounds). This approach has been extended by Burdett and others (as discussed in Section 1.4), and notably by Villars (1983; 1984a; 1984b; 1985) who have been able to rationalize, via the structural-sorting-diagram approach, the structures of most binary and ternary metallic compounds. The link between understanding electrons in a solid and its structure was the topic of a recent meeting held at the Royal Society in September 1990. The contents of this meeting, summarized in the *Bonding and Structure of Solids* (Haydock *et al.* 1991), provide an overview of the theoretical developments in this field.

The outstanding feature of all of the successful approaches to understanding the factors that determine mineral stability has been the identification of the importance of:

1. geometric considerations as determined by relative size of the atomic species involved; and
2. electronic considerations, in terms of the total number of valence electrons, or the relative electronegativity, bonding type, etc.

These two factors are not totally unrelated. Perhaps some of the most interesting results of the structural-sorting-diagram methodology have come from the successful use of so-called *pseudopotential radii*, discussed below in greater detail, to rationalize crystal-structure stability, since it has been shown that this single index scales with both ionic radius and electronegativity. Notwithstanding this correlation, however, initially we will discuss separately the insights obtained from purely geometric and purely electronic considerations before extending our discussion to include the effect that such *atomistic* properties can have on a mineral's *thermodynamic* properties.

1.3 Geometrical constraints on crystal structures

1.3.1 *Topology and eutaxy*

As already discussed, geometrical factors have a direct bearing, not only on the details of crystal structures, but also on the stability of compounds. Geometrical constraints cover both topological aspects that specify the connectedness of each atom, and metric aspects, that cover distances and angles in the structure. The general aim of topological enquiries is to find out what types of structures are possible in principle. In other words, topological constraints provide a necessary but not sufficient condition for the existence of a structure. To pursue topological enquiries we need a suitable language. Hawthorne describes in Chapter 2 how the bond network of a structure can be described in terms of a graphical representation, i.e. a set of vertices (that may be an atom or group of atoms) and edges (that represent linkages or bonds between atoms). Families of such 'nets' (and polyhedra which can be regarded as nets) have been systematically studied by Wells (1984). Wells (1984) further recognized the existence of tightly knit groups of atoms or *complexes* within many crystal structures. He used these complexes as a basis for a broad geometrical description of crystal structures, ranging from one-dimensional complexes (chain structures) to two-dimensional complexes (layer structures), and both finite and infinite three-dimensional complexes (framework structures). An advantage of Wells' classification is that it avoids the question of bond type and division into ionic, covalent, metallic, and molecular crystals, and, we quote from Wells (1984), enables us: 'to discuss the nature of the bonds in a particular crystal without having prejudiced the issue by classifying a crystal as, for example, an ionic crystal'.

Framework structures can be described on the basis of four-connected, three-dimensional nets (e.g. Wells, 1984; Smith, 1977; 1978; 1979). Akporaiye and Price (1989) and Wood and Price (1992), for example, have developed a method to enumerate both known and feasible zeolite frameworks by analysing the frameworks in terms of component sheets, as in a layer structure. Figure 1.1 shows examples of the structural repeat layers of merlinoite and ferrierite in which the tetrahedral interconnections of the net (i.e. nodes) represent Si or Al atoms and the linkages between nodes represent the oxygen bridges. Thus the depiction of the zeolite structure is based on the topology of the constituent nets.

An alternative representation of crystal structures involves description of the structures in terms of periodic packings of equivalent spheres. Wells (1984) describes a sphere packing as one in which the arrangement of spheres makes at least three contacts with other spheres. Table 1.1 summarizes the densities of periodic packing of equal spheres for those packings in which the sphere is in contact with six or more neighbours. The density is defined as the fraction of the total space occupied by spheres. An important factor governing mineral stability is the space-filling postulate of Laves (1955), described in detail by Parthé (1961), which requires the efficient packing of atoms in a solid to give a dense high-symmetry arrangement. This principle is especially important when considering the effect of pressure on the stability of minerals (e.g. Hazen and Finger, 1979; Prewitt, 1982). As shown in Table 1.1, the most economical use of space is made in 'close-packed' structures that are generated by the stacking of triangular nets of atoms at the appropriate spacing, of which the

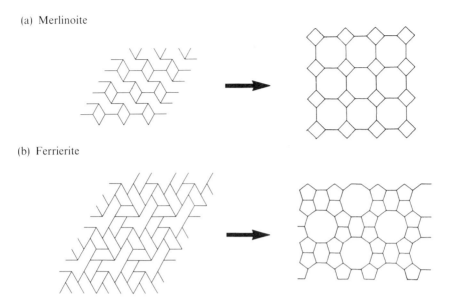

Figure 1.1 Examples of 'generated' and 'relaxed' meshes of (a) merlinoite and (b) ferrierite (Wood and Price, 1992).

cubic close-packed (ccp) and hexagonal close-packed (hcp) are the best-known examples. Spheres cannot fill all space and there are both tetrahedral interstices and octahedral interstices defined by any two close-packed layers. There are two tetrahedral sites and one octahedral site associated with each atom in a close-packed layer. We emphasize that this concept assumes spheres of fixed size in contact and is purely a geometrical approach. O'Keeffe and Hyde (1982) and Burdett (1988) have suggested that a better term defining such an arrangement of atoms, whose centres are at the centres of the spheres of a close-packing, is *eutaxy*.

Table 1.1 Densities of periodic packings of equivalent spheres (Wells, 1984)

Co-ordination number	Type of packing	Density
6	Simple cubic	0.5236
8	Simple hexagonal	0.6046
8	Body-centred cubic	0.6802
10	Body-centred tetragonal	0.6981
11	Tetragonal close-packing	0.7187
12	Closest packings	0.7405

While we find many examples of compounds that crystallize with ccp and hcp structures (e.g. the noble gases and many metals), a large number of additional structures can be developed from eutactic lattices by simple geometrical considerations. For example, some possibilities following the outline given by Adams (1974) are:

1. forming eutactic layers of more than one type and size of sphere;
2. fitting smaller spheres into octahedral and/or tetrahedral sites in eutactic layers composed of one type of sphere only;
3. combining 1. and 2.;
4. omitting a proportion of spheres in eutactic layers and filling a proportion of interstitial sites;
5. distortion of eutactic layers.

More than one structure type may be possible for a given close-packed array even if the stoichiometry of the phase is the same. Therefore the structures of $CaCl_2$, α-PbO_2, and CdI_2, examples of 2, arise from different considerations of how to fill 50% of the available octahedral interstices in an hcp array. Examples of 3. include compounds of the perovskite family with ABX_3 stoichiometry that possess approximately close packed arrays of A and X atoms. The remaining B atoms fill octahedral sites. The closely related structure of ReO_3 may be described in terms of an oxygen-deficient ccp array with rhenium in one quarter of the octahedral sites and therefore is an example of 4. Rutile, TiO_2, is an example of 5., and can be described as an hcp array of oxygens with titanium in one half of the octahedral sites. Upon closer inspection, however, one finds that the anion geometry has relaxed from ideal hcp with the result that the crystal is tetragonal and not orthorhombic. Other well-known mineral structures formed by filling octahedral and tetrahedral interstices in approximate close packed arrays of anions are presented in Table 1.2. There are many more complex variations of this theme and the reader is referred to Adams (1974), Wells (1984), and Hyde and Andersson (1989) for a more extensive discussion of structures described in terms of sphere packings.

The reader should be aware that the compounds cited above can alternatively be described in terms of their bond topology, their component polyhedra, or any other suitable formalism. One such alternative method to geometrically describing and classifying structures was proposed by O'Keeffe and Hyde (1985). In their approach, the oxide structures are described in terms of their *cation* packings and the coordination of the anions by these cations. In other words, instead of regarding a structure as a regular anionic lattice into which cations have been inserted, the alternative is used, namely of inserting anions into a regular cation array. It is interesting to note that the resulting cation packings often correspond to those found in well-known alloy systems. Thus the structure of garnet, which is very complex when considered in terms of cation–anion polyhedra, can be simply described on the basis of its cation

Table 1.2 Examples of some structures formed by filling octahedral or tetrahedral interstices in approximately close-packed lattices

Type and fraction of sites occupied	Cubic close-packing	Hexagonal close-packing
All octahedral	NaCl (rock salt)	NiAs
All tetrahedral	Li_2O (anti-fluorite)	—
1/2 octahedral	TiO_2 (anatase)	TiO_2 (rutile)
1/2 tetrahedral	ZnS (sphalerite)	ZnS (wurtzite)
2/3 octahedral	—	Al_2O_3 (corundum)
1/8 tetrahedral 1/2 octahedral	Al_2MgO_4 (spinel)	Mg_2SiO_4 (olivine)

array which is related to the Cr_3Si (A15) structure type (O'Keeffe and Hyde, 1982). Consideration of cation packing has also shed light on the rationale behind the transition from olivine to spinel in Mg_2SiO_4 at high pressure. This transformation is difficult to understand if the structures are analysed simply in terms of their anion packing. The phase transition occurs with a large decrease in volume although the primary coordination of the cations by anions does not change and the Mg–O and Si–O distances show little change. O'Keeffe and Hyde (1981a), however, have shown that the transformation is understandable when the cation packing in the two structures is considered. One finds that coordination numbers of the cations *by other cations* increases during the transformation. That is, the fairly open packing in olivine with $SiMg_9$ coordination transforms to a more efficiently-packed structure with $SiMg_{12}$ coordination. This is accompanied by a significant increase in the Si–Mg distance, suggesting that Si–Mg repulsion may play a significant role in determining the relative stabilities of the two polymorphs.

Apart from the advantages accruing from high coordination and efficient use of space, little is known of the basis of geometric and topological arrangement as a source of structural ability. Hawthorne explores this area in Chapter 2 and emerges with insights into the energetics and structural consequences of bond topology. Below, we extend our discussion to consider the metric factors associated with structural geometry, with specific reference to the empirical rules refined by Pauling (1929; 1960) to describe the behaviour of *ionic* crystals. The relationship between these metric considerations and bond topology is also addressed in Chapter 2.

1.3.2 Pauling's rules

In Section 1.3, we used the concept of eutaxy, or close packing, to enumerate possible structure types, but did not consider the factors controlling the structural stability of these geometric arrangements. Pauling codified the principles underlying ionic crystal structures into a set of rules that are, for the most part, the expression in formal terms of principles recognized earlier by Goldschmidt and Bragg. In an ionic structure, ions are treated as charged, incompressible, non-polarizable spheres with non-directional, electrostatic (Coulombic) forces holding the ions together. According to Pauling's first rule, structures may be described in terms of coordinated polyhedra of negatively charged anions around positively charged cations. The relative sizes of ions are governed by the insertion of 'small' cations into a lattice of 'large' anions. The type of coordination favoured by the cation is determined by the radius ratio of the anion to the cation. The second rule defines the electrostatic bond strength of a cation–anion bond as equal to the formal charge of the cation divided by the cation coordination number. In a stable ionic structure the formal valence of each anion is equal (but with opposite sign) to the sum of the electrostatic bond strengths of its bonds formed with adjacent cations. The third and fourth rules note that the presence of shared edges and faces between cation-centred polyhedra tend to destabilize the structure, and therefore cations with high valency and small coordination tend not to share polyhedron elements with one another. Thus, for example, a structure based on an hcp array of anions but having all of the available tetrahedral sites filled by cations, would possess face-sharing tetrahedra, and this is consequently not known for any compound. In general, the coordinating ions arrange themselves in such a manner as to minimize the electrostatic energy between them. Of course, the stoichiometry of the system may force the sharing of polyhedral elements as in the rock salt structure.

According to Pauling's second rule, the formal valence of the anion is nearly equal (with opposite sign) to the sum of the bond strengths received by that anion from adjacent cations. Considerable deviations from Pauling's second rule have been observed, however, in minerals such as diopside, melilite, sillimanite, and anorthite, to name a few. Baur (1970) concluded that in structures in which this postulate is not obeyed an effective charge balance is achieved by adjustments in the cation–anion bond distances, in other words by distortion of the cation-centred polyhedra. This led Baur to propose an *extended electrostatic valence rule* to predict bond lengths more accurately from empirically derived parameters. Brown and Shannon (1973) also introduced a method for more accurate prediction of interatomic distances from bond-valence–bond-length relationships. Brown (1981), however, extended bond-valence theory one step further, to a point where the model parameters are associated with ions rather than with bonds. We should emphasize that bond-valence theory is not restricted to ionic solids. Recently, Gibbs (1982) and

Gibbs et al. (1987) have shown that *molecular orbital* calculations on hydroxy-acid molecules containing first- and second-row cations are successful in not only reproducing the observed bond lengths of ionic crystals compiled by Shannon and Prewitt (1969) and Shannon (1976), but also in reproducing the Baur (1970) and Brown and Shannon (1973) bond-strength–bond-length curves. In addition, O'Keeffe and Brese (1991) have shown that a variety of homopolar, heteropolar, and metallic bond lengths can be determined from bond-valence parameters that are, in turn, derived from parameters related to the sizes and electronegativities of the atoms forming the bond. The concept of atomic size and contribution to mineral stability from electronic factors are considered in more detail in the following sections.

1.3.3 Atomic size

According to Pauling's first rule, each ion may be assigned a radius such that the A–C distance in a crystal structure is simply, $R_A + R_C$ (where R_A is the anion radius and R_C is the cation radius). Given a collection of hard spheres in contact, the radius of a tetrahedral interstice in the anion lattice is $0.225 R_A$, that of an octahedral interstice is $0.414 R_A$, and that of a cubically coordinated site is $0.732 R_A$. Thus, for cations to be in contact with six anions in an octahedral arrangement without strong destabilizing repulsions between anions, the radius ratio should be in the range $0.414 < R_C/R_A < 0.732$. If R_C/R_A is less than 0.414 the cation will 'rattle' in the interstice and the structure may be less stable than one with tetrahedral coordination; if the ratio is greater than 0.732, a structure with the cation in larger coordination will be favoured. Thus, the radius ratio can be used to define the boundary between structures with different coordination numbers. Burdett et al. (1981) plotted R_A and R_C pairs for AB binary compounds with the B1 (NaCl), B2 (CsCl), B3 (ZnS), and B4 (ZnO) structures and found that 38 of a total of 98 structure types were incorrectly predicted. Although one might argue that this failure was due to the inclusion of covalent AB compounds, the model incorrectly predicted the structures of ionic compounds such as BaO, SrO, and LiF. One might also argue the correct choice of ionic radii was not made, but notwithstanding these caveats, one must conclude that the radius-ratio rules should be applied with caution.

There are numerous methods for estimating ionic radii from observed internuclear distances involving different criteria for apportioning contributions to each ion (e.g. Goldschmidt, 1927; Pauling, 1927; Ahrens, 1952; Shannon and Prewitt, 1969; Shannon, 1976). The definitive set of ionic radii for use in oxides and fluorides is that of Shannon and Prewitt (1969) as revised by Shannon (1976). In their set of radii, the term 'ionic' refers to compounds containing bonds between atoms of the different *formal valence* charge, and thus applies to many crystals with bond types ranging from ionic to covalent. Interatomic distances are influenced by factors such as the formal valence state

THE STABILITY OF MINERALS: AN INTRODUCTION

of the cation, coordination numbers of cations and anions, and electronic spin state. These factors are taken into account in the Shannon and Prewitt (1969) and Shannon (1976) radii. Additional perturbations of radii sums can result from polyhedral distortion, partial occupancy of cation sites, covalent shortening, and electron delocalization (Shannon, 1976). Thus, whereas oxide structures show relatively small variations in bond length, other compounds such as sulphides show a considerably wider range of distances. Shannon (1981) addresses the problem of determining radii applicable to sulphides. The factor that most influences a particular mean bond length depends on the atoms forming the bonds. We caution readers that while the concept of ionic radius can be useful, there are physical limitations to this concept that can result in erroneous or suspect interpretations which are discussed in more detail by O'Keeffe (1981) and Burdett (1988).

It is doubtful that an unambiguous and useful approach can be devised to divide a bond length into cation and anion parts. Even a 'direct' method, such as measuring the electron density distribution in a compound, is by no means an easy and straightforward task (e.g. Stewart and Spackman, 1981). However, an experimental determination of the electron density distribution in alkali halides (Gourary and Adrian, 1960) and a study of simple oxides (MgO, MnO, CoO, and NiO (Sasaki *et al.*, 1979)), suggest that the ionic radii sets generally defined cations to be too *small* and anions to be too *large*. These findings are consistent with Fumi and Tosi's (1964) analysis of 'basic radii' derived within the framework of Born–Mayer theory. This analysis provided the basis for the 'crystal radii' of Shannon and Prewitt (1969) and Shannon (1976), in which, for example, the crystal radius of oxygen is 1.24 Å while the effective ionic radius for fourfold-coordinated oxygen is defined as 1.38 Å. It is interesting that other radii sets such as Slater's (1964) atomic radii, O'Keeffe and Hyde's (1981b) *non-bonded* or one-angle radii, and Zunger's (1981) *pseudopotential* radii (discussed in more detail on p. 13) show similar periodic trends, increasing to the left and down the periodic table (see Figure 14 in O'Keeffe and Hyde (1981b)). O'Keeffe and Hyde (1981b) further noted that the radii of silicon and aluminum are *larger* than the radius of oxygen in these sets of radii.

Thus far, we have focused on structural features (coordination number, atomic size, etc.) that influence the stability of simple solids and we have attempted to rationalize the existence of some structures in terms of Pauling's rules. These rules are easy to understand and deceptively easy to apply. To progress further, we must now consider the effects of electronic factors on structural stability, which are 'ultimately' founded upon the valence and molecular-orbital theories of bonding.

1.4 Electronic factors and mineral stability

In this section, we address two main electronic factors that have some bearing (or historically have been thought to have some bearing) on structural stability, namely:

1. the number of valence electrons available to form bonds; and
2. the predominant type of chemical bonding within the crystal.

Thus far we have avoided a classification of 'ionic' and 'covalent' structures although the 'billiard ball' description above is consistent with an 'ionic' model in which the attractive force holding two atoms together is the Coulomb interaction between cations and anions. In covalent crystals the attractive force is due to the overlap between atomic orbitals which produces an energy gap between bonding and antibonding states. In Chapter 3, Burdett explains connections between the traditional Pauling viewpoint of inorganic solids with the more rigorous approach of molecular orbital theory.

1.4.1 Number of valence electrons

Most silicates and common inorganic compounds can be considered as having a full 'octet' of valence electrons, so that whatever the bond type (be it 'ionic' or 'covalent'), the species involved can be viewed as having a full valence shell, or more usefully, in terms of molecular orbital or band theory, the valence band is fully occupied. Many materials, however, do not have sufficient valence electrons to fill the valence band, and are metallic. The electronic and geometric factors that determine the stability of metals, alloys, and intermetallic compounds have been extensively analysed by Villars (1983; 1984a; 1984b; 1985), and are also elegantly discussed by Adams (1974). The stability of metallic structures will not be perused further in this volume.

In contrast, some compounds have an excess of valence electrons (e.g. many sulphides such as $CuSbS_2$, Cu_3AsS_4, etc.), which must occupy non-bonding or anti-bonding orbitals. The resulting structures often appear distorted or deformed, when compared with the structures of simple 'octet-compounds'. These distorted structures, however, can frequently be viewed as being derived from the simple octet-compound structures by systematic sequences of bond-breaking, caused by the occupation of anti-bonding orbitals. Thus, Burdett (1980) points out how, by a series of bond-breaking processes (that do not disrupt the topology of the atomic species), one can progress from the wurtzite structure (adopted by ZnS, with 8 valence electrons), to the double-sheet GaSe structure (with 9 valence electrons), to the single-layer As structure (10 valence electrons), to the chain structure of Se (12 valence electrons), to the diatomic molecular structure of crystalline I_2, and finally, to the Van der Waals solid Xe (with 16 valence electrons per atom pair). Interestingly, it is also possible to analyse the degree of polymerization of the Si–O framework in silicate compounds in terms of the number of electrons available to participate in Si–O bonding (Table 1.3).

Table 1.3 Fragmentation of the silicate framework with increasing electron count (Burdett, 1980)

System	Number of electrons per atom	Structural type
SiO_2	5.33	Frameworks
$(Si_2O_5)^{2-}$	5.71	Sheets
$(Si_4O_{11})^{6-}$	5.87	Double chains
$(SiO_3)^{2-}$	6.00	Single chains
$(Si_2O_7)^{6-}$	6.22	Paired tetrahedra
$(SiO_4)^{4-}$	6.4	Isolated tetrahedra

1.4.2 Bonding and structure diagrams

One technique used to systematize the structures of a diverse collection of compounds is *structural mapping*, in which each compound with a given stoichiometry and perhaps total electron count is characterized by two parameters. A plot of one parameter vs. the other will often lead to domains in the map where compounds with the same structure appear to cluster. The goal of structural mapping is to generate a set of atomic parameters which, when used with a judicious choice of indices, will lead to a perfect, or near perfect, structural sorting of a given database that will, in turn, provide insight into why one structure is favoured over another. In the previous section, we noted that utilizing ionic radii as structural parameters was only moderately successful in sorting AB compounds.

The work of Mooser and Pearson (1959) and Phillips and Van Vechten (1969; 1970) has led to semi-quantitative and, in some cases, quantitative, understanding of the relationships between ionicity, f_i, electronegativity, χ, and crystal structure for many of the structures adopted by binary compounds. In an attempt to predict the relative stability of fourfold and sixfold coordinated structures, Mooser and Pearson (1959) recognized two factors that influence the formation of directed bonds. The first is the charge transfer between anion and cation, which they measured in terms of Pauling's electronegativity difference ($\Delta\chi = \chi_A - \chi_B$), and the second is the degree of 'metallization' of s–p bonds through the admixture of d, f, ... states. which increases as the principal quantum number, n, increases. To exhibit the dependence of crystal structures of AB compounds on these two factors, Mooser and Pearson (1959) plotted the mean value of the valence-shell quantum number, $\bar{n}\, (=[n_A + n_B]/2)$, against $\Delta\chi$. They found a sharp boundary between the domains of fourfold and sixfold coordination with 92 out of 100 compounds predicted successfully. Their results suggest that compounds adopt tetrahedrally coordinated structures largely because of the directional (i.e. covalent) character of their bonding rather than for metric reasons.

Phillips and Van Vechten (1969; 1970) further developed a quantum description of the properties of octet compounds that utilized two parameters related to the average band gap, E_g, determined from spectroscopic analysis. For homopolar crystals, E_g is equal to E_h, the homopolar band gap and represents a solely covalent contribution to the bonding. For heteropolar compounds, such as $A^N B^{8-N}$, there is an additional contribution to E_g arising from the differing electronegativities of A and B. This is an ionic contribution to the bonding and is denoted by C. The covalent (E_h) and ionic contributions (C) are related to E_g by,

$$E_g^2 = E_h^2 + C^2 \quad (1)$$

A resulting scale of ionicity, f_i, can therefore be defined,

$$f_i = C^2/E_g^2 \quad (2)$$

where $f_i = 0$ when $C = 0$, that is, when there is no ionic contribution. There is an analogous covalent bond fraction, f_c, equal to E_h^2/E_g^2, such that $f_i + f_c = 1$. It is truly remarkable that Phillip's (1970) structure map for 70 $A^N B^{8-N}$ compounds, using E_h and C as coordinates, *exactly* separates ionic structures from covalent structures with a straight boundary that passes through $C = 0$. The slope of the line corresponds to a 'critical ionicity' of $f_i = 0.79 \pm 0.01$. The important conclusion to be drawn from this work is that the *ionicity*, f_i, of the bond, not the electronegativity difference between the atoms, $\Delta\chi$, is the critical factor in determining which crystal structure is adopted by $A^N B^{8-N}$ compounds. It should be noted that the parameters that Mooser and Pearson used in their structure diagrams, $\Delta\chi$ and $\bar{n}$, are related to C and E_h (Phillips, 1970; Adams, 1974). The Phillips–Van Vechten diagram of the octet compounds, however, was the first *quantum mechanically based* structural sorting diagram.

In recent years, there have been many attempts to find new parameters that successfully describe the various properties of classes of different types of compounds. One new set of quantum coordinates that have been successfully utilized in structural maps of a number of compounds are the *pseudopotential radii* which are derived as follows. A valence electron is attracted to a positively charged nucleus that is screened by other electrons. The attraction increases with decreasing distance from the nucleus, but the electron also feels a strong repulsion which increases with decreasing distance due to the Pauli repulsion forces associated with the other electrons in the atom. The sum of the two gives the pseudopotential, $V(r)$, and the crossing point, the distance at which $V(r) = 0$, defines the pseudopotential radius. There are different pseudopotentials for s, p, and d electrons and they have been derived from first-principles calculations on the free atoms (e.g. St John and Bloch, 1974; Chelikowsky and Phillips, 1978; Zunger and Cohen, 1979; Andreoni et al., 1979; Zunger, 1980; Bloch and Schatteman, 1981; Zunger, 1981). Unlike ionic radii, pseudopotential radii incorporate aspects of both size and electronegativity or bond type, and have been utilized very successfully in structure diagrams. Burdett *et al.* (1981), for example, presented a structure map for AB octets using as indices the sum of the pseudopotential radii, $R_s + R_p$, for A and B, respectively. On the one hand, the simplicity of Phillips' (1970) structure map is lost in the sense that boundaries between crystal structures are no longer linear (which is to be expected since there is no linear relationship between E_g and the pseudopotential radii). On the other hand, there is an improvement in that the new diagram faithfully resolves subtle structural distinctions in the octet compounds, such as that between the B3 (zincblende) and B4 (wurtzite) structure types.

Figure 1.2 displays a spectacular sorting of 187 spinel structures into normal and inverse types using r_σ, the sum of the s and p pseudopotential radii, as indices (Price *et al.*, 1982). Traditionally, crystal field theory has been utilized in rationalizing whether a spinel is a normal one, $A^T[B_2]^oO_4$, or an inverse one, $B^T[AB]^oO_4$. The problem with this approach is that not all spinels contain transition metal ions and among those that do not, the ionic approach

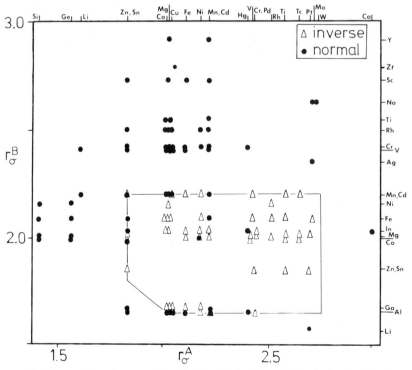

Figure 1.2 Structural sorting map for AB_2X_4 spinels (where X is O, S, Se, Te) using R_σ^A ($= R_S^A + R_P^A$) and R_σ^B ($= R_S^B + R_P^B$) as indices, where R_S and R_P are pseudopotential radii (Price *et al.*, 1982).

is not very successful in resolving the structure type. However, Price and coworkers' (1982) structural sorting map using r_σ as indices (Figure 1.2) has only four errors, of which two are poorly characterized materials. On the boundary, both normal and inverse types are found and it is this region where crystal field arguments come into play and determine the structure. It appears, therefore, that crystal field effects are not the major factor in determining the structure of spinels.

In our discussion thus far, we have attempted to understand the existence of stable structures in terms of geometric parameters and electronic factors. For most mineralogical applications in earth sciences, one wishes to define mineral stability as a function of pressure, temperature, and composition. An alternative description of mineral stability, more suited to answer these questions, is provided through thermodynamics as described in more detail in the next section. One of the challenging goals facing mineralogists today is to establish the link between the atomistic properties of a mineral and its bulk thermodynamic properties.

1.5 Thermodynamics and crystal stability

As we stated above, we may use a *macroscopic* approach involving thermodynamic properties to discuss mineral stability. In the language of thermodynamics, the stable structure of a given composition at a given temperature and pressure is the one with the lowest Gibbs free energy. In particular, we wish to know the possible forms of free energy (G), enthalpy (H), entropy (S), and volume (V) as functions of temperature and pressure, and, in the case of solid solutions, as functions of composition. We desire not only an understanding of atomistic factors controlling mineral stability, but also an ability to *predict* what structures will be stable at a given temperature, pressure, and composition.

Full understanding of the macroscopic behaviour of minerals, and eventually rocks, can only be obtained from a detailed knowledge of their microscopic or atomistic nature. The microscopic understanding is required because it is the response at the atomic level to changes in pressure and temperature that eventually determines the bulk properties of a material. The link between the atomistic properties of a mineral and its bulk thermodynamic properties was the subject of the 1985 Mineralogical Society of America short course (e.g. Kieffer and Navrotsky, 1985). Pauling (1960), in fact, was one of the first to derive a relationship between atomistic properties such as those discussed above and thermodynamics. He formulated an electronegativity scale of the elements by analysis of the single-bond energies and found rough estimates of the enthalpies of formation of compounds could be made by the use of electronegativity values. He also noted that the electronegativity of an element was proportional to the sum of the first ionization energy and the electron affinity. Since electronegativity values vary in a regular way from element to element in the periodic table, Pauling brought a certain amount of systematization into the field of inorganic thermochemistry. Phillips and Van Vechten (1970) further showed that the heats of formation of $A^N B^{8-N}$ compounds can be predicted using the spectroscopic parameters described above. With one free parameter, heats of formation can be fitted to about 10% accuracy.

As Ross describes in Chapter 4, a direct link between the microscopic properties of materials and their macroscopic behaviour can be made through consideration of the lattice vibrations of a crystal. Thermodynamic properties such as heat capacity, entropy, etc. are manifestations of the atomic vibrations within a crystal. The frequencies of atomic vibrations are determined by the strength and nature of the bonding that holds the atoms in that crystal together. More generally, these bonding forces can be described in terms of interatomic potentials, that not only determine the vibrational characteristics of a crystal, but also its structure and physical properties, such as its elastic and dielectric behaviour. Thus, a study of a crystal's lattice vibrations can form a true link between the microscopic and macroscopic properties of the material.

1.5.1 Phase transformations

One application of the study of lattice vibrations has been in the modelling and rationalization of polymorphic phase transitions, that is, between phases with the same composition, such as Mg_2SiO_4, but different structures, such as olivine, modified spinel and spinel (e.g. Akaogi *et al.*, 1984; Price *et al.*, 1987). Using heat capacities, such as those calculated from lattice vibrational models (Chapter 4 this volume), changes in the enthalpy, ΔH_T^0, and entropy, ΔS_T^0, of the transition with temperature can be calculated. The free energy of transition can be then determined from,

$$\Delta G_T^P = \Delta H_T^0 - T\Delta S_T^0 + \int_1^P \Delta V_{T,P}\, dP. \qquad (3)$$

Determination of $\Delta V_{T,P}$, which is the volume change for the transition at a given pressure and temperature, requires knowledge of the thermal expansion and compressibility of both phases at various pressures and temperatures. Generally, such data are lacking, and one has to approximate these effects. Akaogi and co-workers (1984), for example, used a formalism whereby the effect of compression at a desired pressure and 298 K is calculated with a Murnaghan equation and then the effect of expansion at high temperature and one atmosphere is calculated using X-ray and dilatometric data. The delicate balance between the enthalpies, entropies, and volumes of the phases determines which one is stable at a given pressure and temperature.

Transformations among minerals such as the pyroxenes and pyroxenoids typically have small values of ΔH, ΔS, and ΔV, making prediction of their stability on the basis of energetic factors extremely difficult. Pyroxene and pyroxenoid transformations include examples of transitions between *polytypes*, special forms of polymorphs in which the structures can be described in terms of different stacking sequences of structurally similar units. Examples of other compounds that exhibit polytypism include the micas, chlorites, perovskites, and spinelloids, and the classic polytypes with MX and MX_2 stoichiometry, including SiC, ZnS, CdI_2, and MoS_2. Because the polytypes of a given compound are composed of virtually identical structural units, free-energy differences between them are small. Materials that exhibit polytypism have been discussed in terms of their lattice dynamics (Cheng *et al.*, 1990) and other entropy-dependent models, such as the ANNNI model (Price and Yeomans, 1987).

There is a further category of phase transformation that occurs without the coexistence of two phases. These transitions occur with a change of symmetry at a certain thermodynamic state during a continuous structural change through that point. Such transitions may be of three types: order–disorder, displacive, or a combination of both. Such transitions can have a profound effect on equilibrium boundaries of polymorphic phase transitions or heterogeneous reactions as Carpenter points out in Chapter 5. Carpenter also shows how mineralogical examples of such transitions can be analysed in terms of Landau theory. In Chapter 6, McConnell builds on these principles to elucidate the origin and stability of incommensurate phases. Since these phases are generally associated with 'conventional' phase transitions, McConnell extends Landau theory to deal with incommensurate structures, thereby establishing the basis of a theory that links normal and incommensurate phase transformations.

1.5.2 Compositional effects on stability

Stoichiometry has already been identified as one of the key factors in determining the nature of the crystal structure adopted by a compound. Changes in composition can either cause the destabilization of the original crystal structure and the adoption of an alternative form, or these changes can be accommodated via the formation of a solid solution. In this section, we consider briefly the factors that determine the structural relations in so-called *polysomatic series*, and then discuss in greater depth the more classical approach to the behaviour of solid solutions.

As discussed in Section 1.5.1, polytypes may be described in terms of the stacking sequence of structurally compatible, isochemical units. If, however, the constituent structural modules are chemically distinct, it is possible to define a series of structures which has a range of chemical compositions. The resulting structures are known as *polysomes*, and common examples include the biopyriboles (pyroxenes, amphiboles, and sheet silicates), the humite group, pyroxenoids, and phases in the $CeFCO_3$–$CaCO_3$ system.

In principle, an infinite number of structures can be constructed from combinations of two modules. In practice, however, it is found that certain stacking sequences occur much more frequently than others. Recently, Price and Yeomans (1986) analysed the stability and structural characterization of polysomatic series in terms of the ANNNI model, and concluded that the factors that determine the stability of polysomes are (a) the chemical potential, which controls the proportions of the different structural modules, and (b) the competing interactions between first and second neighbour modules within the structure.

In reality, most minerals do not occur as pure phases, rather they can be represented as solid solutions between 'pure' end-member phases. The changes a structure undergoes upon cation substitution may be understood on an atomistic level in terms of structural strains, bonding variations, etc., and, depending on the magnitude of these contributions, the structure may undergo a phase transformation as described by Carpenter in Chapter 5 and McConnell in Chapter 6. On a macroscopic level the microscopic contributions are translated into free energy changes associated with mixing various proportions of the end members of the solid solution. The free energy on mixing is,

$$\Delta G_S = \Delta H_S - T\Delta S_S. \tag{4}$$

The contributions to ΔG_S can be appreciated by considering a simple solution consisting of equal proportions of A and B atoms. The ideal configurational entropy of mixing, ΔS_S, of this solution, assuming random substitution, is

$$\Delta S_S = -R[X_A \ln X_A + X_B \ln X_B] \tag{5}$$

where X_A and X_B are the mole fractions of A and B ($X_A + X_B = 1$) and R is the universal gas constant ($= 8.3144$ J mol^{-1} deg^{-1}). Next consider the enthalpy of forming a solution, ΔH_S. If we assume that the enthalpy of an intermediate composition stems entirely from the interaction between nearest neighbour pairs, then the total energy of interaction varies with $2V_{AB} - V_{AA} - V_{BB}$, where V_{AB}, V_{AA}, V_{BB}, are interaction energies of nearest A–B atoms, nearest A–A atoms, and nearest B–B atoms, respectively. If replacing A–A and B–B bonds by A–B bonds raises the internal energy of the solid solution (i.e. $2V_{AB} > V_{AA} + V_{BB}$), ΔH_S is positive and phase separation will be favoured as temperature is reduced. If replacing A–A and B–B bonds by A–B bonds lowers the internal energy of the solid solution (i.e. $2V_{AB} < V_{AA} + V_{BB}$), ΔH_S is negative and ordering of A and B will be favoured as temperature is reduced. In reality, mineral solid solutions frequently show tendencies to both order and unmix in the same system (e.g. Carpenter, 1985).

The simplest type of solid solution is one in which there is no heat effect upon forming the solution, $\Delta H_S = 0$, and $\Delta G_S = -T\Delta S_S$. This type of solution is called an *ideal* solution. We can describe the process, in thermodynamic terms, of forming a solution of specified composition as a result of two steps

(Fig. 1.3). Continuing with our example of a binary solution, we take proportions of the end members, A and B, having molar free energies, μ_A^0 and μ_B^0, and form a mechanical mixture. This mechanical mixture will have a molar free energy, $\bar{G}_M$, given by

$$\bar{G}_M = X_A \mu_A^0 + X_B \mu_B^0. \tag{6}$$

The second step consists of allowing the mixture of the end members to mutually dissolve in one another, forming a homogeneous solution. The free energy change associated with this process is $\Delta \bar{G}_S = \bar{G}_M - \bar{G}_S$ where $\bar{G}_S$ (for an ideal solution), is

$$\bar{G}_S = X_A \mu_A^0 + X_B \mu_B^0 + RT[X_A \ln X_A + X_B \ln X_B]. \tag{7}$$

In general, the Gibbs free energy of a solution phase is given by,

$$\bar{G}_S = \sum_i X_i \mu_i^0 + RT \sum_i X_i \ln a_i, \tag{8}$$

where a_i is the *activity* of component i. Comparing Equations (7) and (8), it is clear that a_i is equal to X_i^N for an ideal solution in a crystal structure with N sites on which substitution occurs.

More complex expressions for thermodynamic mixing properties arise for

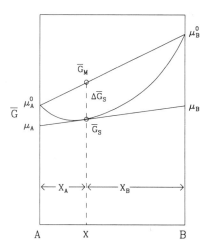

Figure 1.3 Illustration of the Gibbs free energy of a simple binary solution. The process of forming a solution of composition X involves (a) forming a mechanical mixture, $\bar{G}_M$, of the two end members whose molar free energies are μ_A^0 and μ_B^0, and (b) dissolution of two end members to form a homogeneous solution with Gibbs energy, $\bar{G}_S$.

multicomponent solid solutions and when cation substitution occurs on several sublattices, as is found in solid solutions of the fahlore group of minerals as described by Sack in Chapter 7. The behaviour of such complex solutions can be described in terms of thermodynamic 'excess' functions. For example, the excess free energy function may be defined by,

$$\bar{G}_{S,ex} = \bar{G}_S - \bar{G}_{S,ideal}. \qquad (9)$$

These excess variables are related to one another by formulas analogous to those relating standard thermodynamic variables, for example,

$$\bar{G}_{S,ex} = \bar{H}_{S,ex} - T\bar{S}_{S,ex}. \qquad (10)$$

Because of the complexity of describing the microscopic interactions of a disordered system (e.g. Jones et al., 1990; Bertram et al., 1990), we generally lack atomistic models for describing $\bar{G}_{S,ex}$ as a function of composition. As a result, we tend to use a different approach to describe the excess behaviour of solutions. For example, expressions for the excess thermodynamic functions can be derived from Landau theory, as discussed by Carpenter in Chapter 5. A second approach used more commonly in petrology, due in large part to Thompson (1967), describes thermodynamic excess functions in terms of empirical functions with *Margules* parameters. Sack uses this approach to describe the complex behaviour found in the tetrahedrite–tennantite solid solution. Sack also points out that accurate models of activity-composition relations in mineral solid solutions are needed in order to construct internally consistent databases for common rock-forming minerals. Engi assesses the available thermodynamic data bases of minerals in Chapter 8.

1.5.3 Kinetics

As we have emphasized, the driving force for a mineral reaction or transformation is the overall decrease in the free energy of the system. Thermodynamics describes the point towards which these changes are aiming and we can predict what change will occur given appropriate thermodynamic data. The actual behaviour that is observed, however, is governed by both the thermodynamics of the system as well as the *kinetics* of the mechanisms involved in the process. As McConnell points out in Chapter 6, the appearance of some modulated structures can be directly ascribed to kinetic controls. The factors which determine non-equilibrium pathways tend to be diverse and elusive in comparison with the minimization of the free energy which determines equilibrium. Recently, Salje (1988) has shown that kinetics and thermodynamics of structural transformations can be linked via Landau and other theories. Carpenter (Chapter 5) briefly reviews the application of time-dependent Landau theory to phase transitions in minerals. In Chapter 9, Velde explores the interplay of kinetic and thermodynamic factors that control the stability of clay minerals. A full discussion of the complex interplay of thermodynamic and kinetic controls on mineral stability is, however, beyond the scope of this current volume.

1.6 Conclusion

The factors that determine the stability of crystal structures are complex, and difficult to define in detail. There are, however, a number of underlying themes which we have attempted to identify, that enable us to rationalize the reasons for the stability of a given structure. We are still, however, a long way from being able to predict with confidence the structures that will be adopted, under a range of pressure and temperature conditions, by a given compound. The chapters that follow, however, indicate some of the avenues of future research in theoretical crystal chemistry, which, combined with experimental investigations of phase relations, will advance us towards our goal of fully understanding the stability of minerals.

References

Adams, D. M. (1974) *Inorganic Solids; an Introduction to Concepts in Solid-state Structural Chemistry*, John Wiley & Sons, London.

Ahrens, L. H. (1952) The use of ionization potentials; Part 1. Ionic radii of the elements. *Geochimica et Cosmochimica Acta*, **2**, 155–69.

Akaogi, M., Ross, N.L., McMillan, P. *et al.* (1984) The Mg_2SiO_4 polymorphs (olivine, modified spinel and spinel)–thermodynamic properties from oxide melt calorimetry, phase relations, and models of lattice vibrations. *American Mineralogist*, **69**, 499–512.

Akporiaye, D. E. and Price, G. D. (1989) Systematic enumeration of zeolite frameworks. *Zeolites*, **9**, 23–32.

Andreoni, W., Baldereschi, A., Biémont, E. *et al.* (1979) Hard-core pseudopotential radii and structural maps of solids. *Physical Review*, **B20**, 4814–23.

Baur, W. H. (1970) Bond length variation and distorted coordination polyhedra in inorganic crystals. *Transactions of the American Crystallographic Association*, **6**, 129–55.

Bertram, U. C., Heine, V., Jones, I. L. *et al.* (1990) Computer modelling of Al/Si ordering in sillimanite. *Physics and Chemistry of Minerals*, **17**, 326–33.

Bloch, A. N. and Schatteman, G. C. (1981) Quantum-defect orbital radii and the structural chemistry of simple solids, in *Structure and Bonding in Crystals*, vol. 1, (eds M. O'Keeffe and A. Navrotsky) Academic Press, New York, pp. 49–72.

Brown, I. D. (1981) The bond–valence method: an empirical approach to chemical structure and bonding, in *Structure and Bonding in Crystals*, vol. 2, (eds M. O'Keeffe and A. Navrotsky), Academic Press, New York, pp. 1–30.

Brown, I. D. and Shannon, R. D. (1973) Empirical bond-strength–bond-length curves for oxides. *Acta Crystallographica*, **A29**, 266–82.

Burdett, J. K. (1980) *Molecular shapes*, John Wiley & Sons, New York.

Burdett, J. K. (1988) Perspectives in structural chemistry. *Chemical Review*, **88**, 3–30.

Burdett, J.K., Price, G. D., and Price, S. L. (1981) Factors influencing solid-state structure—an analysis using pseudopotential radii structural maps. *Physical Review*, **B24**, 2903–12.

THE STABILITY OF MINERALS: AN INTRODUCTION

Carpenter, M. A. (1985) Order–disorder transformations in mineral solid solutions, in *Reviews in Mineralogy vol. 14: Microscopic to Macroscopic, Atomic Environments to Mineral Thermodynamics*, (eds S. W. Kieffer and A. Navrotsky), Mineralogical Society of America, pp. 187–223.

Catlow, C. R. A. and Price, G. D. (1990) Computer modelling of solid-state inorganic materials. *Nature*, **347**, 243–8.

Chelikowsky, J. R. and Phillips, J. C. (1978) Quantum-defect theory of heats of formation and structural transition energies of liquid and solid simple metal alloys and compounds. *Physical Review*, **B17**, 2453–77.

Cheng, C., Heine, V. and Jones, I. L. (1990) SiC polytypes as equilibrium structures. *Journal of Physics: Condensed Matter*, **2**, 5097–113.

Cohen, M. L. (1989) Novel materials from theory. *Nature*, **338**, 291–2.

Fumi, F. G. and Tosi, M. P. (1964) Ionic sizes and Born repulsive parameters in the NaCl-type alkali halides; I. The Huggins–Mayer and Pauling forms. *Journal of Physics and Chemistry of Solids*, **25**, 31–43.

Gibbs, G. V. (1982) Molecules as models for bonding in silicates. *American Mineralogist*, **67**, 421–50.

Gibbs, G. V., Finger, L. W. and Boisen, M. B. (1987) Molecular mimicry of the bond length–bond strength variations in oxide crystals. *Physics and Chemistry of Minerals*, **14**, 327–31.

Goldschmidt, V. M. (1927) Kristallbau und Chemische Zusammensetzung. *Berichte der Deutschen Chemischen Gesellschaft*, **60**, 1263–96.

Goldschmidt, V. M. (1929) Crystal structure and chemical constitution. *Transactions of the Faraday Society*, **25**, 253–83.

Gourary, B. S. and Adrian, F. J. (1960) Wave functions for electron-excess color centers in alkali halide crystals. *Solid State Physics*, **10**, 127–247.

Haydock, R., Inglesfield, J. E. and Pendry, J. B. (eds) (1991) *Bonding and Structure of Solids*, The Royal Society, London.

Hawthorne, F. C. (1990) Crystals from first principles. *Nature*, **345**, 297.

Hazen, R. M. and Finger, L. W. (1979) Bulk modulus–volume relationship for cation–anion polyhedra. *Journal of Geophysical Research*, **84**, 6723–8.

Hume-Rothery, W. (1963) *Electrons, Atoms, Metals and Alloys*, 3rd edn, Dover, New York.

Hyde, B. G. and Andersson, S. (1989) *Inorganic Crystal Structures*, John Wiley & Sons, New York.

Jones, I. L., Heine, V., Leslie, M. *et al.* (1990) A new approach to simulating disorder in crystals. *Physics and Chemistry of Minerals*, **17**, 238–45.

Kieffer, S. W. and Navrotsky, A. (eds) (1985) *Reviews in Mineralogy, vol. 14: Microscopic to Macroscopic, Atomic Environments to Mineral Thermodynamics*, Mineralogical Society of America.

Laves, F. (1955) Crystal structure and atomic size, in *Theory of Alloy Phases*, American Society for Metals, Cleveland, pp. 124–98.

Maddox, J. (1988) Crystals from first principles. *Nature*, **335**, 201.

Mooser, E. and Pearson, W. B. (1959) On the crystal chemistry of normal valence compounds. *Acta Crystallographica*, **12**, 1015–22.

O'Keeffe, M. (1981) Some aspects of the ionic model of crystals, in *Structure and Bonding in Crystals*, Vol. 1 (eds M. O'Keeffe and A. Navrotsky), Academic Press, New York, pp. 299–322.

O'Keeffe, M. and Brese, N. E. (1991) Atom sizes and bond lengths in molecules and crystals. *Journal of the American Chemical Society*, **113**, 3226–9.

O'Keeffe, M. and Hyde, B. G. (1981a) Why olivine transforms to spinel at high pressure. *Nature*, **293**, 727–8.

O'Keeffe, M. and Hyde, B. G. (1981b) The role of nonbonded forces in crystals, in *Structure and Bonding in Crystals*, vol. 1, (eds M. O'Keeffe and A. Navrotsky), Academic Press, New York, pp. 227–54.

O'Keeffe, M. and Hyde, B. G. (1982) Anion coordination and cation packing in oxides. *Journal of Solid State Chemistry*, **44**, 24–31.

O'Keeffe, M. and Hyde B. G. (1985) An alternative approach to non-molecular crystal structures with emphasis on the arrangements of cations. *Structure and Bonding* (Berlin), **61**, 77–144.

Parthé, E. (1961) Space filling of crystal structures. A contribution to the graphical presentation of geometrical relationships in simple crystal structures. *Zeitschrift für Kristallographie*, **115**, 52–79.

Pauling, L. (1927) The sizes of ions and the structure of ionic crystals. *Journal of the American Chemical Society*, **49**, 765–90.

Pauling, L. (1929) The principles determining the structure of complex ionic crystals. *Journal of the American Chemical Society*, **51**, 1010–26.

Pauling, L. (1960) *The Nature of the Chemical Bond*, 3rd edn, Cornell University Press, Ithaca.

Phillips, J. C. (1970) Ionicity of the chemical bond in crystals. *Reviews of Modern Physics*, **42**, 317–56.

Phillips, J. C. and Van Vechten, J. A. (1969) Dielectric classification of crystal structures, ionization potentials, and band structures. *Physical Review Letters*, **22**, 705–8.

Phillips, J. C. and Van Vechten, J. A. (1970) Spectroscopic analysis of cohesive energies and heats of formation of tetrahedrally coordinated semiconductors. *Physical Review*, **B2**, 2147–60.

Prewitt, C. T. (1982) Size and compressibility of ions at high pressure, in *High-pressure Research in Geophysics*, (eds S. Akimoto and M. H. Manghnani), D. Reidel, Dordrecht, Boston, London, pp. 433–8.

Price, G. D. and Yeomans J. M. (1986) A model for polysomatism. *Mineralogical Magazine*, **50**, 149–56.

Price, G. D. and Yeomans J. M. (1987) Competing interactions and the origins of polytypism, in *Competing Interactions and Microstructures: Statics and Dynamics*, (eds R. LeSar, A. Bishop, and R. Heffner), Springer-Verlag, Berlin, pp. 60–73.

Price, G. D., Parker, S. C. and Leslie, M. (1987) The lattice dynamics and thermodynamics of the Mg_2SiO_4 polymorphs. *Physics and Chemistry of Minerals*, **15**, 181–90.

Price, G. D., Price S. L. and Burdett, J. K. (1982) The factors influencing cation site-preferences in spinels, a new Mendeleyvian approach. *Physics and Chemistry of Minerals*, **8**, 69–76.

Salje, E. (1988) Kinetic rate laws as derived from order parameter theory I: theoretical concepts. *Physics and Chemistry of Minerals*, **15**, 336–48.

Sasaki, S., Fujino, F. and Takéuchi, Y. (1979) X-ray determination of electron-density distributions in oxides, MgO, MnO, CoO, and NiO, and atomic scattering factors of their constituent atoms. *Proceedings of the Japanese Academy*, **55**, 43–8.

Shannon, R. D. (1976) Revised effective ionic radii and systematic studies of interatomic distances in halides and chalcogenides. *Acta Crystallographica*, **A32**, 751–67.

Shannon, R.D. (1981) Bond distances in sulfides and a preliminary table of sulfide crystal radii, in *Structure and Bonding in Crystals*, vol. 2, (eds M. O'Keeffe and A. Navrotsky), Academic Press, New York, pp. 53–70.

Shannon, R. D. and Prewitt, C. T. (1969) Effective ionic radii in oxides and fluorides. *Acta Crystallographica*, **B25**, 925–46.

Slater, J. C. (1964) Atomic radii in crystals. *Journal of Chemical Physics*, **41**, 3199–204.

Smith, J. V. (1977) Enumeration of 4-connected 3-dimensional nets and classification of framework silicates, I. Perpendicular linkage from simple hexagonal net. *American Mineralogist*, **62**, 703–9.

Smith, J. V. (1978) Enumeration of 4-connected 3-dimensional nets and classification of frameworks silicates, II. Perpendicular and near-perpendicular linkages from 4.8^2, 3.12^2, and 4.6.12 nets. *American Mineralogist*, **63**, 960–9.

Smith, J. V. (1979) Enumeration of 4-connected 3-dimensional nets and classification of framework silicates, III. Combination of helix, and zigzag, crankshaft and saw chains with simple 2D nets. *American Mineralogist*, **64**, 551–62.

Stewart, R. F. and Spackman, M. A. (1981) Charge density distributions, in *Structure and Bonding in Crystals*, vol. 1, (eds M. O'Keeffe and A. Navrotsky), Academic Press, New York, 279–298.

St. John, J. and Bloch, A. N. (1974) Quantum-defect electronegativity scale for non-transition elements. *Physical Review Letters*, **33**, 1095–8.

Thompson, A. B. (1967) Thermodynamic properties of simple solutions, in *Researches in Geochemistry*, vol. 2, (ed P. H. Abelson) John Wiley, New York, pp. 340–61.

Villars, P. (1983) A three dimensional structural stability diagram for 998 binary AB intermetallic compounds. *Journal of Less-Common Metals*, **92**, 215–38.

Villars, P. (1984a) A three dimensional structural stability diagram for 1011 binary AB_2 intermetallic compounds. *Journal of Less-Common Metals*, **99**, 33–43.

Villars, P. (1984b) Three dimensional structural stability for 648 binary AB_3 and 389 A_3B_5 intermetallic compounds. *Journal of Less-Common Metals*, **102**, 199–211.

Villars, P. (1985) A semiempirical approach to the prediction of compound formation for 3846 binary alloy systems. *Journal of Less-Common Metals*, **109**, 93–115.

Wells, A. F. (1984) *Structural inorganic chemistry*, 5th edn, Oxford University Press, London.

Wood, I. G. and Price, G. D. (1992) A simple, systematic method for the generation of periodic, 2-dimensional, 3-connected nets for the description of zeolite frameworks. *Zeolites*, **112**, 320–327.

Zunger, A. (1980) Systematization of the stable crystal structure of all AB-type binary compounds: a pseudopotential orbital-radii approach. *Physical Review*, **B22**, 5839–72.

Zunger, A. (1981) A pseudopotential viewpoint of the electronic and structural properties of crystals, in *Structure and Bonding in Crystals*, vol. 1, (eds M. O'Keeffe and A. Navrotsky), Academic Press, New York, pp. 49–72.

Zunger, A. and Cohen, M. L. (1979) First-principles non-local pseudopotential approach in the density-functional formalism: II. Application to electronic and structural properties of solids. *Physical Review*, **B20**, 4082–108.

CHAPTER TWO
Bond topology, bond valence and structure stability

Frank C. Hawthorne

2.1 Introduction

There are approximately 3500 known minerals, varying from the simple (native iron) to the complex (mcgovernite has approximately 1200 atoms in its unit cell) and spanning a wide range in bond type, from metallic (native gold) through 'covalent' (pyrite), to 'ionic' (halite). Most of us unconsciously divide the minerals into two groups: *rock-forming* and *other*. The rock-forming minerals are quantitatively dominant but numerically quite minor, whereas the other minerals are the reverse. Although this may seem a rather frivolous basis for such a division, there are actually some fairly important features to it that bear further examination.

First let us look at the rock-forming minerals. Each of these is stable over a wide range of conditions (pressure, temperature, pH, etc.). In response to changing external conditions, some adjust their structural state (degree of cation and/or anion ordering over non-equivalent structural sites), chemical composition, and the geometrical details of their crystal structure; others retain the same state of internal order and bulk chemistry, adjusting just their structural geometry. The common factor that is characteristic of all these minerals is that the topological details of their bond networks do not change very frequently, i.e. the mineral retains its structural integrity over a wide range of temperature, pressure, and component activities. By and large, the basic bond networks of these structures tend to be quite simple, and we usually follow changes in conditions of equilibration via changes in structural state and/or bulk composition.

Similar considerations for the *other* minerals lead to a totally different set of generalizations. Each of these minerals tends to be stable over a very limited range. In response to changing external conditions, these minerals usually break down and form new phases. Thus, the topological characteristics of the bond networks change very rapidly, and this is complicated by the fact that these structures tend to be very complex. Further difficulties arise from the widespread and varied structural role of (H_2O), a component that has not been well understood in the past.

The Stability of Minerals. Edited by G. D. Price and N. L. Ross.
Published in 1992 by Chapman & Hall, London. ISBN 0 412 44150 0

Progress in understanding the behaviour of the other group has been hindered in the past by the lack of a standard approach to the problem. With the rock-forming minerals, we know what to do. We make field observations, analyse the minerals to determine their bulk composition, determine their crystal structure and structural state, measure the thermodynamic properties, do phase-equilibria studies; the result is a fairly good understanding of their behaviour in geological processes, and a good basis for developing computational methods for structure–property calculations. With the other minerals, the complexity of the problem (e.g. rocks containing 50 or so minerals in an obviously non-equilibrium association) defies complete analysis along standard lines.

Extensive field observations have shown that there is consistency of mineral occurrence in these complex environments. Systematic work has shown that consistent crystallization/alteration sequences of minerals can be recognized (e.g. see Fisher, 1958, for an analysis of pegmatitic phosphate minerals), but the geochemical thread linking these minerals together was not apparent. The increase in speed and power of crystallographic techniques provided new impetus, and work by Moore (1973; 1982) has shown that general features of phosphate paragenesis are paralleled by structural trends in the constituent minerals. Since then, efforts have been made to generalize these ideas, introduce quantitative arguments and provide some sort of theoretical underpinnings to this approach. Here, we will develop the important basic ideas and show how they may be used to understand the behaviour of structurally complex minerals.

2.2 Structures as graphs

One of the problems with thinking about structures is that our normal representations of a crystal come in two forms:

1. a list of atom coordinates with unit cell and symmetry information; this is the representation used by the physicist;
2. a view of the structure based on assumptions as to which atom is bonded to which; this is the representation of the chemist.

Using representation (1), we can do all sorts of structure–property calculations, provided that we have appropriate potentials and sufficient computing power; the problem with this is that it cannot be used for many complex minerals, and offers little or no intuitive feel for the behaviour of minerals in geological processes. Using representation (2), we can make qualitative arguments à la Pauling's rules, but we do not have a quantitative expression of the important features of a structure. It is with these problems in mind that graph theory can help us.

Consider the four atoms shown in Figure 2.1, in which the lines represent

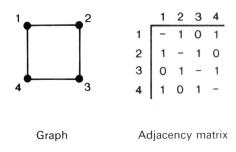

Figure 2.1 A hypothetical molecule consisting of four atoms (•) joined by chemical bonds (———); as drawn, this is a labelled graph (left). An algebraic representation of this graph is the adjacency matrix (right).

chemical bonds between the atoms. This representation, a series of points joined by a series of lines, is the visual representation of a graph. Formally, we may define a graph as a non-empty set of elements, V(G), called vertices, and a non-empty set of unordered pairs of these vertices, E(G), called edges. If we let the vertices of the graph represent atoms (as in Figure 2.1) or groups of atoms, and the edges of the graph represent chemical bonds (or linkages between groups of atoms), then our graph may represent a molecule.

However, we need some sort of digital representation of this graph, something that we can manipulate algebraically. To do this, we introduce an algebraic representation of the graph in the form of a matrix (Fig. 2.1). Each column and row of the matrix is associated with a specific (labelled) vertex, and the corresponding matrix entries denote whether or not two vertices are adjacent, that is joined by an edge. If the edges of the graph are weighted in some form such that the matrix elements denote this weighting, then this matrix is called the *adjacency matrix*. Thus the adjacency matrix is a digital representation of the graph, which is in turn an analogue representation of the structure. The adjacency matrix does not preserve the geometrical features of the structure; information such as bond angles is lost. However, it does preserve information concerning the *topological* features of the bond network, with the possibility of carrying additional information concerning the strengths (or orders) of chemical bonds. Thus, we have a way of quantifying the topological aspects of the bond network of a molecular group. It remains to determine the significance of this information. To do this, we will now examine some of the connections that have recently developed between contemporary theories of chemical bonding and topological (or graphical) aspects of structure. I shall only sketch the outlines of the molecular orbital arguments, except where they serve to emphasize the equivalence or similarity between energetics of bonding and topological aspects of structure. Excellent reviews are given by Burdett (1980), Hoffman (1988), and Albright *et al.* (1985).

2.3 Topological aspects of molecular orbital theory

2.3.1 Molecules

Molecules are built from atoms, and a reasonable first approach to the electronic structure and properties of molecules is to consider a molecule as the sum of the electronic properties of the constituent atoms, as modified by the interaction between these atoms. The most straightforward way of doing this is to construct the molecular orbital wave function from a *Linear Combination of Atomic Orbitals*, the LCAO method of the chemist and the tight-binding method of the physicist. These molecular orbital wave functions are eigenstates of some (unspecified) effective one-electron Hamiltonian, H^{eff}, that we may write as:

$$H^{\text{eff}} \psi = E \psi \tag{1}$$

where E is the energy (eigenvalue) associated with ψ, and the LCAO molecular orbital wave function is written as

$$\psi = \sum_i c_i \phi_i \tag{2}$$

where $\{\phi_i\}$ are the valence orbitals of the atoms of the molecule, and c_i is the contribution of a particular atomic orbital to a particular molecular orbital.

The total electron energy of the state described by this wave function may be written as

$$E = \frac{\int \psi^* H^{\text{eff}} \psi \, d\tau}{\int \psi^* \psi \, d\tau} = \frac{\langle \psi | H^{\text{eff}} | \psi \rangle}{\langle \psi | \psi \rangle} \tag{3}$$

in which the integration is over all space. Substitution of (2) into (3) gives

$$E = \frac{\sum_i \sum_j c_i c_j \langle \phi_i | H^{\text{eff}} | \phi_j \rangle}{\sum_i \sum_j c_i c_j \langle \phi_i | \phi_j \rangle} \tag{4}$$

This equation may be considerably simplified by various substitutions and approximations:

1. The term $\langle \phi_i | \phi_j \rangle$ is the overlap integral between atomic orbitals on different atoms; we will denote this as S_{ij}, and note that it is always ≤ 1; when $i=j$, $\langle \phi_i | \phi_j \rangle = 1$ for a normalized (atomic) basis set of orbitals.
2. We write $\langle \phi_i | H^{\text{eff}} | \phi_i \rangle = H_{ii}$; this is called the Coulomb integral, and represents the energy of an electron in orbital ϕ_i; it can be approximated by the orbital ionization potential.

3. We write $\langle \phi_i | H^{\text{eff}} | \phi_j \rangle = H_{ij}$; it represents the interaction between orbitals ϕ_i and ϕ_j, and is called the *resonance integral*; it can be approximated by the Wolfsberg–Helmholtz relationship $H_{ij} = K S_{ij}(H_{ii} + H_{jj})/2$ (Gibbs et al., 1972).

The molecular orbital energies are obtained from equation (4) via the variational theorem, minimizing the energy with respect to the coefficients c_i. The most familiar form is the following *secular determinant* equation, the eigenvalues (roots) of which give the molecular orbital energy levels:

$$|H_{ij} - S_{ij}E| = 0 \quad (5)$$

Here we will consider the Hückel approximation (Trinajstic, 1983), as this most simply demonstrates the topological content of this approach. In the Hückel approximation, all H_{ii} values for the $p\pi$ orbitals are set equal to α, all H_{ij} are set equal to β, and all $S_{ij}(i \neq j)$ are set equal to zero. As a very simple example, let us consider cyclobutadiene (Fig. 2.2). Writing out the secular determinant equation in full, we get:

$$\begin{vmatrix} \alpha - E & \beta & 0 & \beta \\ \beta & \alpha - E & \beta & 0 \\ 0 & \beta & \alpha - E & \beta \\ \beta & 0 & \beta & \alpha - E \end{vmatrix} = 0 \quad (6)$$

Let us compare the matrix entries in equation (6) with the cyclobutadiene structure of Figure 2.2. The diagonal terms $(\alpha - E)$ can be thought of as the 'self-interaction' terms; in the absence of any off-diagonal β terms, there are no chemical bonds formed, and the roots of the equation are the energies of the electrons in the atomic orbitals themselves. When chemical bonding occurs,

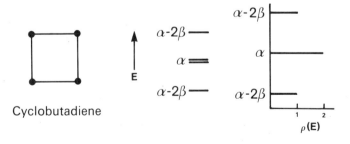

Figure 2.2 The cyclobutadiene molecule (left); to the right are the four roots of equation (6), the electron energy levels expressed in the usual form (centre) and as a density of states form (right).

these energies are modified by the off-diagonal β terms. Thus, when two atoms are bonded together (i.e. atoms 1 and 2 in Figure 2.2), there is non-zero value at this particular (1, 2) entry in the secular determinant; when two atoms are not bonded together (i.e. atoms 1 and 3 in Figure 2.2), then the corresponding determinant entry (1, 3) is zero. Referring back to Figure 2.1, we see that this description is very similar to the adjacency matrix of the corresponding graph. If we use the normalized form of Hückel theory, in which β is taken as the energy unit, and α is taken as the zero-energy reference point (Trinajstic, 1983), then the determinant of equation (6) becomes identical to the corresponding adjacency matrix. The eigenvectors of the adjacency matrix are identical to the Hückel molecular orbitals. Hence it is the *topological (graphical) characteristics* of a molecule, rather than any geometrical details, that determine the form of the Hückel molecular orbitals. For cyclobutadiene, the orbital energies found from the secular determinant (i.e. the four roots of equation (6)) are $E = \alpha + 2\beta, \alpha(\times 2)$, and $\alpha - 2\beta$. These are shown in Figure 2.2 both in a conventional energy representation, and as a *density of states* diagram; the latter shows the 'density' of electrons as a function of electron energy.

2.3.2 Molecular building blocks

When we consider very complicated problems, we like to resolve them into simple (usually additive) components that are easier to deal with. Molecular and crystal structures are no exception; we recognize structural building blocks, and build hierarchies of structures using these 'molecular bricks'. Let us consider this from a graph theoretic point of view.

A graph G' is a subgraph of a graph G if the vertex- and edge-sets $V(G')$ and $E(G')$ are subsets of the vertex- and edge-sets $V(G)$ and $E(G)$; this is illustrated in Figure 2.3. We may express any graph as the sum of a set of subgraphs. The eigenvalues of each subgraph G' are a subset of the eigenvalues of the main graph G, and the eigenvalues of the main graph are the sum of the eigenvalues of all the subgraphs. In the last section, we saw that the eigenvalues of an adjacency matrix are identical to the Hückel molecular orbitals. Now let us consider the construction of large molecules from smaller building blocks. This provides us with a convenient visual way of analysing the connectivity of our molecule, and of relating molecules together. But this is not all. The fact that the eigenvalues of the graphs of our building blocks are contained in the eigenvalues of the graph of the complete molecule indicates that we may consider our building blocks as *orbital* or *energetic building blocks*. Thus, there is an *energetic basis* for the use of fundamental building blocks in the representation and hierarchical analysis of complex structures.

2.3.3 Crystals

So far, we have been considering molecules; however, crystals are far more interesting, particularly if they are minerals. We can envisage constructing a

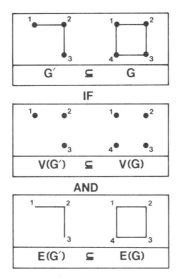

Figure 2.3 The relationship between a graph G and a subgraph G' expressed in terms of the relevant vertex and edge sets.

crystal from constituent molecular building blocks, in this way considering the crystal as a giant molecule. However, it is not clear what influence translational periodicity will have on the energetics of this conceptual building process. Consequently, we will now examine the energetic differences between a molecule and a crystal.

Consider what would happen if we were able to solve the secular determinant equation (6) for a giant molecule; the results are sketched in Figure 2.4. Solution of the secular determinant will give a very large number of molecular orbital energies, and obviously their conventional representation solely as a function of energy is not very useful; such results are more usefully expressed as a density of states diagram (Fig. 2.4), in which the electron occupation of a specific energy interval (band) is expressed as a function of orbital energy.

So what happens in a crystal which has translational symmetry? Obviously, we cannot deal with a crystal using exactly the same sort of calculation, as

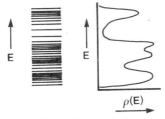

Figure 2.4 The electron energy levels for a giant molecule expressed in the usual way (left) and as a density of states (right).

there are approximately Avogadro's number of atoms in a (macroscopic) crystal, far beyond any foreseeable computational capability. Instead, we must make use of the translational symmetry to reduce the problem to a manageable size. We do this by using what are called Bloch orbitals (Ziman, 1965), in which the orbital content of the unit cell is constrained to the periodicity of the crystal. The secular determinant is solved at a representative set of points within the Brillouin zone (the *special points method*), giving a (hopefully) representative sampling of the orbital energy levels that may be used as the basis of a density of states diagram; this may be smoothed to give the usual density of states diagram. The total orbital energy can then be calculated by integrating the electronic energy density of states up to the Fermi level.

We may summarize the differences between a molecule and a crystal as follows: in a molecule, there is a discrete set of orbital energy levels; in a crystal, these levels are broadened into bands whose occupancies as a function of energy is represented by the corresponding electronic energy density of states.

2.3.4 *The method of moments*

The traditional method for generating the electronic energy density of states has little intuitive connection to what we usually think of as the essential features of a crystal structure, the relative positions of the atoms and the disposition of the chemical bonds. In this regard, Burdett *et al.* (1984) have come up with a very important method of deriving the electronic energy density of states using the method of moments. Here, I will give a brief outline of the method; interested readers should consult the original paper for mathematical details, and are also referred to Burdett (1986; 1987) for a series of applications in solid-state chemistry.

When we solve the secular determinant (equation (6)), we diagonalize the Hamiltonian matrix. The trace of this matrix may be expressed as follows:

$$\text{Tr}(H^n) = \sum_i \sum_{j,k...n} H_{ij} H_{jk} ... H_{ni}. \tag{7}$$

A topological (graphical) interpretation of one term in this sum is shown in Figure 2.5. Each H_{ij} term is the interaction integral between orbitals i and j, and hence is equal to β (if the atoms are bonded) or zero (if the atoms are not bonded, or if $i=j$ when $\alpha=0$). Thus a single term $\{H_{ij}H_{ik}...H_{1n}\}$ in equation (7) is non-zero only if all H_{ij} terms are non-zero. As the last H_{ij} term is the interaction between the nth orbital and the first orbital, the $\{H_{ij}H_{jk}...H_{ni}\}$ term represents a closed path of length n in the graph of the orbitals (molecule). Thus in Figure 2.5, the term $\{H_{ij}H_{jk}H_{kl}H_{li}\}$ represents the clockwise path of length 4 around the cyclobutadiene pπ orbitals. Thus the complete sum of equation (7) represents all circuits of length n through the graph of the (orbital structure of) the molecule.

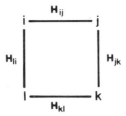

Figure 2.5 Topological interpretation of a single term in the sum of equation 7; for each orbital i, the non-zero terms are a series of circuits of length n with orbital i as the origin; the term shown here has n=4 (for cyclobutadiene).

The trace of the Hückel matrix remains invariant under diagonalization, and thus

$$\mathrm{Tr}(H^n) = \mathrm{Tr}(E^n) = \mu_n \qquad (8)$$

where E is the diagonal matrix of eigenvalues (energy levels) and μ_n is the nth *moment* of E, formally denoted by

$$\mu_n = \sum_i E_i^n. \qquad (9)$$

The collection of moments $\{\mu_n\}$ may be inverted (see Burdett *et al.*, 1984 for mathematical details) to give the density of states. As we can evaluate $\mathrm{Tr}(H^n)$ directly from the topology of the orbital interactions (bond topology), we thus derive the electronic energy density of states *directly* from the bond topology. Of course, we have already shown that this is the case by demonstrating the equivalence of the secular determinant and the adjacency matrix of the molecule. However, this method of moments generalizes quite readily to infinite systems (i.e. crystals).

For an infinite system, we can define the nth moment of E as

$$\mu_n = \int E^n \rho(E) \, dE \qquad (10)$$

where $\rho(E)$ is the density of states. In principle, the moments may be evaluated as before, and inverted to give the electronic energy density of states. Thus, we see in principle the topological content of the electronic energy density of states in an infinite system, which in turn emphasizes the energetic content of a topological (graphical) representation of (periodic) structure. However, we can go further than this. Burdett (1986) has shown that the energy difference between two structures can be expressed in terms of the first few disparate moments of their respective electronic energy density of states. Thus, when comparing two structures, the important energetic terms are the most *local topological differences* between the structures. Putting this in structural terms, the important energetic terms involve differences in coordination number and differences in local polyhedral linkage. Furthermore, in structures with bonds of different strengths, each edge of each path (walk) that contributes to each moment will be weighted according to the value of the strength (resonance integral) of the bond defining that edge. Thus, strongly bonded paths through the structure will contribute most to the moments of the electronic energy density of states. The most important energetic features of a structure are thus not only the local connectivity, but the local connectivity of the strongly bonded coordination polyhedra in the structure. This provides an energetic justification for the hypothesis that will be introduced later on, that structures may be ordered according to the polymerization of the more strongly bonded coordination polyhedra (Hawthorne, 1983).

2.4 Topological aspects of crystal chemistry

There are some empirical rules that (sometimes weakly) govern the behaviour of minerals and inorganic crystals, rules that date back to early work on the modern electronic theory of valence and the structure of crystals. The most rigorous rule is that of electroneutrality: *the sum of the formal charges of all the ions in a crystal is zero*. Although we tend to take this rule for granted, it is an extremely powerful constraint on possible chemical variations in crystals. The other rules grew out of observations on a few mineral and inorganic structures. In 1920 the idea was first put forward that atoms have a certain size, and a table of atomic radii was produced; in 1927 the idea of coordination number was introduced, and silicate minerals were considered as polymerizations of coordination polyhedra. These ideas were refined by Pauling (1929; 1960), who systematized them into his well-known rules for the behaviour of 'complex ionic crystals':

1. A coordination polyhedron of anions is formed about each cation, the cation–anion distance being determined by the radius sum, and the ligancy (coordination number) of the cation being determined by the radius ratio.
2. The strength of a bond from a cation to an anion is equal to the cation charge divided by the cation coordination number; in a stable (ionic) structure, the formal valence of each anion is approximately equal to the sum of the incident bond-strengths.
3. The presence of shared faces and edges between coordination polyhedra decreases the stability of a structure; this effect is large for cations of large valence and small ligancy.
4. In a crystal containing different cations, those with large valence and small coordination number tend not to share polyhedral elements with each other.

These rules put some less rigorous constraints on the behaviour of mineral structures, constraints that are traditionally associated with the ionic model of the chemical bond; they allow us to make the following statements about the structure and chemistry of minerals:

1. the formula is electrically neutral;
2. we may make (weak) predictions of likely coordination numbers from the radius ratio rules;
3. we can make fairly good (<0.02Å) predictions of mean bond lengths in crystals.

Compared with the enormous amount of structural and chemical data available for minerals, our predictive capabilities concerning this information are limited in the extreme. The following questions are pertinent in this regard:

1. Within the constraint of electroneutrality, why do some stoichiometries occur while others do not?
2. Given a specific stoichiometry, what is its bond connectivity (bond topology)?
3. Given a specific stoichiometry and bond connectivity, what controls the site occupancies?
4. What is the role of 'water' of hydration in minerals?

These are some of the basic questions that need answering if we are going to understand the stability of minerals and their role in geological processes from a mechanistic point of view.

2.4.1 Pauling's rules and bond topology

Here I will briefly consider each of Pauling's rules, and indicate how they each relate to the topology of the bond connectivity in structures.

1. The mean interatomic distance in a coordination polyhedron can be determined by the radius sum. This point has been extensively developed up to the present time (Shannon, 1976; Baur, 1987), together with consideration of additional factors that also affect mean bond lengths in crystals (Shannon, 1975). The first rule also states that the coordination number is determined by the radius ratio. This works reasonably well for small high-valence cations, but does not work well for large, low-valence cations. For example, inspection of Shannon's (1976) table of ionic radii shows Na radii listed for coordination numbers from [4] to [12] with oxygen ligands, whereas a radius ratio criterion would indicate that any cation can have (at the most) two coordination numbers for a specific anion. It is important to note that the coordination number of an atom is one of the lowest moments of the electronic energy density of states.
2. This is also known (rather unfortunately) as the electrostatic valence rule.

It has been further extended by Baur (1970; 1971), who developed a scheme for predicting individual bond lengths in crystals given the bond connectivity, and by Brown and Shannon (1973), who quantitatively related the length of a bond to its strength (bond-valence). The latter scheme has proved a powerful *a posteriori* method of examining crystal structures for crystal chemical purposes. This rule relates strongly to the local connectivity of strong bonds in a structure, and again involves significant low-order moments of the electronic energy density of states.

3.,4. Both of these rules again relate to the local connectivity in a structure, and strongly affect the important low-order moments, both by different short paths resulting from different local bond topologies, and from differences in anion coordination numbers.

The bottom line is that Pauling's rules can all be intuitively related to bond topology and its effect on the low-order moments of the electronic energy density of states.

2.4.2 *Ionicity, covalency, and bond topology*

Pauling's rules were initially presented as *ad hoc* generalizations, rationalized by qualitative arguments based on an electrostatic model of the chemical bond. This led to an association of these rules with the ionic model, and there has been considerable criticism of the second rule as an 'unrealistic' model for bonding in most solids. Nevertheless, these rules have been too useful to disgard, and in various modifications, continue to be used to the present day. Clearly, their proof is in their applicability to real structures rather than in the details of somewhat vague ionic arguments (Burdett and McLarnan, 1984). In this regard, I will use the terms cation and anion to denote atoms that are of lesser or greater electronegativity, respectively; here, these terms carry no connotation as to models of chemical bonding.

For the past 15 years, Gibbs, Tossell, and coworkers (e.g. Gibbs, 1982; Tossell and Gibbs, 1977) have approached the structure of minerals from a molecular-orbital viewpoint, and have made significant progress in both rationalizing and predicting geometrical aspects of structures. In particular, they have shown that many of the geometrical predictions of Pauling's rules can also be explained by molecular-orbital calculations on small structural fragments. Burdett and McLarnan (1984) show how the same predictions of Pauling's rules can be rationalized in terms of band-structure calculations, again focusing on the covalent interactions, but doing so for an infinite structure. It is interesting to note how these two approaches parallel the arguments given previously concerning the relationship between bond topology and energetics:

1. The energy of a molecular fragment is a function of its graphical/topological characteristics via the form of the secular determinant.
2. The electronic energy density of states of a continuous structure can be

expressed in terms of the sum of the moments of the energy density of states, which is related to the topological properties of its bond network.

The underlying thread that links these ideas together is the topology of the bond network via its effect on the energy of the system. This conclusion also parallels our earlier conclusion that all of Pauling's rules relate to the graphical/topological characteristics of the bond network of a crystal.

We can take this argument a little further. Consider two (dimorphic) structures of the same stoichiometry but different atomic arrangement. As the chemical formulae of the two structures are the same, the atomic components of the energy of each structure must be the same, and the difference in energy between the two structures must relate *completely* to the difference in bond connectivity. This 'general principles' argument emphasizes the importance of bond topology in structural stability, and finds more specific expression in the method of moments developed by Burdett *et al.* (1984). Thus we come to the general conclusion that *arguments of ionicity and/or covalency in structure are secondary to the overriding influence of bond topology on the stability and energetics of structure.*

2.5 Bond-valence theory

From Pauling's second rule and its more quantitative generalization by Brown and Shannon (1973), Brown (1981) has developed a coherent approach to chemical bonding in inorganic structures. Although the empirical bond-valence curves are now widely used, the general ideas of bond-valence theory have not yet seen the use that they deserve. Consequently, I shall review these ideas in detail here, particularly as they can be developed further to deal in a very simple way with the complex hydroxy-hydrated oxysalt minerals.

2.5.1 Bond-valence relationships

According to Pauling's second rule, bond-strength, p, is defined as

$$p = \text{cation valence/cation coordination number} = Z/\text{cn}. \tag{11}$$

If we sum the bond-strengths around the anions, the second rule states that the sum should be approximately equal to the magnitude of the anion valence:

$$\sum_{\text{anion}} p \sim |Z_{\text{anion}}|. \tag{12}$$

Table 2.1 shows the results of this procedure for forsterite and diopside; as is apparent, the rule works well for forsterite but poorly for diopside, with deviations of 0.40 v.u. (valence units ($\sim 20\%$ for the O(2) anion in diopside). Many people have noticed the correlation between deviations from Pauling's

Table 2.1 Pauling bond–strength tables for forsterite and diopside

	Forsterite			
	Mg(1)	Mg(2)	Si	Sum
O(1)	$(\frac{1}{3})^{\times 2}$	$(\frac{1}{3})^{\times 2}$	1	2.00
O(2)	$(\frac{1}{3})^{\times 2}$	$(\frac{1}{3})^{\times 2}$	1	2.00
O(3)	$(\frac{1}{3})^{\times 2}$	$(\frac{1}{3})^{\times 2}$	$1^{\times 2}$	2.00
Sum	2	2	4	

	Diopside			
	Ca	Mg	Si	Sum
O(1)	$(\frac{1}{4})^{\times 2}$	$(\frac{1}{3})^{\times 2}$	1	1.92
O(2)	$(\frac{1}{4})^{\times 2}$	$(\frac{1}{3})^{\times 2}$	1	1.58
O(3)	$(\frac{1}{4})^{\times 2}$		$1^{\times 2}$	2.50
Sum	2	2	4	

second rule and bond-length variations in crystals (see Allmann, 1975), and have parameterized this variation for specific cation–anion bonds. For such schemes, I use the term *bond-valence*, in contrast to the Pauling scheme for which I use the term *bond-strength*; this is merely a convenient nomenclature without any further significance.

The most useful bond-valence–bond-length relationship was introduced by Brown and Shannon (1973). They expressed bond-valence, s, as a function of bond-length, R, in the following way:

$$s = s_0 |R/R_0|^{-N} \tag{13}$$

where s_0, R_0 and N are constants characteristic of each cation–anion pair, and were derived by fitting such equations to a large number of well-refined crystal structures under the constraint that the valence-sum rule work as closely as possible. Values of s_0, R_0 and N are given in Table 2.2. Table 2.3 shows the application of this relationship to the forsterite and diopside structures. Although the valence-sum rule is not obeyed exactly, comparison with Table 2.1 shows great improvement over the simple Pauling bond-strength model.

In equation (13), R_0 is nominally a refined parameter, but is obviously equal to the grand mean bond-length for the particular bond pair and cation coordination number under consideration; also s_0 is equal to the Pauling bond-strength. Thus $(R/R_0) \sim 1$, and s_0 is actually a scaling factor that ensures that the sum of the bond-valences around an atom is approximately equal to the magnitude of its valence.

Table 2.2 Individual bond-valence parameters for some geologically relevant cation–anion pairs

Cation	s_0(v.u.)	R_0(Å)	N
H^+	0.50	1.184	2.2
Li^+	0.25	1.954	3.9
Be^{2+}	0.50	1.639	4.3
B^{3+}	1.00	1.375	3.9
Na^+	0.166	2.449	5.6
Mg^{2+}	0.333	2.098	5.0
Al^{3+}	0.50	1.909	5.0
Si^{4+}	1.00	1.625	4.5
P^{5+}	1.25	1.534	3.2
S^{6+}	1.50	1.466	4.0
K^+	0.125	2.833	5.0
Ca^{2+}	0.25	2.468	6.0
Sc^{3+}	0.50	2.121	6.0
Ti^{4+}	0.666	1.952	4.0
V^{5+}	1.25	1.714	5.1
Cr^{6+}	1.50	1.648	4.9
Mn^{2+}	0.333	2.186	5.5
Fe^{3+}	0.50	2.012	5.3
Fe^{2+}	0.333	2.155	5.5
Co^{2+}	0.333	2.118	5.0
Cu^{2+}	0.333	2.084	5.3
Zn^{2+}	0.50	1.947	5.0
Ga^{3+}	0.75	1.837	4.8
Ge^{4+}	1.0	1.750	5.4
As^{5+}	1.25	1.681	4.1

Table 2.3 Bond-valence tables for forsterite and diopside

	Forsterite			
	Mg(1)	Mg(2)	Si	Sum
O(1)	$0.341^{\times 2}$	$0.289^{\times 2}$	1.019	1.938
O(2)	$0.348^{\times 2}$	$0.370^{\times 2}$	0.954	2.042
O(3)	$0.302^{\times 2}$	0.266 0.356	$0.917^{\times 2}$	1.895
Sum	1.982	1.940	3.915	

	Diopside			
	Ca	Mg	Si	Sum
O(1)	$0.316^{\times 2}$	$0.318^{\times 2}$ $0.359^{\times 2}$	1.055	2.048
O(2)	$0.329^{\times 2}$	$0.365^{\times 2}$	1.092	1.786
O(3)	$0.205^{\times 2}$ $0.234^{\times 2}$		0.845	2.245
Sum	2.168	2.084	3.953	

Let us suppose that there is a delocalization of charge into the bonds, together with a reduction in the charge on each atom. For an A–B bond, let the residual charges change by $Z_A p_A$ and $Z_B p_B$ respectively. The (Pauling) bond-strength [=scaling constant s_0 in equation (13)] is given by $Z_A p_A / cn$, where cn is the coordination number of atom A. Inserting these values into equation (13) and summing over the bonds around B gives:

$$\sum_B = p_A \sum s_0 |R/R_0|^{-N} = p_B |Z_B|. \quad (14)$$

If $p_A \sim p_B$, these terms cancel and the bond-valence equation works, provided the relative delocalization of charge from each formally ionized atom is not radically different. Thus the bond-valence equation will apply from 'very ionic' to 'very covalent' situations.

Brown and Shannon (1973) also introduce what they call *universal curves* for bond-valence relationships. These are parameterized such that a single curve applies to all the atoms of an isoelectronic series in the periodic table, and are of the form

$$s = |R/R_1|^{-n}. \quad (15)$$

The constants R_1 and n are given in Table 2.4, and the curves are illustrated in Figure 2.6. These work as well as the individual curves of equation (13), and are often more convenient to use as it is not necessary to distinguish between cations of the same isoelectronic series (e.g. Si and Al, or Mg and Al). A new form of the bond-valence relationship is also given by Brown (1981), but this does not affect the basic ideas of the theory itself.

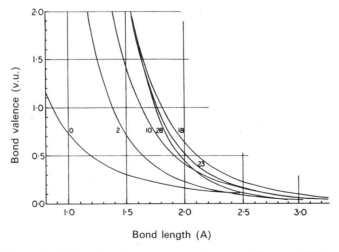

Figure 2.6 Universal bond-valence curves (Brown and Shannon, 1973).

Table 2.4 Universal bond-valence curves for isoelectronic series

Cations	# of core electrons	R_1(Å)	N_1
H^+	0	0.86	2.17
$Li^+ Be^{2+} B^{3+}$	2	1.378	4.065
$Na^+ Mg^{2+} Al^{3+} Si^{4+} P^{5+} S^{6+}$	10	1.622	4.290
$K^+ Ca^{2+} Sc^{3+} Ti^{4+} V^{5+} Cr^{6+}$	18	1.799	4.483
$Mn^{2+} Fe^{3+}$	23	1.760	5.117
$Zn^{2+} Ga^{3+} Ge^{4+} As^{5+}$	28	1.746	6.050

2.5.2 *The conceptual basis of bond-valence theory*

We start by defining a crystal, liquid, or molecule as a network of atoms connected by chemical bonds. For the materials in which we are interested (i.e. minerals), any path through this network contains alternating cations and anions, and the total network is subject to the *law of electroneutrality*: the total valence of the cations is equal to the total valence of the anions. A bond-valence can be assigned to each bond such that the *valence sum rule* is obeyed: *the sum of the bond-valences at each atom is equal to the magnitude of the atomic valence*. If the interatomic distances are known, then the bond-valences can be calculated from the curves of Brown (1981); if the interatomic distances are not known, then the bond-valences can be approximated by the Pauling bond-strengths.

This far, we are dealing just with formalizations from and extensions of Pauling's rules. Although these ideas are important, they essentially involve *a posteriori* analysis: the structure must be known in detail before we can apply these ideas. This is obviously not satisfactory. We need an *a priori* approach to structure stability if we are to develop any predictive capability. In this regard, Brown (1981) introduced a very important idea that abstracts the basic ideas of bond-valence theory, and associates the resulting quantitative parameters with *ions* rather than with bonds (between specific atom pairs). This means that we can examine what would happen if atoms were to bond together in a specific configuration.

If we examine the bond-valences around a specific cation in a wide range of crystal structures, we find that the values lie within ~20% of the mean value; this mean value is thus characteristic of that particular cation. If the cation only occurs in one type of coordination, then the mean bond-valence for that cation will be equal to the Pauling bond-strength; thus P (phosphorus) always occurs in tetrahedral coordination in minerals, and will hence have a mean bond-valence of 1.25 v.u. (Table 2.2). If the cation occurs in more than one coordination number, then the mean bond-valence will be equal to the weighted mean of the bond-valences in all the observed structures. Thus Fe^{2+}

occurs in various coordinations from [4] to [8]; the tendency is for [4] and [5] to be more common than [7] and [8], and hence the mean bond-valence is 0.40 v.u. As the mean bond-valence correlates with formal charge and cation size, then it should vary systematically through the periodic table; this is in fact the case. Table 2.5 shows these characteristic values, smoothed across the periods and down the groups of the periodic table.

The mean bond-valence of a cation correlates quite well with the *electronegativity*, as shown in Figure 2.7. Conceptually this is not surprising. The electronegativity is a measure of the electrophilic strength (electron-accepting capacity) of the cation, and the correlation with the characteristic bond-valence (Fig. 2.7) indicates that the latter is a measure of the *Lewis acid strength* of the cation. Thus we have the following definition (Brown, 1981):

The Lewis acid strength of a cation = characteristic (bond-)valence
= atomic (formal)valence/mean coordination number

We can define the Lewis base strength of an anion in exactly the same way— as the characteristic valence of the bonds formed by the anion. However, it is notable that the bond-valence variations around anions are much greater than those around cations. For example, the valences of the bonds to O^{2-} vary between nearly zero and 2.0 v.u.; thus in sodium alum $(Na[Al(SO_4)_2(H_2O)_6](H_2O)_6$, Cromer et al., 1967), Na is in [12]-coordination, and the oxygen to which it is bonded receives 0.08 v.u. from the Na–O bond; conversely in CrO_3 (a pyroxene-like structure without any [6]-coordinated cations), one oxygen is bonded only to Cr^{6+} and receives 2.00 v.u. from the Cr–O bond. With this kind of variation, it is not particularly useful to define a Lewis base strength for a simple anion such as O^{2-}. However, the situation is entirely different if we consider *complex oxyanions*. Consider the

Table 2.5 Lewis acid strengths (vu) for cations

Li	0.22	Sc	0.50	Cu^{2+}	0.45
Be	0.50	Ti^{3+}	0.50	Zn	0.36
B	0.88	Ti^{4+}	0.75	Ga	0.50
C	1.30	V^{3+}	0.50	Ge	0.75
N	1.75	V^{5+}	1.20	As	1.02
Na	0.16	Cr^{3+}	0.50	Se	1.30
Mg	0.36	Cr^{6+}	1.50	Rb	0.10
Al	0.63	Mn^{2+}	0.36	Sr	0.24
Si	0.95	Mn^{3+}	0.50	Sn	0.66
P	1.30	Mn^{4+}	0.67	Sb	0.86
S	1.65	Fe^{2+}	0.36	Te	1.06
Cl	2.00	Fe^{3+}	0.50	Cs	0.08
K	0.13	Co^{2+}	0.40	Ba	0.20
Ca	0.29	Ni^{2+}	0.50	Pb^{2+}	0.20

Values taken from Brown (1981), except Pb^{2+} which was estimated from several oxysalt mineral structures.

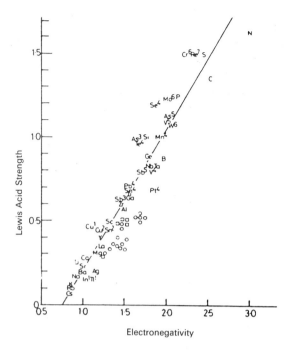

Figure 2.7 Lewis acid strength (mean bond-valence for a specific cation) as a function of cation electronegativity (Brown, 1981).

$(SO_4)^{2-}$ oxyanion shown in Figure 2.8. Each oxygen receives 1.5 v.u. from the central S^{6+} cation, and hence each oxygen of the group needs an additional 0.5 v.u. to be supplied by additional cations. If the oxygen coordination number is $[n]$, then the average valence of the bonds to O^{2-} (exclusive of the S–O bond) is $0.5/(n-1)$ v.u., thus if $n = 2, 3, 4$, or 5, then the mean bond-

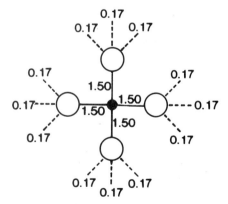

Figure 2.8 Bond-valence structure of the $(SO_4)^{2-}$ oxyanion, with the individual bond-valences shown in vu; • = sulphur, 0 = oxygen.

valencies to the oxygen are 0.50, 0.25, 0.17, or 0.11 v.u., respectively. As all of the oxygens in the $(SO_4)^{2-}$ oxyanion have the same environment, then the average bond-valence received by the oxyanion is the same as the average bond-valence received by the individual oxygens. In this way, we can define the Lewis basicity of an oxyanion. Note that for the $(SO_4)^{2-}$ oxyanion discussed above, the possible average bond-valences are quite tightly constrained (0.50–0.11 v.u.), and we can easily calculate a useful Lewis basicity. Table 2.6 lists Lewis basicities for geologically relevant oxyanions.

These definitions of Lewis acid and base strengths lead to a specific criterion for chemical bonding, the *valence-matching principle*:

The most stable structures will form when the Lewis acid strength of the cation closely matches the Lewis base strength of the anion. We can consider this as the chemical analogue of the handshaking principle in combinatorial mathematics, and the 'kissing' principle in social relationships. As a chemical bond contains two ends, then the ends must match up for a stable configuration to form.

2.5.3 Simple applications of the valence-matching principle

Thenardite (Na_2SO_4 (Hawthorne and Ferguson, 1975a)) illustrates both the utility of defining a Lewis base-strength for an oxyanion, and the working of the valence-matching principle (Fig. 2.8). As outlined above, the bond-valences to O^{2-} vary between 0.17 and 1.50 v.u. Assuming a mean oxygen coordination number of [4], the Lewis base strength of the $(SO_4)^{2-}$ oxyanion is 0.17 v.u., which matches up very well with the Lewis acidity of 0.16 v.u. for Na given in Table 2.5. Thus the Na–(SO_4) bond accords with the valence-matching principle, and thenardite is a stable mineral.

Let us consider the compound Na_4SiO_4. The Lewis basicity of the $(SiO_4)^{4-}$ oxyanion is 0.33 v.u. (Table 2.6); the Lewis acidity of Na is 0.16 v.u. These values do not match up, and thus a stable bond cannot form; consequently Na_4SiO_4 is not a stable mineral.

Let us consider Ca_2SiO_4. The Lewis basicity of $(SiO_4)^{2-}$ is 0.33 v.u. and the Lewis acidity of Ca is 0.29 v.u.; these values match up reasonably well, and Ca_2SiO_4 is the mineral larnite.

Consider $CaSO_4$. The relevant Lewis basicity and acidity are 0.17 and

Table 2.6 Lewis basicities for oxyanions of geological interest

$(BO_3)^{3-}$	0.33	$(CO_3)^{2-}$	0.25
$(SiO_4)^{4-}$	0.33	$(NO_3)^{3-}$	0.12
$(AlO_4)^{3-}$	0.42	$(VO_4)^{3-}$	0.25
$(PO_4)^{3-}$	0.25	$(SO_4)^{2-}$	0.17
$(AsO_4)^{3-}$	0.25	$(CrO_4)^{2-}$	0.17

0.29 v.u., respectively; thus according to the valence-matching principle, we do not expect a stable structure to form. However, the mineral anhydrite is stable, the cation and anion coordination numbers both reducing to allow the structure to satisfy the valence-sum rule (Hawthorne and Ferguson, 1975b). However, anhydrite is not very stable, as it hydrates in the presence of water to produce gypsum, $CaSO_4.2H_2O$; this instability (and as we will show later, the hydration mechanism) is suggested by the violation of the valence-matching principle as considered with Lewis acidity/basicity parameters.

Thus we see the power of the valence-matching principle as a simple way in which we can consider the possibility of cation–anion interactions of interest. It is important to recognize that this is an *a priori* analysis, rather than the *a posteriori* analysis of Pauling's second rule and its various modifications.

2.5.4 *Bond-valence theory as a molecular orbital model*

As noted above, there has been considerable criticism of Pauling's second rule and its more recent extensions, criticisms based on its perception as a description of ionic bonding. This viewpoint is totally wrong. In their original work, Brown and Shannon (1973) emphasize the difference between bond-valence theory and the ionic model. In bond-valence theory, the structure consists of a series of atomic cores held together by valence electrons that are associated with the chemical bonds between atoms; they also explicitly state that the valence electrons may be associated with chemical bonds in a symmetric (covalent) or asymmetric (ionic) position. However, *a priori* knowledge of the electron distribution is not necessary, as it is quantitatively derived from the application of the bond-valence curves to the observed structure. Indeed, Burdett and Hawthorne (1992) show how the bond-valence bond-length relationship may be derived algebraically from a molecular orbital description of a solid in which there is a significant energy gap between the interacting orbitals on adjacent atoms. *Thus we may consider bond-valence theory as a very simple form of molecular orbital theory, parameterized via interatomic distance rather than electronegativity or ionization potential*, and (arbitrarily) scaled via the valence-sum rule.

2.6 Structural hierarchies

The need to organize crystal structures into hierarchical sequences has long been recognized. Bragg (1930) classified the silicate minerals according to the way in which the (SiO_4) tetrahedra polymerize, and this was developed further by Zoltai (1960) and Liebau (1985). The paragenetic implications of this are immediately apparent by comparing this scheme with Bowen's discontinuous reaction series in a fractionating basaltic magma (Fig. 2.9). This suggests that structure has a major influence on mineral paragenesis. Further developments along similar lines are the classification of the aluminium hexafluoride minerals

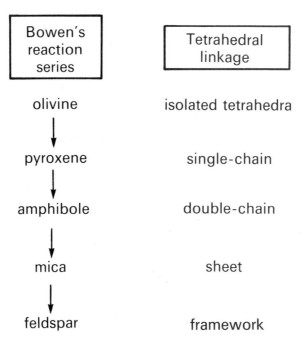

Figure 2.9 Bowen's reaction series shown as a function of the polymerization characteristics of the structures involved.

(Pabst, 1950; Hawthorne, 1984a) and the borate minerals (Christ, 1960; Christ and Clark, 1977), both of which focus on the modes of polymerization of the principal oxyanions. Such an approach to hierarchical organization is of little use in minerals such as the phosphates or the sulphates, in which the principal oxyanion does not self-polymerize. Moore (1984) developed a very successful classification of phosphate minerals, based on the polymerization of divalent and trivalent metal octahedra, and again showed the influence of structure on mineral paragenesis (Moore, 1973). However, all these hierarchical schemes focus on specific chemical classes of minerals. From a paragenetic point of view, this introduces divisions between different chemical classes of minerals, divisions that the natural parageneses indicate to be totally artificial.

2.6.1 *A general hypothesis*

Hawthorne (1983) has proposed the following hypothesis: *structures may be (hierarchically) ordered according to the polymerization of the coordination polyhedra with higher bond-valences.* There are three important points to be made with regard to this idea:

1. We are defining the structural elements by bond-valences rather than by chemistry; consequently, there is no division of structures into different chemical groups except as occurs naturally via the different 'strengths' of the chemical bonding.
2. We can rationalize this hypothesis from the viewpoint of bond-valence theory. First, let us consider the cations in a structure. The cation bond-valence requirements are satisfied by the formation of anion co-ordination polyhedra around them. Thus, we can think of the structure as an array of complex anions that polymerize in order to satisfy their (simple) anion bond–valence requirements according to the valence-sum rule. Let the bond-valences in an array of coordination polyhedra be represented by s_0^i ($i=1, n$) where $s_0^i > s_0^{i+1}$. The valence-sum rule indicates that polymerization can occur when

$$s_0^1 + s_0^i < |V_{\text{anion}}| \qquad (16)$$

and the valence-sum rule is most easily satisfied when

$$s_0^1 + s_0^i = |V_{\text{anion}}| \qquad (17)$$

This suggests that the most important polymerizations involve those coordination polyhedra with higher bond-valences, subject to the constraint of equation (16), as these linkages most easily satisfy the valence-sum rule (under the constraint of maximum volume).

3. Earlier we argued that the topology of the bond network is a major feature controlling the energy of a structure. The polymerization of the principal coordination polyhedra is merely another way of expressing the topology of the bond network, and at the intuitive level, we can recognize an energetic basis for the hierarchical organization of structures according to the details of their polyhedral polymerization.

2.6.2 Structural specifics

Many classifications of complex structures recognize families of structures based on different arrangements of a *fundamental building block* (FBB). This is a tightly bonded unit within the structure, and can be envisaged as the inorganic analogue of a molecule in an organic structure. The FBB is usually a *homo-* or *heteropolyhedral cluster* of coordination polyhedra that have the strongest bond-valence linkages in the structure. The FBB is repeated, usually polymerized, to form the *structural unit*, a complex anionic polyhedral array whose charge is balanced by the presence of large low-valence *interstitial* cations (usually alkalis or alkaline earths). These definitions are illustrated for the mineral tornebohmite in Figure 2.10. The various structural units can be

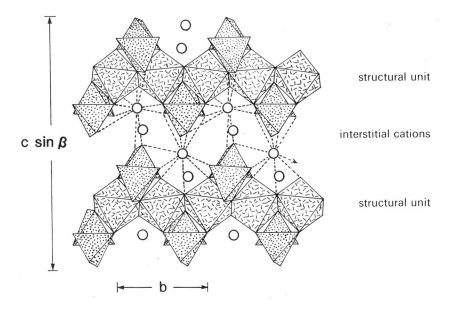

Figure 2.10 The structural unit (as shaded polyhedra) and interstitial cations (as circles) for tornebohmite, $(RE)_2[Al(SiO_4)_2(OH)]$; for clarity, not all bonds from the interstitial cations to the structural unit are shown.

arranged according to the mode of polymerization:

1. unconnected polyhedra
2. finite clusters
3. infinite chains
4. infinite sheets
5. infinite frameworks

Most work has focused on minerals with tetrahedra, triangles, and octahedra as principal components of the structural unit (Hawthorne, 1979; 1984a; 1985a; 1986; 1990; Moore, 1970a, b; 1973; 1974; 1975; 1984), although there has been some notable work (Moore, 1981) on structures with important higher coordinations. Here we will examine minerals based on tetrahedra and octahedra, concentrating specifically on the general stoichiometries $M(T\phi_4)\phi_n$ and $M(T\phi_4)_2\phi_n$ (M = [6]-coordinate, T = [4]-coordinate, ϕ = unspecified anion); these are quantitatively and petrologically most important, as well as showing the most structural diversity.

2.6.3 Unconnected polyhedra

Minerals of this class are listed in Table 2.A1 (Appendix p. 77). The tetrahedra and octahedra are linked together by large low-valence interstitial cations and by hydrogen bonding; thus the (H_2O) group plays a major role in the structural chemistry of this particular class of minerals. The tetrahedral cations are coordinated by oxygens, and the octahedral cations are coordinated predominantly by (H_2O) groups; the exceptions to the latter are khademite and the minerals of the fleischerite group (Table 2.A1), in which the octahedral groups are (Al(H_2O)$_5$F) and (Ge(OH)$_6$), respectively. It is notable that khademite is the only $M(T\phi_4)\phi_n$ structure (in this class) with a trivalent octahedral cation; all other minerals have divalent octahedral cations. Similarly, the fleischerite group minerals are the only $M(T\phi_4)_2\phi_n$ minerals (in this class) with tetravalent octahedral cations; all other minerals of this stoichiometry have trivalent octahedral cations.

There is one very notable generalization that comes from an inspection of Table 2.A1. All $M(T\phi_4)_2\phi_n$ structures have interstitial cations, whereas virtually none of the $M(T\phi_4)\phi_n$ structures have interstitial cations. The one exception to this is struvite, $NH_4[Mg(H_2O)_6][PO_4]$, in which the cation is the complex group $(NH_4)^+$ which links the isolated polyhedra together via hydrogen bonds, as is the case for the rest of the $M(T\phi_4)\phi_n$ unconnected polyhedra structures. For the other cases in which the stoichiometry would suggest that an interstitial cation is needed for electroneutrality reasons (e.g. M^{2+} = Mg, T^{5+} = P, As), we get an acid phosphate or arsenate group instead (e.g. as in phosphorroesslerite and roesslerite, Table 2.A1). Thus for reasons that are as yet unclear, the $M(T\phi_4)\phi_n$ isolated polyhedra structures seemingly will not accept interstitial cations, in contrast to the $M(T\phi_4)_2\phi_n$ structures which always have interstitial cations.

Although this is not the place to go into the details of the hydrogen-bonding schemes in these structures, it should be emphasized that in all structures, the (H_2O) groups participate in an ordered network of hydrogen bonds, and hence are a fixed and essential part of the structure. There has been a tendency in mineralogy to regard (H_2O) as an unimportant component of minerals. Nothing could be further from the truth. Non-occluded (H_2O) is just as important a component as $(SiO_4)^{4-}$ or $(PO_4)^{3-}$ in these minerals, and is the 'glue' that often holds them together.

2.6.4 Finite cluster structures

Minerals of this class are given in Table 2.A2 (Appendix). The different types of clusters found in these minerals are illustrated in Figure 2.11.

In jurbanite, the cluster consists of an octahedral edge-sharing dimer of the form $[Al_2(OH)_2(H_2O)_8]$ and an isolated (SO_4) tetrahedron (Figure 2.11(a)). These two fragments are bound together by hydrogen bonding from the octahedral dimer (donor) to the tetrahedron (acceptor), and hence jurbanite is actually transitional between the unconnected polyhedra structures and the finite cluster structures.

In the $M(T\phi_4)\phi_n$ minerals, the structures of the members of the rozenite group are based on the $[M_2(T\phi_4)_2\phi_8]$ cluster (Fig. 2.11(b)), linked solely by

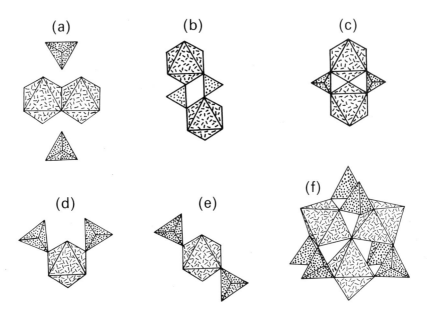

Figure 2.11 Finite polyhedral clusters in $[M(T\phi_4)\phi_n]$ and $[M(T\phi_4)_2\phi_n]$ structures: (a) the $[M_2(T\phi_4)_2\phi_{10}]$ cluster in jurbanite; (b) the $[M_2(T\phi_4)_2\phi_8]$ cluster in the rozenite group minerals; (c) the $[M_2(T\phi_4)_2\phi_7]$ cluster in morinite; (d) the *cis* $[M(T\phi_4)_2\phi_4]$ cluster in roemerite; (e) the *trans* $[M(T\phi_4)_2\phi_4]$ cluster in anapaite, bloedite, leonite, and schertelite; (f) the $[M_3(T\phi_4)_6\phi_4]$ cluster in metavoltine.

hydrogen bonding between adjacent clusters. The morinite structure is based on the $[M_2(T\phi_4)_2\phi_7]$ cluster (Fig. 2.11(c)), linked by interstitial cations as well as inter-unit hydrogen bonds. Hawthorne (1983) derived all possible finite clusters of the form $[M_2(T\phi_4)_2\phi_n]$ with no linkage between tetrahedra and with only corner-sharing between tetrahedra and octahedra. Based on the conjecture that the more stable clusters are those in which the maximum number of anions have their bond-valences most nearly satisfied, four clusters were predicted to be the most stable; two of these are the clusters of Figures 2.11(b),(c).

There is far more structural variety in the $M(T\phi_4)_2\phi_n$ minerals (Table 2.A2). Anapaite, bloedite, leonite and schertelite are based on the simple $[M(T\phi_4)_2\phi_4]$ cluster of Figure 2.11(d), linked by a variety of interstitial cations and hydrogen-bond arrangements. Roemerite is also based on an $[M(T\phi_4)_2\phi_4]$ cluster, but in the *cis* rather than in the *trans* arrangement (Fig. 2.11(e)). Metavoltine is built from a complex but elegant $[M_3(T\phi_4)_6\phi_4]$ cluster (Fig. 2.11(f)) that is also found in a series of synthetic compounds investigated by Scordari (1980; 1981). Again it is notable that the $M(T\phi_4)_2\phi_n$ minerals in this class are characterized by interstitial cations, whereas the bulk of the

$M(T\phi_4)\phi_n$ minerals are not, as was the case for the unconnected polyhedra structures.

The energetic considerations outlined previously suggest that the stability of these finite-cluster structures will be dominated by the topological aspects of their connectivity. Nevertheless, it is apparent from the structures of Table 2.A2 that this is not the only significant aspect of their stability. Figure 2.12 shows the structures of most of the minerals of Table 2.A2. It is very striking that these clusters are packed in essentially the same fashion, irrespective of their nature, and irrespective of their interstitial species. Although a more detailed examination of this point is desirable, its very observation indicates that not only does Nature choose a very small number of fundamental building blocks, but she also is very economical in her ways of linking them together.

2.6.5 Infinite chain structures

A large number of possible $[M(T\phi_4)\phi_n]$ and $[M(T\phi_4)_2\phi_n]$ chains can be constructed from fundamental building blocks involving one or two octahedra and one, two or four tetrahedra. Only a few of these possible chains have actually been found in mineral structures, and these are shown in Figure 2.13. Minerals with structures based on these chains are listed in Table 2.A3 (Appendix).

First let us consider the $M(T\phi_4)\phi_n$ minerals, focusing in particular on the first three chains (Figures 2.13(a),(b),(c)). These are the more important of the chains in this group, and it is notable that:

1. they all have a fairly simple connectivity;
2. there is just one particular chain for each type of connectivity between octahedra.

Thus in the first chain, there is no direct linkage between octahedra; in the second chain, there is corner-sharing between octahedra; in the third chain, there is edge-sharing between octahedra.

The more complex chains of Figures 13(d),(e),(f) are found in a smaller number of (far less common) minerals. In addition, there seems to be a trend emerging, that the most complex structural units tend to occur for the ferric iron sulphates.

If we examine the structural units of the $M(T\phi_4)_2\phi_n$ minerals, the same sort of hierarchy is apparent, perhaps even more so than for the $M(T\phi_4)\phi_n$ minerals. Again the three most common and important types of $[M(T\phi_4)_2\phi_n]$ chains are those which show the simplest connectivity. It is significant that they also show the same distribution of connectivities: there is one chain for each of the possible octahedral–octahedral linkages (i.e. no linkage, corner-sharing, and edge-sharing). As indicated in Table 2.A3, these dominate as structural units in the $M(T\phi_4)_2\phi_n$ minerals, with the remaining more complex chains just being found in a few very rare and complicated ferric iron sulphates.

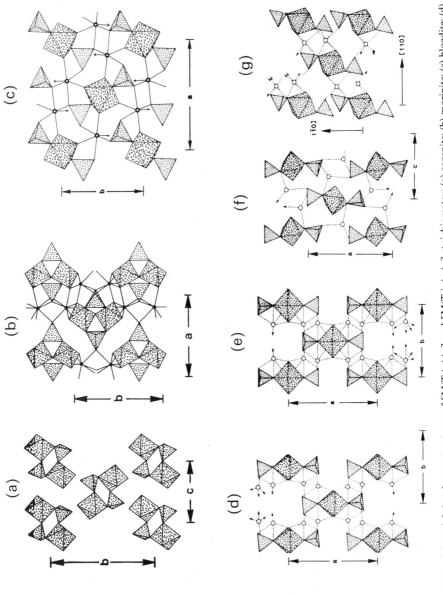

Figure 2.12 Selected finite cluster structures of $[M(T\phi_4)\phi_n]$ and $[M(T\phi_4)_2\phi_n]$ stoichiometry: (a) rozenite; (b) morinite; (c) bloedite; (d), (e) leonite; (f) schertelite; (g) anapaite.

2.6.6 Infinite sheet structures

Minerals of this class are given in Table 2.A4 (Appendix). As the degree of polymerization of the structural unit increases, the number of possible bond-connectivities becomes enormous. However, Nature still seems to favour only a fairly small number of them; these are illustrated in Figures 2.14, 2.15, and 2.16.

There is far more variety in the sheet structures of the $M(T\phi_4)\phi_n$ minerals. Notable in the less-connected structural units is that of minyulite (Fig. 14(b)), which is built by condensation (via corner-linkage between octahedra and tetrahedra) of $[M_2(T\phi_4)_2\phi_7]$ clusters that are the structural unit in morinite (Fig. 2.11(c)). The structures of the laueite, stewartite, pseudolaueite, strunzite, and metavauxite groups (Fig. 2.14(c)–(f)) are based on sheets formed from condensation of the vertex-sharing, octahedral–tetrahedral chains of the sort shown in Figures 2.13(b),(h). The tetrahedra cross-link the chains into sheets, and there is much possible variation in this type of linkage; for more details, see Moore (1975). The five structural groups of these minerals are based on the four sheets of Figures 2.14(c)–(f). These sheets are linked through insular divalent-metal octahedra, either by direct corner-linkage to phosphate tetrahedra plus hydrogen bonding, or by hydrogen bonding alone. There is great potential for stereoisomerism in the ligand arrangement of these linking octahedra, but only the *trans* corner-linkages occur in these groups.

More condensed sheets from the $M(T\phi_4)\phi_n$ minerals are shown in Figure 2.15. Again it is notable that we can identify fragments of more primitive (less condensed) structural units in these sheets. In the whitmoreite group sheet (Fig. 2.15a), we can see both the $[M_2(T\phi_4)_2\phi_7]$ cluster of the morinite structure and the $[M(T\phi_4)_2\phi_8]$ cluster of the rozenite group structures (Fig 2.11(c),(b)). Similarly in the $[M(T\phi_4)\phi]$ sheet of the bermanite and tsumcorite structures (Fig. 2.15(b)), we can see the $[M(T\phi_4)\phi_2]$ chain that is the structural unit in the minerals of the linarite group (Fig. 2.13(e)).

The structural sheet units found in the $M(T\phi_4)_2\phi_n$ minerals are shown in Figure 2.16. Again we see this structural building process, whereby our structural units of more primitive connectivities act as fundamental building blocks for the more condensed structural units of corresponding composition. Thus the $[M(T\phi_4)_2\phi_2]$ sheet found in rhomboclase (Fig. 2.16(a)) is constructed from the *cis* $[M(T\phi_4)_2\phi_4]$ cluster that is the structural unit of roemerite (Fig. 12(e)). Similarly, the $[M(T\phi_4)_2\phi_2]$ sheet of olmsteadite (Fig. 2.16(b)) is based on the *trans* $[M(T\phi_4)_2\phi_4]$ cluster (Fig. 2.12(d)) found in anapaite, bloedite, leonite and schertelite. Note that the rhomboclase and olmsteadite sheets are actually *geometrical isomers* (Hawthorne, 1983).

Analogous relationships are obvious for the $[M(T\phi_4)_2]$ merwinite-type sheet and the $[M(T_2\phi_7)\phi_2]$, bafertisite-type sheet (Fig. 2.16(c),(d)). Both are based on the $[M(T\phi_4)_2\phi_2]$ krohnkite chain of Figure 2.13(g), but in each sheet, the chains are cross-linked in a different fashion. In the merwinite sheet,

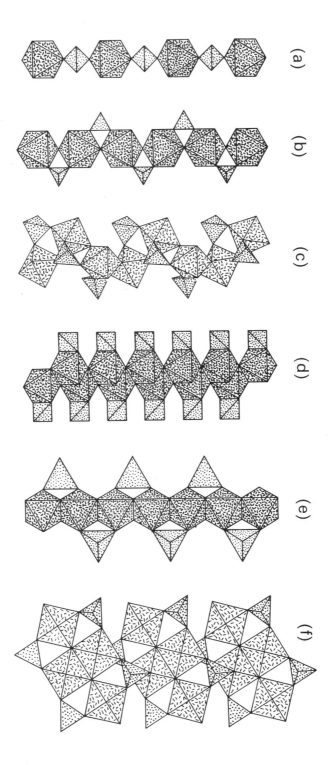

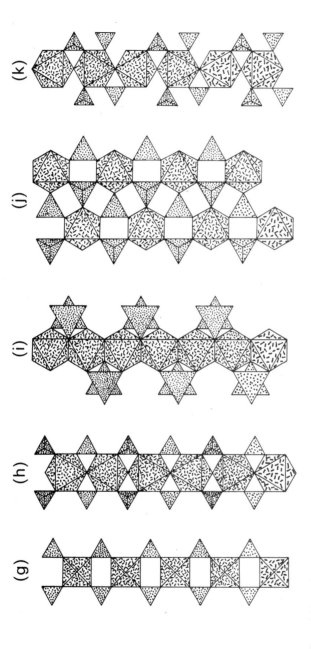

Figure 2.13 Infinite chains in $[M(T\phi_4)\phi_n]$ and $[M(T\phi_4)_2\phi_n]$ structures: (a) the $[M(T\phi_4)\phi_4]$ chain in the chalcanthite group minerals, liroconite and brassite; (b) the $[M(T\phi_4)\phi_3]$ chain found in butlerite, parabutlerite, the childrenite group, and uklonskovite; (c) the $[M(T\phi_4)\phi_3]$ chain in fibbroferrite; (d) the $[M(T\phi_4)\phi]$ chain in chlorotionite; (e) the $[M(T\phi_4)\phi_2]$ chain in the linarite group minerals; (f) the $[M_2(T\phi_4)_4\phi_5]$ chain found in amarantite and hohmannite; (g) the $[M(T\phi_4)\phi_2]$ chain in the krohnkite, talmessite, and fairfieldite groups; (h) the $[M(T\phi_4)_2\phi]$ chain found in tancoite, sideronatrite, the jahnsite and segelerite groups, guildite, and yftisite; (i) the $[M(T\phi_4)_2\phi]$ chain found in the brackebuschite, fornacite, and vauquelinite groups; (j) the $[M(T\phi_4)_2\phi]$ chain in ransomite and krausite; (k) the $[M_2(T\phi_4)_4\phi_5]$ chain found in botryogen.

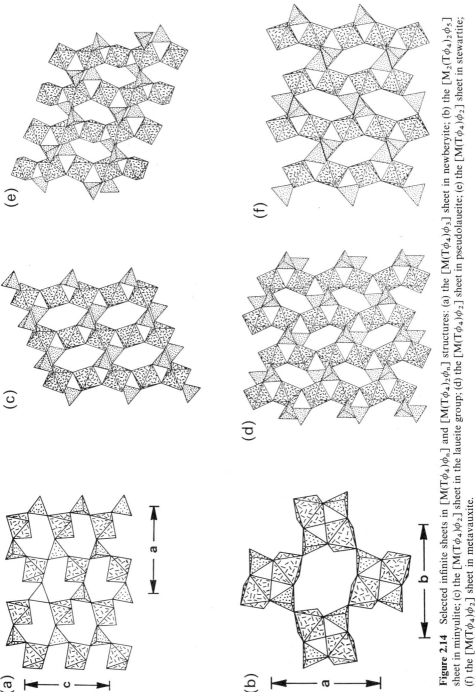

Figure 2.14 Selected infinite sheets in $[M(T\phi_4)\phi_n]$ and $[M(T\phi_4)_2\phi_n]$ structures: (a) the $[M(T\phi_4)\phi_3]$ sheet in newberyite; (b) the $[M_2(T\phi_4)_2\phi_5]$ sheet in minyulite; (c) the $[M(T\phi_4)\phi_2]$ sheet in the laueite group; (d) the $[M(T\phi_4)\phi_2]$ sheet in pseudolaueite; (e) the $[M(T\phi_4)\phi_2]$ sheet in stewartite; (f) the $[M(T\phi_4)\phi_2]$ sheet in metavauxite.

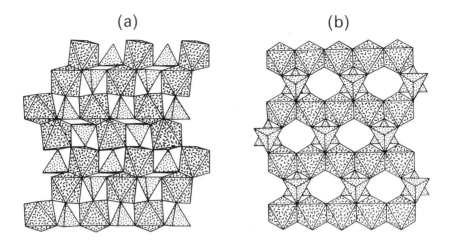

Figure 2.15 Selected infinite sheets in $[M(T\phi_4)\phi_n]$ and $[M(T\phi_4)_2\phi_n]$ structures: (a) the $[M_2(T\phi_4)_2\phi_7]$ sheet in whitmoreite; (b) the $[M(T\phi_4)\phi]$ sheet in tsumcorite and bermanite.

tetrahedra from one chain share corners with octahedra of adjacent chains, with neighbouring tetrahedra pointing in opposite directions relative to the plane of the sheet. In the bafertisite sheet, the $[M(T\phi_4)_2\phi_2]$ chains link by sharing corners between tetrahedra. Thus both sheets are 'built' from the same more primitive structural unit, and these two sheets are in fact *graphical isomers* (Hawthorne, 1983).

2.6.7 Framework structures

Minerals of this class are listed in Table A5 (Appendix). Unfortunately, the topological aspects of the framework structures cannot be easily summarized in a graphical fashion, partly because of their number, and partly because of the complexity that results from polymerization in all three spatial dimensions. Consequently, we will consider just a few examples that show particularly clearly the different types of linkages that can occur.

The structure of bonattite is shown in Figure 2.17(a) Now bonattite is quite hydrated (Table 2.A5), and comparison with the minerals of Table 2.A4 suggests that it should be a sheet structure (cf. newberyite, Table 2.A4). Prominent in the structure are the $[M(T\phi_4)\phi_4]$ chains (Fig. 2.13(a)) that also occur as fragments of the newberyite sheet (Fig. 2.14(a)). In bonattite, adjacent chains are skew and link to form a framework; in newberyite, the chains are parallel, and with the same number of interchain linkages, they link to form sheets rather than a framework. Thus bonattite and newberyite are graphical isomers, and provide a good illustration of how different modes of linking the same fundamental building block can lead to structures of very different connectivities and properties.

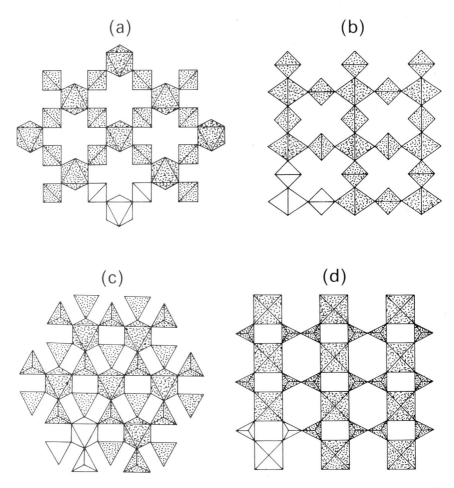

Figure 2.16 Selected infinite sheets in $[M(T\phi_4)\phi_n]$ and $[M(T\phi_4)_2\phi_n]$ structures: (a) the $[M(T\phi_4)_2\phi_2]$ sheet found in rhomboclase; (b) the $[M(T\phi_4)_2\phi_2]$ sheet found in olmsteadite; (c) the $[M(T\phi_4)_2]$ sheet found in the merwinite group and yavapaite; (d) the $[M(T_2\phi_7)\phi_2]$ sheet found in bafertisite.

The structure of titanite is shown in Figure 2.17b; this basic arrangement is found in a considerable number of minerals (Table 2.A5) of widely differing chemistries. The $[M(T\phi_4)\phi]$ framework can be constructed from $[M(T\phi_4)\phi]$ vertex-sharing chains of the sort found in butlerite, parabutlerite, the childrenite group, and uklonskovite (Table 2.A3, Fig. 2.13(b)). The chains pack in a C-centered array and cross-link by sharing corners between octahedra and tetrahedra of adjacent chains. It is notable that this chain is also a fundamental building block of the sheets (Fig. 2.14(c)–(f)) in the laueite, stewartite, pseudolaueite, strunzite, and metavauxite groups (Table 2.A4).

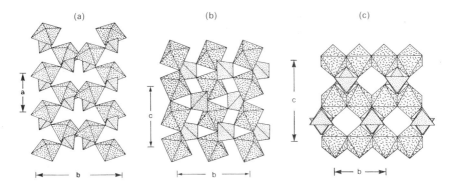

Figure 2.17 Selected framework structures in $[M(T\phi_4)\phi_n]$ and $[M(T\phi_4)_2\phi_n]$ minerals: (a) the $[M(T\phi_4)\phi_3]$ framework structure of bonattite; (b) the $[M(T\phi_4)\phi]$ framework structure of titanite; (c) the $[M(T\phi_4)\phi]$ framework structure of descloizite.

The structure of descloizite is shown in Figure 2.17(c); again this is a popular structural arrangement (Table 2.A5). Prominent features of the tetrahedral–octahedral framework are the edge-sharing chains of octahedra flanked by staggered tetrahedra that link along the chain. This $[M(T\phi_4)\phi]$ chain is found in the structures of the minerals of the linarite group (Figure 2.13(e)), and is also a fundamental building block for the $[M(T\phi_4)\phi]$ sheet (Figure 2.15(b)) that is the structural unit in tsumcorite and bermanite (Table 2.A4).

These three examples show the type of structural variability we find in the framework structures, and also the small number of polyhedral linkage patterns (fundamental building blocks) that occur and seem common to a wide range of structural types. This suggests that these patterns of bond connectivity are very stable, and hence tend to persist from one structure to another. In addition, the incorporation of relatively primitive fragments into more highly condensed structural units tends to support the conceptual approach of considering a large structure both topologically and energetically as an assemblage of smaller structural fragments.

2.7 (OH) and (H$_2$O) in oxysalt structures

In inorganic minerals, the hydrogen cation H^+ most commonly has a coordination number of [2]; higher coordination numbers are not rare, but for simplicity we will consider the former, as the arguments presented here can easily be generalized to higher coordination numbers. Usually this arrangement undergoes a spontaneous distortion, with the hydrogen ion moving off-centre towards one of the two coordinating anions. The geometry of this arrangement has been very well-characterized by neutron diffraction (Ferraris and Franchini-Angela, 1972); the typical arrangement is shown in Figure 2.18. Brown (1976) has shown that the most common bond-valence distribution is

Figure 2.18 Typical geometry of hydrogen coordination: the hydrogen is [2]-coordinated, and spontaneously moves off-centre to form two bonds of the approximate valence shown, and a bent O-H-O angle; the anion closer to the hydrogen is called the 'donor' anion, and the anion further from the hydrogen is called the 'acceptor' anion.

about 0.80 v.u. to the closer oxygen, and approximately 0.20 v.u. to the further oxygen; this generally leads to the stronger bond being subsumed within $(H_2O)^0$ or $(OH)^-$ groups that now become complex anions, and the longer (weaker) bond being referred to as a *hydrogen bond*. The oxygen closer to the hydrogen is called the (hydrogen bond) *donor*, and the oxygen further from the hydrogen is called the (hydrogen bond) *acceptor* (Fig. 2.18).

There are four different hydrogen-bearing groups in minerals: $(OH)^-$, $(H_2O)^0$, $(H_3O)^+$, and $(H_5O_2)^+$; sketches of typical bond-valence distributions for these groups are shown in Figure 2.19. The positively charged groups act as cations and are extremely uncommon, although they have been identified in such minerals as hydronium jarosite ($\{H_3O\}[Fe_3^{2+}(SO_4)_2(OH)_6]$, Ripmeester et al., 1986) and rhomboclase ($\{H_5O_2\}[Fe^{3+}(SO_4)_2(H_2O)_2]$, Mereiter, 1974).

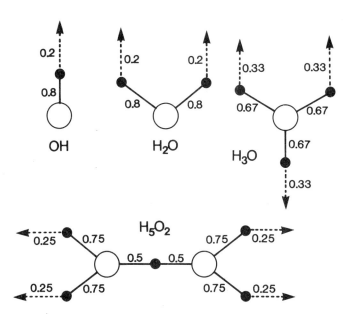

Figure 2.19 Typical bond-valence distributions for the hydrogen-bearing groups found in minerals: $(OH)^{1-}$, $(H_2O)^0$, $(H_3O)^{1+}$ and $(H_5O_2)^{1+}$.

On the other hand, the (OH)⁻ and $(H_2O)^0$ groups play a very important role in the structures of the oxysalt minerals, particularly with regard to the topological properties of their bond networks. The reason for this stems from the extremely directional nature of the bonding associated with these two groups. On the oxygen side of each group, they function as an anion, whereas on the hydrogen side of each group, they function as a cation (Fig. 2.18); it is because of this unusual property that they play such a unique role in the structure and chemistry of minerals.

2.7.1 (OH) and (H_2O) as components of the structural unit

The very important role of these groups in the constitution of the structural unit of a mineral stems from their bond-valence distribution (Fig. 2.19). On the anionic side of each group, the bond-valence is relatively strong, approximately 1.2 v.u. for (OH) and 0.4 v.u. for (H_2O); the remainder of the bond-valence requirements of the central oxygen is satisfied by the hydrogens, and on the cationic side of the group, the bond-valence is relatively weak, about 0.2 v.u. for each group (Fig. 2.19). Thus, on the anionic side of the group, the strong bonding constitutes part of the bonding network of the structural unit; conversely, on the cationic side of the group, the hydrogen bond is too weak to form a part of the bonding network of the structural unit. The role of both (OH) and (H_2O) is thus to '*tie off*' the polymerization of the structural unit in specific directions. Consequently, these groups play a crucial role in controlling the class of the polymerization of the structural unit (Hawthorne, 1985a), and hence control many of the physical and chemical properties of a mineral.

An excellent example of this is the structure of newberyite (Sutor, 1967), [Mg(PO_3OH)(H_2O)$_3$]. The structural unit is a sheet of corner-sharing (MgO_6) octahedra and (PO_4) tetrahedra, with the polyhedra arranged at the vertices of a 6_3 net, as illustrated in Figure 2.14(a); the bond-valence structure is shown in Table 2.7. In the (PO_4) tetrahedra, three of the ligands link to (MgO_6) octahedra within the sheet. The other ligand is 'tied off' orthogonal to the sheet

Table 2.7 Bond-valence table for newberyite

	Mg	P	H(6)	H(71)	H(72)	H(81)	H(82)	H(91)	H(92)	Sum
O(3)	0.389	1.399								1.788
O(4)	0.349	1.242	0.20	0.20						1.891
O(5)	0.364	1.232			0.20	0.20				1.996
O(6)		1.095	0.80					0.20		2.095
O(7)	0.326			0.80	0.80			0.20		2.126
O(8)	0.316					0.80	0.80		0.20	2.116
O(9)	0.313							0.80	0.80	1.913
Sum	2.057	4.968	1.0	1.0	1.0	1.0	1.0	1.0	1.0	

by the fact that the oxygen is strongly bonded to a hydrogen atom (i.e. it is a hydroxyl group); the long P−O bond of 1.59 Å contributes a bond-valence of 1.10 v.u. to the oxygen, and the remaining 0.90 v.u. is contributed by the hydrogen atom which then weakly hydrogen-bonds (bond-valence of about 0.10 v.u.) to the neighbouring sheet in the Y-direction. In the (MgO_6) octahedra, three of the ligands link to (PO_4) tetrahedra within the sheets. The other ligands are 'tied off' by the fact that they are (H_2O) groups; the Mg−O bonds of 2.11, 2.12 and 2.13 Å contribute a bond-valence of approximately 0.32 v.u. to each oxygen, and the remaining 1.68 v.u. is contributed by the two hydrogen atoms which then weakly hydrogen-bond (bond-valences of about 0.16 v.u. for each bond) to the neighbouring sheets in the Y direction. The chemical formula of the structural unit is also the chemical formula of the mineral, and the sheet-like nature of the structural unit is controlled by the number and distribution of the hydrogen atoms in the structure.

In newberyite, all intra-unit linkage was stopped at the (OH) and (H_2O) groups. This is not necessarily the case; for specific topologies, both (OH) and (H_2O) can allow intra-unit linkage in some directions and prevent it in others. A good example of this is the mineral artinite (Akao and Awai, 1977), $[Mg_2(CO_3)(OH)_2(H_2O)_3]$, the structure of which is shown in Figure 2.20; the bond-valence structure is shown in Table 2.8. The structural unit is a ribbon (chain) of edge-sharing (MgO_6) octahedra, flanked by (CO_3) triangles linked to alternate outer octahedral vertices of the ribbon, and occurring in a staggered arrangement on either side of the ribbon. The anions bonded to Mg and running down the centre of the ribbon are bonded to three Mg cations; they receive about $0.36 \times 3 = 1.08$ v.u. from the Mg cations, and thus receive 0.92 v.u. from their associated hydrogen atoms which then weakly hydrogen-bond (bond-valence approximately 0.08 v.u.) to an adjacent ribbon. The $(OH)^-$ group thus allows linkage in the X and Y directions but prevents linkage in the Z direction.

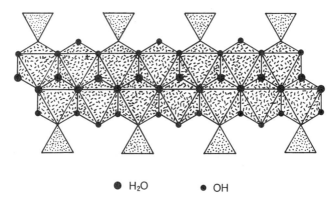

● H_2O ● OH

Figure 2.20 The structural unit in artinite, a ribbon of (MgO_6) octahedra and (CO_3) triangles; all simple anions not bonded to carbon are either (OH) or (H_2O).

Table 2.8 Bond-valence table for artinite

	Mg	C	H(1)	H(2)	H(3)	H(4)	Sum
*O(1)	0.391		0.08			0.80$^{\times 2}$	2.071
*O(1)'	0.391	1.678	0.08				2.149
O(2)		1.264$^{\times 2}$		0.30	0.30		1.864
OH	0.372$^{\times 2}$ 0.350		0.92				2.018
OW	0.283$^{\times 2}$			0.70	0.70		1.966
	2.051	4.206	1.00	1.00	1.00	1.00	

*O(1) and O(1)' are disordered, and are both half-occupied.

The anions bonded to Mg and running along the edge of the ribbon are bonded to either one Mg, two Mg, or one Mg and one C, with bond-valence contributions of about 0.3, 0.6, and 1.7 v.u., respectively. The former two ligands are therefore (H_2O) groups which hydrogen bond fairly strongly to anions in the same structural unit and in adjacent structural units. Thus the (H_2O) group bonded to one Mg prevents further unit polymerization in all three directions, whereas the (H_2O) group bonded to two Mg atoms allows polymerization in the Y direction but prevents polymerization in the other two directions. The bond-valence requirements of the two anions just bonded to C are satisfied by hydrogen bonding involving donor atoms both in the same structural unit and in different structural units. Thus in artinite, all linkage between structural units is through hydrogen bonding via (OH) and (H_2O) groups of the structural units; in addition, the (OH) groups allow polymerization in two directions within the structural unit, whereas the two types of (H_2O) groups allow polymerization in one and no directions, respectively, within the structural unit.

The (OH) and (H_2O) groups play a crucial role in controlling the polymerization of the structural unit in oxysalt minerals. Because of its very asymmetric distribution of bond-valences, the hydrogen atom can link to any strongly bonded unit, essentially preventing any further polymerization in that direction. Thus the *dimensionality* of the structural unit in a mineral is primarily controlled by the amount and role of hydrogen in the structure.

2.7.2 (H_2O) groups bonded to interstitial cations

Interstitial cations are usually large and of low charge. Generally, they are alkali or alkaline earth cations with Lewis acidities significantly less than those of the cations belonging to the structural unit. Consequently, (H_2O) can function as a ligand for these cations whereas (OH) cannot, as the cation to which it must bond cannot contribute enough bond-valence (i.e. about 1.0 v.u.)

for its bond-valence requirements to be satisfied. There are (at least) three possible reasons for (H$_2$O) groups to act as ligands for interstitial *cations*:

1. to satisfy the bond-valence requirements around the interstitial cation in cases where there are insufficient anions available from adjacent structural units;
2. to carry the bond-valence from the interstitial cation to a distant unsatisfied anion of an adjacent structural unit;
3. to act as bond-valence transformers between the interstitial cation and the anions of the structural unit; this is a mechanism of particular importance, and will be discussed separately later on.

A good example of (H$_2$O) of this kind is found in the structure of stringhamite (Hawthorne, 1985b), [CaCu(SiO$_4$)](H$_2$O), the structure of which is illustrated in Figure 2.21. The structural unit is a sheet of corner-sharing (SiO$_4$) tetrahedra and square-planar (CuO$_4$) polyhedra, arranged parallel to (010). These sheets are linked together by interstitial Ca atoms; each Ca links to four anions from one sheet and one anion from the adjacent sheet. Presumably the Ca coordination number of [5], a value that is rare for Ca, is not adequate with regard to the satisfaction of local bond-valence requirements, and two (H$_2$O) groups complete the Ca coordination polyhedron. As shown in Figure 2.21, each

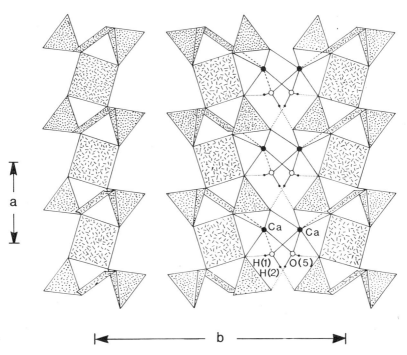

Figure 2.21 The crystal structure of stringhamite projected on to (001); interstitial species are omitted to the left of the diagram to emphasize the sheet-like nature of the structural unit.

(H_2O) group bonds to two Ca atoms, and also hydrogen bonds to anions in adjacent sheets, carrying the Ca bond-valence to anions which otherwise it could not reach. Thus the (H_2O) groups of this type, that is bonded only to interstitial cations, play a very different role from those (H_2O) groups that form part of the structural unit.

2.7.3 Hydrogen-bonded interstitial (H_2O) groups

In some structures, there are interstitial (H_2O) groups that are not bonded to any interstitial cations and yet participate in a well-defined hydrogen-bonding network. The (H_2O) groups of this sort act as both hydrogen-bond donors *and* hydrogen-bond acceptors. Any hydrogen-containing group (both (OH) and (H_2O) of the structural unit, interstitial (H_2O) bonded to interstitial cations, and interstitial (H_2O) groups not bonded to the structural unit or interstitial cations) can act as a hydrogen-bond donor to (H_2O) groups of this sort, and any anion or (H_2O) group can act as hydrogen-bond acceptor for such (H_2O) groups. Minerals with such hydrogen-bonding networks can be thought of as intermediate between anhydrous structures and clathrate structures.

A good example of such a structure is the mineral mandarinoite (Hawthorne, 1984b), [$Fe_2^{3+}(SeO_3)_3(H_2O)_3$]($H_2O$)$_3$, the structure of which is illustrated in Figure 2.22. The structural unit is a heteropolyhedral framework of corner-linking (SeO_3) triangular pyramids and (FeO_6) octahedra, with large cavities that are occupied by hydrogen-bonded (H_2O) groups in well-defined positions. Thus of the six (H_2O) groups in the formula unit, three are bonded to Fe^{3+} and are part of the structural unit; the three remaining (H_2O) groups are interstitial and not bonded to any cation at all, but held in place solely by a network of hydrogen bonds.

2.7.4 Occluded (H_2O) groups

Some structures contain (H_2O) groups that are not bonded to any cation and are not associated with any hydrogen-bonding scheme; normally such (H_2O) groups are located in holes within or between structural units. Such groups can occupy well-defined crystallographic positions, but their interaction with the rest of the structure is solely through a Van der Waals interaction.

A good example of such (H_2O) groups occurs in the structure of beryl. Alkali-free beryl can have non-bonded (H_2O) groups occurring in the channels of the framework structure. Most natural beryls contain alkali cations partly occupying sites within these channels, and these cations are bonded to channel (H_2O) groups. However, Hawthorne and Cerny (1977) have shown that most natural beryls contain (H_2O) groups in excess of that required to coordinate the channel cations, and hence some of the (H_2O) groups must be occluded rather than occurring as bonded components of the structure. Although such (H_2O) does not play a significant structural role, it can have important effects

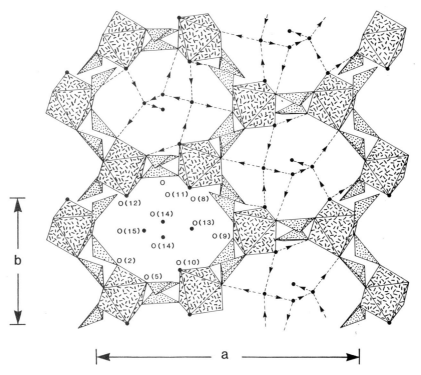

Figure 2.22 The crystal structure of mandarinoite projected on to (001); note the two different types of (H$_2$O) groups, one bonded to cations of the structural unit, and the other held in the structure by hydrogen-bonding only.

on such physical properties as specific gravity, optical properties (Cerny and Hawthorne, 1976) and dielectric behaviour.

2.8 (H$_2$O) as a bond-valence transformer

Consider a cation, M, that bonds to an anion X (Fig. 2.23(a)); the anion X receives a bond-valence of v valence units from the cation M. Consider a cation, M, that bonds to an (H$_2$O) group which in turn bonds to an anion X (Fig. 2.23(b)). In the second case, the oxygen receives a bond-valence of v valence units from the cation M, and its bond-valence requirements are satisfied by two short O−H bonds of valence $(1-v/2)$ valence units. To satisfy the bond-valence requirements around each hydrogen atom, each hydrogen forms at least one hydrogen bond with its neighbouring anions. In Figure 2.23(b), one of these hydrogen bonds is to the X anion which thus receives a bond-valence of one half what it received when it was bonded directly to the M cation. Thus the (H$_2$O) group has acted as a *bond-valence transformer*, causing one bond (bond-valence = v v.u.) to be split into two weaker bonds

Figure 2.23 The transformer effect of (H_2O) groups: a cation bonds to an oxygen of an (H_2O) group, and the strong bond is split into two weaker bonds (hydrogen bonds) via the bond-valence requirements of the constituent H^+ and O^{2-} ions; this is shown for both M^{2+} and M^{3+} cations.

(bond-valence $= v/2$ v.u.). It is this *transformer effect* that is the key to understanding the role of interstitial (H_2O) in minerals.

2.9 Binary structural representation

It has been proposed (Hawthorne, 1985a; 1986) that the structural unit be treated as a (very) complex oxyanion. Within the framework of bond-valence theory, we can thus define a Lewis basicity for the structural unit in exactly the same way as we do for a more conventional oxyanion. We may then use the valence-matching principle to examine the interaction of the structural unit with the interstitial cations. In this way, we can get some quantitative insight into the weak bonding in minerals. It is worth emphasizing here that we have developed a *binary representation* that gives us a simple quantitative model of even the most complicated structure. We consider structures in this way not to convey the most complete picture of the bond topology, but to *express structure* in such a way that we may apply bond-valence arguments in an *a priori* fashion to problems of structural chemistry.

Let us look at goedkenite, $Sr_2[Al(PO_4)_2(OH)]$, the structural unit of which is shown in Figure 2.13(i). The bond network in the structural unit is shown in Figure 2.24 as a sketch of the smallest repeat fragment in the structural unit. There are 9 oxygens in this fragment (as indicated by the general $[M(T\phi_4)_2\phi]$ form of the structural unit), and the residual anionic charge is 4^-. In order to calculate the basicity of this structural unit, we must assign simple anion coordination numbers to the unit. Obviously, we must have an objective process for doing this, as the calculation of structural unit basicity hinges on this assignment. Fortunately, this assignment is fairly well-constrained by the general observation that most minerals have oxygen in [3]- or [4]-coordination; of course, it is easy to think of exceptions, quartz for example, but the fact that these exceptions are few 'proves the rule'. Normally it is adequate to use the coordination number [4]; however, there are the following exceptions:

1. Minerals with $M = 3^+$ and $T = 6^+$, for which the coordination number [3] is more appropriate.
2. A coordination number of [3] (including H atoms) is more appropriate for (H_2O), and is also used for (OH) when it is bonded to M^{3+} cations.

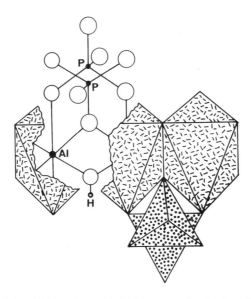

Figure 2.24 The bond network in the structural unit of goedkenite.

To attain an oxygen coordination number of [4], the cluster shown in Figure 2.24 needs an additional number of bonds from the interstitial cations. From the connectivity of the structural unit, the cluster of Figure 2.24 needs an additional 20 bonds; however, it will receive one (hydrogen) bond from an adjacent chain, which leaves 19 bonds to be received from the interstitial cations. These 19 bonds must come from 4^+ charges, and thus the average bond-valence required by the cluster is $4/19 = 0.22$ v.u.; this is the basicity of the structural unit in goedkenite. Examination of the table of Lewis acid strengths (Table 2.5) shows that the cations of appropriate Lewis acidity are Pb(0.20 v.u.), Sr(0.24 v.u.), and Ba(0.20 v.u.); in agreement with this, Sr is the interstitial cation in goedkenite. Note that Ca, with a Lewis basicity of 0.29 v.u., does not match with the Lewis basicity of the structural unit, and thus the valence-matching principle accounts for the fact that goedkenite has Sr rather than (the more common) Ca as the interstitial cation.

In this way of treating minerals, we have a simple binary interaction: the structural unit bonds with the interstitial cation(s). We may evaluate the stability of this interaction via the valence-matching principle, using the Lewis basicity of the structural unit and the Lewis acidity of the interstitial cation(s) as measures of their interaction. This reduces the most complex structure to a fairly simple representation, and looks at the interaction of its component features in a very simple but quantitative fashion.

2.10 Interstitial (H_2O) in minerals

So far, we have seen that (H_2O) plays a major role in controlling the character (dimensionality) of the structural unit; such (H_2O) is part of the structural unit and is stoichiometrically fixed by the topology of the unit connectivity. Interstitial (H_2O) is obviously very different in character. As we discussed previously, it may coordinate interstitial cations or it may occur solely as a component of a hydrogen-bonded network. Whichever is the case, the (H_2O) occupies fixed atomic positions and must play a role in the stability of the structure. The key to understanding this role is found in two distinct ideas of bond-valence theory:

1. the role of (H_2O) as a bond-valence transformer;
2. application of the valence-matching principle to the interaction between the structural unit and the interstitial cations.

Ideally, the valence of the bonds from the interstitial cations to the structural unit must match the Lewis basicity of that structural unit; if they do not match, then there cannot be a stable interaction and that particular structural arrangement will not occur. However, if the Lewis acidity of the interstitial cation is too large, the cation may bond to an interstitial (H_2O) group which acts as a bond-valence transformer (see Section 2.9), taking the strong bond and transforming it into two weaker bonds (Fig. 2.23). In this way, incorporation of interstitial (H_2O) into the structure can moderate the Lewis acidity of the interstitial cations such that the valence-matching principle is satisfied.

A good example of this is the ferric iron sulphate mineral botryogen, $Mg_2[Fe_2^{3+}(SO_4)_4(OH)_2(H_2O)_2](H_2O)_{10}$; why does this mineral have 10 interstitial (H_2O) groups per structural formula? The structural unit of botryogen is illustrated in Figure 2.13(k), and the coordinations of the various anions in the structural unit are shown in Table 2.9. Using the ideal coordination numbers discussed earlier (=[3] for all the simple anions in botryogen), the structural unit of botryogen needs an additional 26 bonds to achieve ideal coordination of all its simple anions. Six of these bonds will be hydrogen bonds from (OH) and (H_2O) groups within or in adjacent structural units, leaving 20 bonds needed from interstitial cations. Thus the Lewis basicity of the structural unit in botryogen is the charge divided by the number of required bonds: $4/20 = 0.20$ v.u. The interstitial cations in botryogen are Mg, with a Lewis acidity of 0.36 v.u. The valence-matching principle is violated, and a stable structure should not form. However, the interstitial Mg atoms are coordinated by $\{5(H_2O)+O\}$, and this will moderate the effective Lewis acidity of the cation via the transformer effect of (H_2O). Thus, the effective Lewis acidity of the 'complex cation' $\{Mg(H_2O)_5O\}$ is the charge divided by the number of bonds: $2/(5 \times 2+1) = 0.19$ v.u. The moderated Lewis acidity of the complex interstitial cation thus matches the Lewis basicity of the structural unit, and a stable mineral is formed.

Table 2.9 Details of H_2O 'of hydration' in botryogen

Botryogen: $Mg_2[Fe_2^{3+}(SO_4)_4(OH)_2(H_2O)_2](H_2O)_{10}$

Bonded atoms	Number of anions	Ideal coord. no.	Bonds needed for ideal coord.
S	10	3	2×10
$S + Fe^{3+}$	6	3	1×6
$2Fe^{3+} + H$	2	3	0
$Fe^{3+} + 2H$	2	3	0

Bonds needed to structural unit $= 2 \times 10 + 1 \times 6 = 26$
No. of H bonds to structural unit $= 2 \times 2 + 2 \times 1 = 6$
No. of additional bonds needed $= 26 - 6 = 20$
Charge on structural unit $= 4^-$
Lewis basicity of structural unit $= 4/20 = 0.20$ v.u.

Interstitial cation(s) is Mg
Mg coordination $= \{5(H_2O) + 0\}$
Bonds from Mg to structural unit $= 5 \times 2 + 1 = 11$
Effective Lewis acidity of $Mg = 2/\{5 \times 2 + 1\} = 0.19$ v.u.
The interstitial (H_2O) has moderated the Lewis acidity of the interstitial cation such that the valence-matching principle is satisfied

2.11 Bond-valence controls on interstitial cations

The structural unit is (usually) of anionic character, and thus has negative charge; this is neutralized by the presence of interstitial cations. Apart from the requirement of electroneutrality, the factors that govern the identity of the interstitial cations have been obscure. However, inspection of Tables 2.A2, 2.A3, and 2.A4 indicates that there must be controls on the identity of the interstitial cation. It is immediately apparent that different structural units are associated with different interstitial cations; thus $[M^{2+}(T^{5+}O_4)_2(H_2O)_2]$ chains (Fig. 2.13(a), Table 2.A3) always have Ca as the interstitial cation, whereas $[M^{2+}(T^{5+}O_4)_2(H_2O)]$ chains (Fig. 2.13(b), Table 2.A3) always have Pb^{2+} as the interstitial cation. Why is this so? If we were dealing with one or two very rare minerals, we might suspect that the difference is geochemically controlled; however, these are reasonably common minerals with significantly variable chemistry and paragenesis. We are forced to conclude that the control on interstitial cation type is crystal chemical rather than geochemical.

We find the answer to this problem in the application of the valence-matching principle to our binary representation of structure. The Lewis acidity of the interstitial cation must match up with the basicity of the structural unit. Thus it is not enough that the interstitial cation has the right valence; it must also have the right Lewis acidity. Let us examine the example outlined in the previous paragraph, that is the identity of the interstitial cations in the

$[M^{2+}(T^{5+}O_4)_2(H_2O)_2]$ and $[M^{2+}(T^{5+}O_4)_2(H_2O)]$ structures, using brandtite and brackebuschite as examples.

The situation for brandtite is shown in Table 2.10; counting the bonds within the structural unit indicates that an additional 20 bonds to the structural unit are needed to attain the requisite simple anion coordination numbers. Four of these bonds are hydrogen bonds from other structural units, leaving 16 bonds to be contributed by the interstitial cations. The residual charge on the structural unit is 4^- (per $[Mn^{2+}(AsO_4)_2(H_2O)_2]$ unit), and hence the basicity of the structural unit is $4/16 = 0.25$ v.u. Inspection of the Lewis acidity table (Table 2.5) shows that Ca has a Lewis acidity of 0.29 v.u., matching up with the Lewis basicity of the structural unit. Hence the valence-matching principle is satisfied, and $Ca_2[Mn^{2+}(AsO_4)_2(H_2O)_2]$ is a stable structure.

The situation for brackebuschite is also shown in Table 2.10; an additional 20 bonds are needed to satisfy the requisite simple anion coordination requirements. Two of these bonds are hydrogen bonds from adjacent structural units, leaving 18 bonds to be satisfied by the interstitial cations. The residual charge on the structural unit is 4^-, and hence the basicity of the structural unit is $4/18 = 0.22$ v.u. This value matches up quite well with the Lewis basicity of Pb^{2+} (0.20 v.u., see Table 2.5), the valence-matching principle is satisfied, and $Pb^{2+}[Mn^{2+}(V^{5+}O_4)_2(H_2O)]$ is a stable structure.

Table 2.10 Calculation of structural unit basicity for brandtite

---

Brandtite $= Ca_2[Mn^{2+}(AsO_4)_2(H_2O)_2]$ Structural unit $= [Mn^{[6]}(As^{[4]}O_4)_2(H^{[2]}_2O)_2]$
Number of bonds in structural unit $= 1 \times [6] + 2 \times [4] + 2 \times [2] = 18$
Number of bonds needed for [4]-coordination of all simple anions (except (H_2O) for which [3]-coordination is assigned) $= 8 \times [4] + 2 \times [3] = 38$
Number of additional bonds to structural unit to achieve this coordination $= 20$
Number of hydrogen bonds to structural unit $= 2 \times 2 = 4$
Therefore the number of bonds required from interstitial cations $= 20 - 4 = 16$
Charge on the structural unit $[Mn^{2+}(AsO_4)_2(H_2O)_2]$ in brandtite $= 4^-$
Lewis basicity of structural unit $=$ charge/bonds $= 4/16 = 0.25$ v.u.
This basicity matches most closely with the Lewis acidity of Ca at 0.27 v.u.
Thus the formula of brandtite is $Ca_2[Mn(AsO_4)_2(H_2O)_2]$

Brackebuschite $= Pb_2[Mn^{2+}(VO_4)_2(H_2O)]$ Structural unit $= [Mn^{[6]}(V^{[4]}O_4)_2(H^{[2]}_2O)]$
Number of bonds in structural unit $= 1 \times [6] + 2 \times [4] + 2 \times [1] = 16$
Number of bonds needed for [4]-coordination of all simple anions (including (H_2O) which is [4]-coordinated in this structural unit) $= 9 \times [4] = 36$
Number of additional bonds to structural unit to achieve this coordination $= 20$
Number of hydrogen bonds to structural unit $= 2$
Number of bonds required from interstitial cations $= 18$
Charge on the structural unit $[Mn^{2+}(VO_4)_2(H_2O)]$ in brackebuschite $= 4^-$
Lewis basicity of structural unit $=$ charge/bonds $= 4/18 = 0.22$ v.u.
This basicity matches most closely with the Lewis acidity of Pb at 0.20 v.u.
Thus the formula of brackebuschite is $Pb_2[Mn(VO_4)_2(H_2O)]$

Now that we understand the basis of this selectivity of interstitial cations, some very interesting geological questions become apparent. Does the form of the structural unit dictate the identity of the interstitial cations, or does the availability of a particular interstitial cation dictate the form of the structural unit? Does the pH of the environment affect the form of the structural unit or the amount of interstitial (H_2O) incorporated into the structure? Are there synergetic interactions between these factors? Using bond-valence theory in conjunction with the topological characteristics of the structural unit, we can begin to investigate some of these questions that previously we have had no conceptual basis to think about.

2.12 Structural controls on mineral paragenesis

2.12.1 Magnesium sulphate minerals

The magnesium sulphate minerals are important phases in many marine salt deposits, and these parageneses can be very complicated. Common minerals in these deposits are given in Table 2.11, together with an indication of the character of the structural unit and its Lewis basicity. The paragenetic scheme shown in Figure 2.25 was constructed from an examination of natural occurrences, together with consideration of their phase relations in aqueous solutions. The arrows in Figure 2.25 indicate a change in the crystallizing phase from an aqueous solution of the bulk composition of the previously crystallizing phase, and/or an alteration sequence. Obviously, these equilibria will be specifically dependent on temperature, bulk composition and pH, but the natural assemblages suggest that the scheme of Figure 2.25 corresponds reasonably well to the general case.

Table 2.11 Common magnesium sulphate minerals

Mineral	Formula	Structural unit	Unit basicity (v.u.)
Langbeinite	$K_2Mg_2(SO_4)_3$	Infinite framework	0.11
Loeweite	$Na_{12}Mg_7(SO_4)_{13} \cdot 15H_2O$	Infinite framework	0.14
Vanthoffite	$Na_6Mg(SO_4)_4$	Infinite sheet	0.14
Polyhalite	$K_2Ca_2Mg(SO_4)_4 \cdot 2H_2O$	Finite cluster	0.21
Kieserite	$Mg(SO_4) \cdot H_2O$	Infinite framework	0.00
Leonite	$K_2Mg(SO_4)_2 \cdot 4H_2O$	Finite cluster	0.14
Bloedite	$Na_2Mg(SO_4)_2 \cdot 4H_2O$	Finite cluster	0.14
Pentahydrite	$Mg(SO_4) \cdot 5H_2O$	Infinite chain	0.00
Hexahydrite	$Mg(SO_4) \cdot 6H_2O$	Isolated polyhedra	0.00
Picromerite	$K_2Mg(SO_4)_2 \cdot 6H_2O$	Isolated polyhedra	0.13
Epsomite	$Mg(SO_4) \cdot 7H_2O$	Isolated polyhedra	0.00

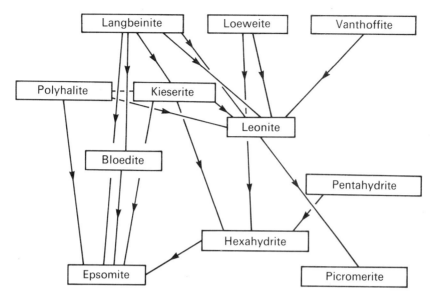

Figure 2.25 Approximate paragenetic scheme for $A_v[Mg_x(SO_4)_y\phi_z]$ salt minerals. The arrows denote progressive crystallization and/or alteration.

It is apparent from inspection of Table 2.11 that there is a gradual depolymerization of the structural unit down through the paragenetic sequence of Figure 2.25. This change is effected by the incorporation of increasing amounts of (H_2O) into the structural units, the (H_2O) groups essentially blocking further polymerization where they attach to the structural unit. This decrease in the polymerization of the structural unit is also accompanied by a decrease in the basicity of the structural unit (Table 2.11, Fig. 2.26). This has the effect of changing the character of the interstitial cations, which are gradually decreasing in Lewis acidity, until at the last stages of the crystalliz-

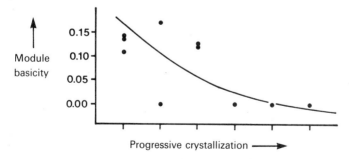

Figure 2.26 Change in character of structural unit with progressive crystallization for the minerals of Figure 2.25.

ation sequence, there are no interstitial cations left, and the structural units are neutral.

Figure 2.27 shows the evolution of the character of the structural units. For a specific Lewis basicity, there is a gradual de-polymerization of the structural unit with progressive crystallization. The pattern exhibited by Figure 2.27 suggests a specific control on the character of the crystallizing mineral(s). Crystallization begins with the formation of a structural unit of maximum connectivity and the highest possible Lewis basicity consonant with the Lewis acidity of the most acid cation available. With continued crystallization, the dimension of polymerization of the structural unit gradually decreases while the Lewis basicity of the structural units is maintained. As the most acid cation

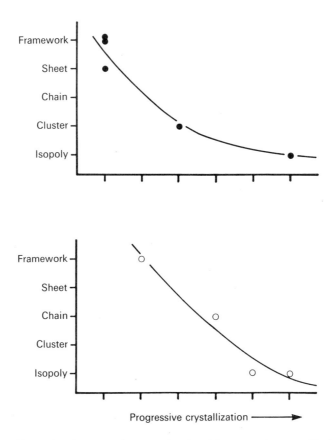

Figure 2.27 Change in character of structural unit with progressive crystallization for the minerals of Figure 2.25; the upper figure is for minerals with a Lewis basicity of ~ 0.15 v.u. for their structural units, the lower figure is for minerals with a Lewis basicity of zero for their structural units.

becomes depleted, there is the beginning of crystallization of structural units of lower basicity and the incorporation of less acid cations into the structure. Again the degree of polymerization of the structural units decreases with progressive crystallization, but the basicity of the structural units is maintained until the cations of corresponding acidity are gone. This process repeats itself with structural units and interstitial cations of progressively decreasing basicity and acidity respectively, until we reach the final stage when the structural units have zero basicity and there are no interstitial cations in the resulting minerals. Note that at any stage in the crystallization process structural units of differing basicities may be crystallizing simultaneously, the resulting minerals incorporating very different interstitial cations; however, the connectivities (dimensions of polymerization) of the structural units are very different. This scheme seems to work quite well for these minerals, and one can make all sorts of inferences about possible complexing in solution and crystallization mechanisms. It will be interesting to see how well this scheme works in more complicated systems.

2.13 Summary

Here I have tried to develop a global approach to questions of stability and paragenesis of oxide and oxysalt minerals. The fundamental basis of the approach relates to the energetic content of the bond topology of a structure. Bond topology has a major effect on the energetics of a structure, suggesting that major trends in structure stability, properties, and behaviour should be systematically related to the coordination geometry and polyhedral linkage of a structure. Combination of these ideas with bond-valence theory (a very simple form of molecular orbital theory) allows a simple binary representation of even the most complex structure: a (usually anionic) structural unit that interacts with (usually cationic or neutral) interstitial species to form the complete structure. This interaction can be quantitatively examined in terms of the Lewis basicities and acidities of the binary components, and such chemical variables as interstitial cation chemistry and 'water' of hydration can be quantitatively explained. Examination of a few complex mineral parageneses, together with the topological character of the structural unit, indicates that the latter has a major control on the sequence of crystallization of the constituent minerals. In addition, the persistence of particular fundamental building blocks in series of associated minerals suggests that such clusters may occur as complex species in associated hydrothermal and saline fluids.

The principal idea behind this work is to develop an approach that is reasonably transparent to chemical and physical intuition, and that can be applied to large numbers of very complex structures in geological environments. Mineralogy is currently absorbing a large number of 'theoretical' techniques from physics and chemistry, examining aspects of (fairly simple) minerals. One of the dangers with this is that it is easy to become lost in the technical complexities of the computational approach, and the basic physics and chemistry of what is going on can become lost; one is merely doing a numerical experiment. By and large, mineralogists have not fallen into this trap; in particular, the work of G. V. Gibbs and colleagues has continually pursued the important energetic mechanisms operative in minerals. However, I feel that there is also room for a much simpler approach that addresses more global aspects of complex minerals, working well below the 'Pauling point' rather than above it. These ideas tend to be more intuitive and only semi-quantitative, but seem to be capable of organizing a large amount of factual information into a coherent framework, and also provide a basis for thinking about many questions that were intractable to previous approaches.

Acknowledgements

I have been strongly influenced by the ideas of I. David Brown, Jeremy K. Burdett, and Paul B. Moore; as a result, they can take the blame for any shortcomings of the present work. This work was supported by the Natural Sciences and Engineering Council of Canada in the form of an Operating Grant to the author.

Appendix

Table 2.A1 $M(T\phi_4)\phi_n$ and $M(T\phi_4)_2\phi_n$ minerals based on isolated $M\phi_6$ octahedra and $T\phi_4$ tetrahedra

$M(T\phi_4)\phi_n$ mineral	Formula	$M(T\phi_4)_2\phi_n$ mineral	Formula
Bianchite	$[Zn(H_2O)_6][SO_4]$	Amarillite	$Na[Fe^{3+}(SO_4)_2(H_2O)_6]$
Ferrohexahydrite	$[Fe^{2+}(H_2O)_6][SO_4]$	Tamarugite	$Na[Al(SO_4)_2(H_2O)_6]$
Hexahydrite	$[Mg(H_2O)_6][SO_4]$		
Moorhouseite	$[Co(H_2O)_6][SO_4]$	Mendozite	$Na[Al(SO_4)_2(H_2O)_6](H_2O)_5$
Nickel-hexahydrite	$[Ni(H_2O)_6][SO_4]$	Kalinite	$K[Al(SO_4)_2(H_2O)_6](H_2O)_5$
Retgersite	$[Ni(H_2O)_6][SO_4]$	Sodium alum	$Na[Al(SO_4)_2(H_2O)_6](H_2O)_6$
Khademite	$[Al(H_2O)_5F][SO_4]$	Potassium alum	$K[Al(SO_4)_2(H_2O)_6](H_2O)_6$
Epsomite	$[Mg(H_2O)_6][SO_4](H_2O)$	Tschermigite	$NH_4[Al(SO_4)_2(H_2O)_6](H_2O)_6$
Goslarite	$[Zn(H_2O)_6][SO_4](H_2O)$	Apjohnite	$Mn[Al(SO_4)_2(H_2O)_6](H_2O)_{10}$
Morenosite	$[Ni(H_2O)_6][SO_4](H_2O)$	Bílinite	$Fe^{2+}[Fe^{3+}(SO_4)_2(H_2O)_6](H_2O)_{10}$
		Dietrichite	$Zn[Al(SO_4)_2(H_2O)_6](H_2O)_{10}$
		Halotrichite	$Fe^{2+}[Al(SO_4)_2(H_2O)_6](H_2O)_{10}$
Bierberite	$[Co(H_2O)_6][SO_4](H_2O)$	Pickeringite	$Mg[Al(SO_4)_2(H_2O)_6](H_2O)_{10}$
Boothite	$[Cu(H_2O)_6][SO_4](H_2O)$	Redingtonite	$Fe^{2+}[Cr(SO_4)_2(H_2O)_6](H_2O)_{10}$
Mallardite	$[Mn(H_2O)_6][SO_4](H_2O)$		
Melanterite	$[Fe^{2+}(H_2O)_6][SO_4](H_2O)$	Aubertite	$Cu^{2+}[Al(SO_4)_2(H_2O)_6]Cl(H_2O)_8$
Zinc-melanterite	$[Zn(H_2O)_6][SO_4](H_2O)$	Boussingaultite	$(NH_4)_2[Mg(SO_4)_2(H_2O)_6]$
		Cyanochroite	$K_2[Cu^{2+}(SO_4)_2(H_2O)_6]$
Phosphorroesslerite	$[Mg(H_2O)_6][PO_3(OH)](H_2O)$	Mohrite	$(NH_4)_2[Fe^{2+}(SO_4)_2(H_2O)_6]$
Roesslerite	$[Mg(H_2O)_6][AsO_3(OH)](H_2O)$	Picromerite	$K_2[Mg(SO_4)_2(H_2O)_6]$
Struvite	$NH_4[MgH_2O)_6][PO_4]$	Despujolsite	$Ca_3[Mn^{4+}(SO_4)_2(OH)_6](H_2O)_3$
		Fleischerite	$Pb_3[Ge(SO_4)_2(OH)_6](H_2O)_3$
		Schauertite	$Ca_3[Ge(SO_4)_2(OH)_6](H_2O)_3$

Table 2.A2 $M(T\phi_4)\phi_n$ and $M(T\phi_4)_2\phi_n$ minerals based on finite clusters of $M\phi_6$ octahedra and $T\phi_4$ tetrahedra

$M(T\phi_4)\phi_n$ mineral	Formula	$M(T\phi_4)_2\phi_n$ mineral	Formula
Aplowite	$[Co(SO_4)(H_2O)_4]$	Anapaite	$Ca_2[Fe^{2+}(PO_4)_2(H_2O)_4]$
Boyleite	$[Zn(SO_4)(H_2O)_4]$	Bloedite	$Na_2[Mg(SO_4)_2(H_2O)_4]$
Ilesite	$[Mn(SO_4)(H_2O)_4]$	Leonite	$K_2[Mg(SO_4)_2(H_2O)_4]$
Rozenite	$[Fe^{2+}(SO_4)(H_2O)_4]$	Schertelite	$(NH_4)_2[Mg(PO_3OH)_2(H_2O)_4]$
Starkeyite	$[Mg(SO_4)(H_2O)_4]$	Roemerite	$Fe^{2+}[Fe^{3+}(SO_4)_2(H_2O)_4]_2(H_2O)_6$
Morinite	$Ca_2Na[Al_2(PO_4)_2F_4(OH)(H_2O)_2]$	Metavoltine	$K_2Na_6Fe^{2+}[Fe_3^{3+}(SO_4)_6O(H_2O)_3]_2(H_2O)_{12}$

Table 2.A3 $M(T\phi_4)\phi_n$ and $M(T\phi_4)_2\phi_n$ minerals based on infinite chains of $M\phi_6$ octahedra and $T\phi_4$ tetrahedra

$M(T\phi_4)\phi_n$ mineral	Formula	$M(T\phi_4)_2\phi_n$ mineral	Formula
Chalcanthite	$[Cu(SO_4)(H_2O)_4](H_2O)$	Brandtite	$Ca_2[Mn(AsO_4)_2(H_2O)_2]$
Jokokuite	$[Mn(SO_4)(H_2O)_4](H_2O)$	Krohnkite	$Na_2[Cu(SO_4)_2(H_2O)_2]$
Pentahydrite	$[Mg(SO_4)(H_2O)_4](H_2O)$	Roselite	$Ca_2[Co(AsO_4)_2(H_2O)_2]$
Siderotil	$[Fe^{2+}(SO_4)(H_2O)_4](H_2O)$		
		Cassidyite	$Ca_2[Ni(PO_4)_2(H_2O)_2]$
Liroconite	$Cu_2[Al](AsO_4)(OH)_4](H_2O)_4$	Collinsite	$Ca_2[Mg(PO_4)_2(H_2O)_2]$
		Gaitite	$Ca_2[Zn(AsO_4)_2(H_2O)_2]$
Brassite	$[Mg(AsO_3(OH))(H_2O)_4]$		
		Talmessite	$Ca_2[Mg(AsO_4)_2(H_2O)_2]$

Mineral	Formula	Mineral	Formula
Butlerite	$[Fe^{3+}(SO_4)(OH)(H_2O)_2]$	Fairfieldite	$Ca_2[Mn(PO_4)_2(H_2O)_2]$
Parabutlerite	$[Fe^{3+}(SO_4)(OH)(H_2O)_2]$	Messelite	$Ca_2[Fe^{2+}(PO_4)_2(H_2O)_2]$
Childrenite	$Mn^{2+}[Al(PO_4)(OH)_2(H_2O)]$	Tancoite	$Na_2LiH[Al(PO_4)_2(OH)]$
Eosphorite	$Fe^{2+}[Al(PO_4)(OH)_2(H_2O)]$		
		Sideronatrite	$Na_2[Fe^{3+}(SO_4)_2(OH)](H_2O)_3$
Uklonskovite	$Na[Mg(SO_4)(OH)(H_2O)_2]$		
		Jahnsite	$CaMnMg_2[Fe^{3+}(PO_4)_2(OH)]_2(H_2O)_8$
Fibroferrite	$[Fe^{3+}(SO_4)(OH)(H_2O)_2](H_2O)_4$	Whiteite	$CaFe^{2+}Mg_2[Al(PO_4)_2(OH)]_2(H_2O)_8$
Chlorothionite	$K_2[Cu(SO_4)Cl_2]$	Lun'okite	$Mn_2Mg_2[Al(PO_4)_2(OH)]_2(H_2O)_8$
		Overite	$Ca_2Mg_2[Al(PO_4)_2(OH)]_2(H_2O)_8$
Linarite	$Pb[Cu(SO_4)(OH)_2]$		
		Segelerite	$Ca_2Mg_2[Fe^{3+}(PO_4)_2(OH)]_2(H_2O)_8$
Schmiederite	$Pb_2[Cu_2(SeO_3)(SeO_4)(OH)_4]$	Wilhelmvierlingite	$Ca_2Mn_2[Fe^{3+}(PO_4)_2(OH)]_2(H_2O)_8$
Amarantite	$[Fe_2^{3+}(SO_4)_2OH_2O)_4](H_2O)_3$	Guildite	$Cu^{2+}[Fe^{3+}(SO_4)_2(OH)](H_2O)_4$
Hohmannite	$[Fe_2^{3+}(SO_4)_2OH_2O)_4](H_2O)_4$	Yftisite	$Y_4[Ti(SiO_4)_2O](F,OH)_6$
		Arsenbrackebuschite	$Pb_2[Fe^{2+}(AsO_4)_2(H_2O)]$
		Arsentsumebite	$Pb_2[Cu(SO_4)(AsO_4)(OH)]$
Fornacite	$Pb_2[Cu(AsO_4)(CrO_4)(OH)]$		
Molybdofornacite	$Pb_2[Cu(AsO_4)(MoO_4)(OH)]$	Brackebuschite	$Pb_2[Mn(VO_4)_2(H_2O)]$
		Gamagarite	$Ba_2[(Fe^{3+}, Mn)(VO_4)_2(OH, H_2O)]$
Tornebohmite	$(RE)_2[Al(SiO_4)_2(OH)]$		
		Goedkenite	$Sr_2[Al(PO_4)_2(OH)]$
Ransomite	$Cu[Fe^{3+}(SO_4)_2(H_2O)]_2(H_2O)_4$	Tsumebite	$Pb_2[Cu(PO_4)(SO_4)(OH)]$
Krausite	$K[Fe^{3+}(SO_4)_2(H_2O)]$	Vauquelinite	$Pb_2[Cu(PO_4)(CrO_4)(OH)]$
Botryogen	$Mg_2[Fe_2^{3+}(SO_4)_4(OH)_2(H_2O)_2](H_2O)_{10}$		
Zincbotryogen	$Zn_2[Fe_2^{3+}(SO_4)_4(OH)_2(H_2O)_2](H_2O)_{10}$		

Table 2.A4 $M(T\phi_4)\phi_n$ and $M(T\phi_4)_2\phi_n$ minerals based on infinite sheets of $M\phi_6$ octahedra and $T\phi_4$ tetrahedra

$M(T\phi_4)\phi_n$ mineral	Formula	$M(T\phi_4)_2\phi_n$ mineral	Formula
Tsumcorite	$Pb[(Zn, Fe^{3+})(AsO_4)(H_2O, OH)]_2$	Rhomboclase	$(H_5O_2)[Fe^{3+}(SO_4)_2(H_2O)_2]$
Bermanite	$Mn^{2+}[Mn^{3+}(PO_4)(OH)]_2(H_2O)_4$	Olmsteadite	$KFe_2^{2+}[Nb(PO_4)_2O_2](H_2O)_2$
Foggite	$Ca[Al(PO_4)(OH)_2](H_2O)$	Brianite	$Na_2Ca[Mg(PO_4)_2]$
		Merwinite	$Ca_3[Mg(SiO_4)_2]$
Arthurite	$Cu[Fe^{3+}(AsO_4)(OH)]_2(H_2O)_4$		
Earlshannonite	$Mn^{2+}[Fe^{3+}(PO_4)(OH)]_2(H_2O)_4$	Yavapaite	$K[Fe^{3+}(SO_4)_2]$
Ojuelaite	$Zn[Fe^{3+}(AsO_4)(OH)]_2(H_2O)_4$		
Whitmoreite	$Fe^{2+}[Fe^{3+}(PO_4)(OH)]_2(H_2O)_4$	Bafertisite	$BaFe^{2+}[Ti(Si_2O_7)O_2]$
Krautite	$[Mn^{2+}(AsO_3OH)(H_2O)]$	Pyrophyllite	$[AlSi_2O_5(OH)]$
		Dioctahedral micas	(M^+, M^{2+}) $[(M^{3+}, M^{2+})(Si, Al)_2O_5(OH)]_2$
Fluckite	$[CaMn^{2+}(AsO_3OH)_2(H_2O)_2]$	Ephesite	$NaLi[Al(Si, Al)_2O_5(OH)]_2$
Co-Koritnigite	$[Co(AsO_3OH)(H_2O)]$	Taenolite	$KLi[MgSi_2O_5(OH)]_2$
Koritnigite	$[Zn(AsO_3OH)(H_2O)]$	Dioctahedral smectites	(M^+, H_2O) $[(M^{3+}, M^{2+})(Si, Al)_2O_5(OH)]_2$
Kaolinite	$[Al_2Si_2O_5(OH)_4]$	Bramallite	$(M^+, H_2O)_x$ $[(Al, Mg, Fe)(Si, Al)_2O_5(OH)]_2$
Dickite	$[Al_2Si_2O_5(OH)_4]$		

Nacrite	$[Al_2Si_2O_5(OH)_4]$	
Endellite	$[Al_2Si_2O_5(OH)_4] \cdot 2(H_2O)$	Hydromica $(M^+, H_2O)_x[Al(Si, Al)_2O_5(OH)]_2$
		Illite $(M^+, H_2O)_x$
		$[(Al, Mg, Fe)(Si, Al)_2O_5(OH)]_2$
Halloysite	$[Al_2Si_2O_5(OH)_4]$	Goldichite $K_2[Fe_2^{3+}(SO_4)_4(H_2O)_4](H_2O)_4$
Arseniosiderite	$Ca_2[Fe_3^{3+}(AsO_4)_3O_2](H_2O)_3$	
Kolfanite	$Ca_2[Fe_3^{3+}(AsO_4)_3O_2](H_2O)_2$	
Mitridatite	$Ca_2[Fe_3^{3+}(PO_4)_3O_2](H_2O)_3$	
Robertsite	$Ca_2[Mn_3^{3+}(PO_4)_3O_2](H_2O)_3$	
Newberyite	$[Mg(PO_3OH)(H_2O)_3]$	
Minyulite	$K[Al_2(PO_4)_2F(H_2O)_4]$	
Gordonite	$Mg[Al_2(PO_4)_2(OH)_2(H_2O)_2](H_2O)_4 \cdot 2(H_2O)$	
Laueite	$Mn^{2+}[Fe_2^{3+}(PO_4)_2(OH)_2(H_2O)_2](H_2O)_4 \cdot 2(H_2O)$	
Paravauxite	$Fe^{2+}[Al_2(PO_4)_2(OH)_2(H_2O)_2](H_2O)_4 \cdot 2(H_2O)$	
Sigloite	$(Fe^{3+}, Fe^{2+})[Al_2(PO_4)_2(OH)_2(H_2O)_2](H_2O, OH)_4 \cdot 2(H_2O)$	
Ushkovite	$Mg[Fe_2^{3+}(PO_4)_2(OH)_2 H_2O)_2](H_2O)_4 \cdot 2(H_2O)$	
Stewartite	$Mn^{2+}[Fe_2^{3+}(PO_4)_2(OH)_2(H_2O)_2](H_2O)_4 \cdot 2(H_2O)$	
Pseudolaueite	$Mn^{2+}[Fe^{3+}(PO_4)(OH)(H_2O)]_2(H_2O)_4 \cdot 2(H_2O)$	
Strunzite	$Mn^{2+}[Fe^{3+}(PO_4)(OH)(H_2O)]_2(H_2O)_4$	
Ferrostrunzite	$Fe^{2+}[Fe^{3+}(PO_4)(OH)(H_2O)]_2(H_2O)_4$	
Metavauxite	$Fe^{2+}[Al(PO_4)(OH)(H_2O)]_2(H_2O)_6$	

Table 2.A5 $M(T\phi_4)\phi_n$ and $M(T\phi_4)_2\phi_n$ minerals based on infinite frameworks of $M\phi_6$ octahedra and $T\phi_4$ tetrahedra

$M(T\phi_4)\phi_n$ mineral	Formula	$M(T\phi_4)_2\phi_n$ mineral	Formula
Bonattite	$[Cu(SO_4)(H_2O)_3]$	Keldyshite	$(Na,H_3O)[Zr(Si_2O_7)]$
		Parakeldyshite	$Na[Zr(Si_2O_7)]$
Kolbeckite	$[Sc(PO_4)(H_2O)_2]$		
Metavariscite	$[Al(PO_4)(H_2O)_2]$	Nenadkevichite	$Na_2[Nb(Si_2O_6)OH](H_2O)_2$
Phosphosiderite	$[Fe^{3+}(PO_4)(H_2O)_2]$	Labuntsovite	$K_2[Ti(Si_2O_6)OH](H_2O)_2$
Mansfieldite	$[Al(AsO_4)(H_2O)_2]$	Batisite	$Na_2Ba[Ti(Si_2O_6)O]_2$
Scorodite	$[Fe^{3+}(AsO_4)(H_2O)_2]$	Shcherbakovite	$K_2Ba[(Ti, Nb)(Si_2O_6)O]_2$
Strengite	$[Fe^{3+}(PO_4)(H_2O)_2]$	Alkali pyroxenes	$M^+[M^{3+}(Si_2O_6)]$
Variscite	$[Al(PO_4)(H_2O)_2]$	Calcic pyroxenes	$Ca[M^{2+}(Si_2O_6)]$
Kainite	$K_4[Mg_4(SO_4)_4(H_2O)_{11}]Cl_4$	Lavenite	$(Na, Ca)_3[Zr(Si_2O_7)O]F$
Amblygonite	$Li[Al(PO_4)F]]$	Wohlerite	$Na_2Ca_4[ZrNb(Si_2O_7)_2O_2]F$
Montebrasite	$Li[Al(PO_4)OH]$		
Tavorite	$Li[Fe^{3+}(PO_4)OH]$	Rosenbuschite	$(Ca, Na)_2[Zr_2Ti_2(Si_2O_7)_4O_2F_2]F_4$
		$M(T\phi_4)\phi_n$ Mineral	Formula
Durangite	$Na[Al(AsO_4)F]$	Barbosalite	$Fe^{2+}[Fe^{3+}(PO_4)OH]_2$
Isokite	$Ca[Mg(PO_4)F]$	Lazulite	$Mg[Al(PO_4)OH]_2$
Lacroixite	$Na[Al(PO_4)F]$	Scorzalite	$Fe^{2+}[Al(PO_4)OH]_2$
Malayaite	$Ca[Sn^{4+}(SiO_4)O]$		
Panasquereite	$Ca[Mg(PO_4)OH]$		
Tilasite	$Ca[Mg(AsO_4)F]$	Lawsonite	$Ca[Al_2(Si_2O_7)(OH)_2](H_2O)$

Mineral	Formula	Mineral	Formula
Titanite-P2$_1$/c	Ca[Ti(SiO$_4$)O]	Dioptase	[Cu$_6$(Si$_6$O$_{18}$)(H$_2$O)$_6$]
Titanite-C2/c	Ca[(Ti, Al, Fe(SiO$_4$)O]	Veszelyite	[Cu$_2$(ZnPO$_4$OH)(OH)$_2$(H$_2$O)$_2$]
Dwornikite	[Ni(SO$_4$)(H$_2$O)]	Holdenite	[Mn$_6^{3+}$(Zn$_3$As$_2$SiO$_{12}$(OH)$_8$)]
Gunningite	[Zn(SO$_4$)(H$_2$O)]	Alluaudite	(Na, Ca)[Fe^{2+}(Mn, Fe, Mg)$_2$(PO$_4$)$_3$]
Kieserite	[Mg(SO$_4$)(H$_2$O)]	Hagendorfite	(Na, Ca)[Mn^{2+}(Fe, Mg)$_2$(PO$_4$)$_3$]
Poitevinite	[Cu(SO$_4$)(H$_2$O)]	Maghagendorfite	Na[Mn^{2+}(Mg, Fe)$_2$(PO$_4$)$_3$]
Szmikite	[Mn^{2+}(SO$_4$)(H$_2$O)]	Varulite	(Na, Ca)[Mn^{2+}(Mn, Fe)$_2$(PO$_4$)$_3$]
Szomolnokite	[Fe^{2+}(SO$_4$)(H$_2$O)]	Plancheite	[Cu$_8$(Si$_8$O$_{22}$)(OH)$_4$(H$_2$O)]
Adelite	Ca[Mg(AsO$_4$)OH]	Tiragalloite	[Mn$_4$(AsSi$_3$O$_{12}$(OH)]
Austinite	Ca[Zn(AsO$_4$)OH]		
Conichalcite	Ca[Cu(AsO$_4$)OH]	Medaite	[Mn$_6$(VSi$_5$O$_{18}$)(OH)]
Duftite	Pb[Cu(AsO$_4$)OH]		
Gabrielsonite	Pb[Fe^{2+}(AsO$_4$)OH]	Zoisite	Ca$_2$[Al$_3$(SiO$_4$)(Si$_2$O$_7$)O(OH)]
Vuagnatite	Ca[Al(SiO$_4$)OH]	Allanite	(Ce, Ca)$_2$[Al$_3$(SiO$_4$)(Si$_2$O$_7$)O(OH)]
Arsendescloizite	Pb[Zn(AsO$_4$)OH]	Allanite-(Y)	(Y, Ce, Ca)$_2$[Al$_3$(SiO$_4$)(Si$_2$O$_7$)O(OH)]
Calciovolborthite	Ca[Cu(VO$_4$)OH]	Clinozoisite	Ca$_2$[Al$_3$(SiO$_4$)(Si$_2$O$_7$)O(OH)]
Cechite	Pb[Fe^{2+}(VO$_4$)OH]	Epidote	Ca$_2$[Fe^{3+}(SiO$_4$)(Si$_2$O$_7$)O(OH)]
Descloizite	Pb[Zn(VO$_4$)OH]	Hancockite	(Pb, Ca)$_2$[Al$_3$(SiO$_4$)(Si$_2$O$_7$)O(OH)]
Mottramite	Pb[Cu(VO$_4$)OH]	Piemontite	Ca$_2$[Mn$_3^{3+}$(SiO$_4$)(Si$_2$O$_7$)O(OH)]

$M(T\phi_4)\phi_n$ mineral	Formula	$M(T\phi_4)_2\phi_n$ mineral	Formula
Pyrobelonite	Pb[Mn(VO$_4$)OH]		
Jagowerite	Ba[Al(PO$_4$)OH]$_2$	Orthoenstatite	[Mg(SiO$_3$)]
Melonjosephite	Ca[Fe^{2+}Fe^{3+}(PO$_4$)$_2$OH]	Hypersthene	[(Mg, Fe^{2+})(SiO$_3$)]
Bertossaite	CaLi$_2$[Al(PO$_4$)OH]$_4$	Orthoferrosilite	[Fe^{2+}(SiO$_3$)]
Palermoite	SrLi$_2$[Al(PO$_4$)OH]$_4$	Clinoenstatite	[Mg(SiO$_3$)]
		Clinohypersthene	[(Mg, Fe^{2+})(SiO$_3$)]
Carminite	Pb[Fe^{3+}(AsO$_4$)OH]$_2$	Clinoferrosilite	[Fe^{2+}(SiO$_3$)]
Leucophosphite	K[Fe$_2^{3+}$(PO$_4$)$_2$(OH)(H$_2$O)](H$_2$O)$_2$		
Ferropumpellyite	Ca$_2$[Al$_2$(Al, Fe^{2+})(SiO$_4$)(Si$_2$O$_7$)(OH)$_2$(OH, H$_2$O)]		
Jugoldite	Ca$_2$[Fe$_2^{3+}$(Fe^{3+}, Fe^{2+})(SiO$_4$)(Si$_2$O$_7$)(OH)$_2$(OH, H$_2$O)]		
Pumpellyite	Ca$_2$[Al$_2$(Al, Mg)(SiO$_4$)(Si$_2$O$_7$)(OH)$_2$(OH, H$_2$O)]		
Shuiskite	Ca$_2$[Cr$_2$(Al, Mg)(SiO$_4$)(Si$_2$O$_7$)(OH)$_2$(OH, H$_2$O)]		
Ardennite	(Mn, Ca)$_4$[Al$_4$(Mg, Al)$_2$(SiO$_4$)$_2$(Si$_3$O$_{10}$)(AsO$_4$, VO$_4$)(OH)$_6$]		

References

Akao, M. and Iwai S. (1977) The hydrogen bonding of artinite. *Acta Crystallographica* **B33**, 3951–3.

Albright, T. A., Burdett, J. K., and Whangbo, M. H. (1985) *Orbital Interactions in Chemistry*, Wiley-Interscience, New York.

Allmann, R. (1975) Beziehungen zwischen Bindungslangen und Bindungsstarken in Oxidstrukturen. *Monatschefte für Chemie* **106**, 779–93.

Baur, W. H. (1970) Bond length variation and distorted coordination polyhedra in inorganic crystals. *Transactions of the American Crystallographic Association*, **6**, 129–55.

Baur, W. H. (1971) The prediction of bond length variations in silicon-oxygen bonds. *American Mineralogist*, **56**, 1573–99.

Baur, W. H. (1987) Effective ionic radii in nitrides. *Crystallography Reviews*, **1**, 59–83.

Brown, I. D. (1976) On the geometry of O-H...O hydrogen bonds. *Acta Crystallographica*, **A32**, 24–31.

Brown, I. D. (1981) The bond-valence method: an empirical approach to chemical structure and bonding, in *Structure and Bonding in Crystals II*, (eds M. O'Keeffe and A. Navrotsky), Academic Press, New York, pp. 1–30.

Brown, I. D. and Shannon, R. D. (1973) Empirical bond-strength—bond-length curves for oxides. *Acta Crystallographica*, **A29**, 266–82.

Burdett, J. K. (1980) *Molecular Shapes*, John Wiley, New York.

Burdett, J. K. (1986) Structural–electronic relationships in the solid state, in *Molecular Structure and Energetics*, (eds A. Greenberg and J. F. Liebman), VCH Publishers, Boca Raton, pp. 209–75.

Burdett, J. K. (1987) Some structural problems examined using the method of moments. *Structure and Bonding*, **65**, 29–89.

Burdett, J. K. and Hawthorne, F. C. (1992) An orbital approach to the theory of bond-valence. *American Mineralogist*, (submitted).

Burdett, J. K. and McLarnan, T. M. (1984) An orbital interpretation of Pauling's rules. *American Mineralogist*, **69**, 601–21.

Burdett, J. K., Lee, S., and Sha, W. C. (1984) The method of moments and the energy levels of molecules and solids. *Croatia Chemica Acta*, **57**, 1193–216.

Cerny, P. and Hawthorne, F. C. (1976) Refractive indices versus alkali contents in beryl: general limitations and applications to some pegmatite types. *Canadian Mineralogist*, **14**, 491–7.

Christ, C. L. (1960) Crystal chemistry and systematic classification of hydrated borate minerals. *American Mineralogist*, **45**, 334–40.

Christ, C. L. and Clark, J. R. (1977) A crystal-chemical classification of borate structures with emphasis on hydrated borates. *Physics and Chemistry of Minerals*, **2**, 59–87.

Cromer, D. T., Kay, M. I., and Larsen, A. C. (1967) Refinement of the alum structures, II. X-ray and neutron diffraction of $NaAl(SO_4)_2 \cdot 12H_2O$. *Acta Crystallographica*, **22**, 182–7.

Ferraris, G. and Franchini-Angela, M. (1972) Survey of the geometry and environment of water molecules in crystalline hydrates studied by neutron diffraction. *Acta Crystallographica*, **B28**, 3572–83.

Fisher, D. J. (1958) Pegmatite phosphates and their problems. *American Mineralogist*, **43**, 181–207.

Gibbs, G. V. (1982) Molecules as models for bonding in silicates. *American Mineralogist*, **67**, 421–50.

Gibbs, G. V., Hamil, M. M., and Louisnathan, S. J. *et al.* (1972) Correlations between Si-O bond length, Si–O–Si angle and bond-overlap populations calculated using extended Huckel molecular orbital theory. *American Mineralogist*, **57**, 1578–613.

Hawthorne, F. C. (1979) The crystal structure of morinite. *Canadian Mineralogist*, **17**, 93–102.

Hawthorne, F. C. (1983) Graphical enumeration of polyhedral clusters. *Acta Crystallographica*, **A39**, 724–36.

Hawthorne, F. C. (1984a) The crystal structure of stenonite and the classification of the aluminofluoride minerals. *Canadian Mineralogist*, **22**, 245–51.

Hawthorne, F. C. (1984b) The crystal structure of mandarinoite, $Fe_2^{3+}Se_3O_9 \cdot 6H_2O$. *Canadian Mineralogist*, **22**, 475–80.

Hawthorne, F. C. (1985a) Towards a structural classification of minerals: the $^{vi}M^{iv}T_2O_n$ minerals. *American Mineralogist*, **70**, 455–73.

Hawthorne, F. C. (1985b) The crystal structure of stringhamite. *Tschermaks Mineralogische und Petrographische Mitteilungen*, **34**, 15–24.

Hawthorne, F. C. (1986) Structural hierarchy in $^{vi}M_x^{iii}T_yO_z$ minerals. *Canadian Mineralogist*, **24**, 625–42.

Hawthorne, F. C. (1990) Structural hierarchy in $M^{[6]}T^{[4]}O_n$ minerals. *Zeitschrifte für Kristallographie*, **192**, 1–52.

Hawthorne, F. C. and Cerny, P. (1977) The alkali-metal positions in Cs-Li beryl. *Canadian Mineralogist*, **15**, 414–21.

Hawthorne, F. C. and Ferguson, R. B. (1975a) Anhydrous sulphates: I. Refinement of the crystal structure of celestite, with an appendix on the structure of thenardite. *Canadian Mineralogist*, **13**, 181–7.

Hawthorne, F. C. and Ferguson, R. B. (1975b) Anhydrous sulphates: II. Refinement of the crystal structure of anhydrite. *Canadian Mineralogist*, **13**, 181–7.

Hoffman, R. (1988) *Solids and Surfaces: A Chemist's view of Bonding in Extended Structures*, VCH Publishers, New York.

Liebau, F. (1985) *Structural Chemistry of Silicates*, Springer-Verlag, Berlin.

Mereiter, K. (1974) Die Kristallstruktur von Rhomboklas, $H_5O_2\{Fe[SO_4]_2 \cdot 2H_2O\}^-$. *Tschermaks Mineralogische und Petrographische Mitteilungen*, **21**, 216–32.

Moore, P. B. (1970a) Structural hierarchies among minerals containing octahedrally coordinating oxygen: I. Stereoisomerism among corner-sharing octahedral and tetrahedral chains. *Neues Jahrbuch fur Mineralogie Monatschefte*, pp. 163–73.

Moore, P. B. (1970b) Crystal chemistry of the basic iron phosphates. *American Mineralogist*, **55**, 135–69.

Moore, P. B. (1973) Pegmatite phosphates: mineralogy and crystal chemistry. *Mineralogical, Record*, **4**, 103–30.

Moore, P. B. (1974) Structural hierarchies among minerals containing octahedrally coordinating oxygen: II. Systematic retrieval and classification of octahedral edge-sharing clusters: an epistemological approach. *Neues Jahrbuch für Mineralogie Abhandlungen* **120**, 205–27.

Moore, P. B. (1975) Laueite, pseudolaueite, stewartite and metavauxite: a study in combinatorial polymorphism. *Neues Jahrbuch für Mineralogie Abhandlungen* **123**, 148–59.

Moore, P. B. (1981) Complex crystal structures related to glaserite, $K_3Na(SO_4)_2$: evidence for very dense packings among oxysalts. *Bulletin de la Societe francaise Mineralogie et Cristallographie* **104**, 536–47.

Moore, P. B. (1982) Pegmatite minerals of P(V) and B(III). *Mineralogical Association of Canada Short Course*, **8**, 267–91.

Moore, P. B. (1984) Crystallochemical aspects of the phosphate minerals, in *Phosphate Minerals*, (eds J. O. Niagru and P. B. Moore), Springer-Verlag, Berlin, pp. 155–170.

Pabst, A. (1950) A structural classification of fluoaluminates. *American Mineralogist*, **35**, 149–65.

Pauling, L. (1929) The principles determining the structure of complex ionic crystals. *Journal of the American Chemical Society*, **51**, 1010–26.

Pauling, L. (1960) The Nature of the Chemical Bond, 3rd edn, Cornell University Press, Ithaca, New York.

Ripmeester, J. A., Ratcliffe, C. I., and Dutrizac, J. E. *et al.* (1986) Hydronium ion in the alunite-jarosite group. *Canadian Mineralogist*, **24**, 435–47.

Scordari, F. (1980) Structural considerations of some natural and artificial iron hydrated sulphates. *Mineralogical Magazine*, **43**, 669–73.

Scordari, F. (1981) Crystal chemical implications on some alkali hydrated sulphates. *Tschermaks Mineralogische und Petrographische Mitteilungen*, **28**, 207–22.

Shannon, R. D. (1975) Systematic studies of interatomic distances in oxides, in *The Physics and Chemistry of Minerals and Rocks*, (ed. R. G. J. Sterns), John Wiley & Sons, London, pp. 403–31.

Shannon, R. D. (1976) Revised effective ionic radii and systematic studies of interatomic distances in halides and chalcogenides. *Acta Crystallographica*, **A32**, 751–67.

Sutor, D. J. (1967) The crystal and molecular structure of newberyite, $MgHPO_4 \cdot 3H_2O$. *Acta Crystallographica*, **23**, 418–22.

Tossell, J. A. and Gibbs, G. V. (1977) Molecular orbital studies of geometries and spectra of minerals and inorganic compounds. *Physics and Chemistry of Minerals*, **2**, 21–57.

Trinajstic, N. (1983) *Chemical Graph Theory*, vol. I, CRC Press, Boca Raton.

Ziman, J. (1965) *Principles of the Theory of Solids*, Cambridge University Press, Cambridge.

Zoltai, T. (1960) Classification of silicates and other minerals with tetrahedral structures. *American Mineralogist*, **45**, 960–73.

CHAPTER THREE
Electronic paradoxes in the structures of minerals

Jeremy K. Burdett

3.1 Introduction

For the majority of readers of this chapter, perhaps the traditional view of solids, and especially of minerals, is the one usually described as the 'ionic' model. On this scheme solids are assembled by packing together charged billiard ball-like ions of different sizes, and described by the rules collected together by Pauling (1929). Such a viewpoint has the advantage that it leads to an easy visualization based upon packing concepts learned since infancy. In more quantitative language the 'size' of the ion concerned is set by the form of the repulsive part of the potential between it and other ions, which prohibits a mutual approach less than a certain distance. In recent years this conceptually simple model has been substantially improved by the inclusion of more sophisticated potentials between atoms, derived either empirically or from quantum chemical calculations on small molecules. Some of the results which have been obtained are quite impressive and have led to a useful route with which to model solids, from the study of structural stability fields to the determination of the compressibility of quartz, for example. In spite of the increase in sophistication it is still, of course, a ball-and-spring model. The basic ideas behind such a model are very different from those behind the quantum-mechanical-based models. These dominate chemical theory and are employed by the physicist who wishes to predict parameters such as the critical pressure for the NaCl→CsCl transition in rocksalt, or the critical superconducting temperature in $NbGe_3$. The orbital model is usually introduced early on in chemical and physical education. Every beginning chemist learns about the octet rule and the covalent origin of the two-electron bond, localized between adjacent bonded atoms, which holds together the atoms in diamond and in much of the organic chemical world. Introduced a little later is the concept of the delocalized interaction that we must use to view the existence of metals and conjugated organic molecules such as benzene. Tight-binding theory is a universal topic in beginning solid-state physics courses.

The idea behind this chapter is to make some connections between the very different ways of viewing solids. In the world of minerals our current thinking is still enormously influenced by the Pauling viewpoint, but we will show how orbital ideas often give the same result. This is the paradox to which we refer

The Stability of Minerals. Edited by G. D. Price and N. L. Ross.
Published in 1992 by Chapman & Hall, London. ISBN 0 412 44150 0

in the title. Often more than one theory leads to the same prediction. Which is correct? By and large most of the models we use to explain experimental observations are just that – models, invariably gross simplifications of what can be called 'the quantum mechanical truth'. Surely as long as the scheme is useful, able to produce rationalizations and predictions, then it can take its place as a useful model, and is judged on what it can accomplish in terms of breadth and detail. The molecular-orbital approach as used qualitatively and quantitatively in molecules and solids is a particularly fine example of a model, and we shall use it widely in this chapter. The approach will be a personal one, in that almost all of the examples we will use to make links with structural questions either initially, or even currently, viewed in a different language, come from our own research.

3.2 Orbitals of molecules

The important thrust of this section will be to tie together localized and delocalized viewpoints of chemical bonding (for a review see Albright et al., 1985). We shall see how the orbital picture is related to traditional ways to look at chemical bonds. In the next section we will show how analogous ideas work for solids. Before we develop the energy bands of solids we will work through three simple molecular examples to show how the energy levels of molecules are generated using a simple form of molecular orbital theory — the one-electron model. Symmetry is invariably important and allows not only a simplification of what could be messy arithmetic but for the purposes of this chapter will allow us a useful entry into the solid state in the next section. Take a molecule which is described by the point group G and has a collection of symmetry elements $g \in G$, and whose atoms hold a set of basis orbitals $\{\phi_i\}$. Initially we choose G to be a cyclic group. This would describe the benzene molecule, or the equilateral triangular structure of H_3^+. It has the property that there are the same number of representations as the dimension of the group. The symmetry elements g of this cyclic group simply take one orbital to another one in the ring. Thus in H_3^+ a C_3 operation takes ϕ_1 to ϕ_2. There is a standard way to construct the molecular orbitals of the system by evaluating the form of the orbitals of different symmetry. For the jth irreducible representation of G, the resulting molecular orbital, ψ_j, is

$$\psi(j) = \sum_g \chi(j,g) g \phi_1 \qquad (1)$$

where $\chi(j,g)$ is the character of the jth irreducible representation corresponding to the operation g. If there is only one possible $\psi(j)$ then the energy of the molecular orbital is found by substitution into the wave equation

$$E_j = \langle \psi(j) | \mathcal{H} | \psi(j) \rangle / \langle \psi(j) | \psi(j) \rangle \qquad (2)$$

For cyclic molecules, such as the H_n series, a general result is that each of the n representations are included just once in the collection of molecular orbitals. For the H_3^+ molecule (the point group for our purposes is C_3) the three H 1s orbitals transform as $a_1 + e$. (Two of the representations are related by time-reversal symmetry and lead to the degenerate e representation.) Use of standard group theoretical methods gives from the relevant character table

$$\psi(a_1) = \phi_1 + \phi_2 + \phi_3 \qquad (3)$$

$$\psi(e)_a = 2\phi_1 - \phi_2 - \phi_3$$
$$\psi(e)_b = \phi_2 - \phi_3 \qquad (4)$$

where the a,b subscripts apply to the two components of the degenerate e orbital. These wave functions are not normalized but this does not matter if we evaluate the energy correctly. For the a_1 representation the energy is

$$E(a_1) = \langle \phi_1 + \phi_2 + \phi_3 | \mathcal{H} | \phi_1 + \phi_2 + \phi_3 \rangle / \langle \phi_1 + \phi_2 + \phi_3 | \phi_1 + \phi_2 + \phi_3 \rangle$$
$$= (3\langle \phi_i | \mathcal{H} | \phi_i \rangle + 6\langle \phi_i | \mathcal{H} | \phi_j \rangle)/3$$
$$= \alpha_{HH} + 2\beta_{HH} \qquad (5)$$

where α_{HH} is the energy of an electron in an isolated 1s orbital ($\langle \phi_i | \mathcal{H} | \phi_i \rangle$ or H_{ii}), related to the ionization energy from that orbital, and β_{HH} the interaction or resonance energy between two adjacent orbitals ($\langle \phi_i | \mathcal{H} | \phi_j \rangle$ or H_{ij}). α_{HH} and β_{HH} are negative. This is the one-electron model at its most basic – the Hückel model. For the orbitals of symmetry e we can use the same recipe to give the complete picture of Figure 3.1. On climbing the stack of orbitals the

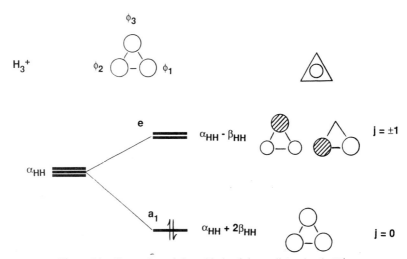

Figure 3.1 Generation of the orbitals of the cyclic molecule H_3^+.

number of nodes increases and we find one bonding and two anti-bonding orbitals. Notice the form of the two components of the e orbitals.

There is a general equation which gives the energy levels of such cyclic systems. For a ring containing n atoms it is just

$$E_j = \alpha + 2\beta \cos(2j\pi/n), \quad j = 0, \pm 1, \pm 2, \pm (n-1)/2 \ldots \qquad (6)$$

where we have dropped the H subscripts for generality. The wave functions are given by equation (1), which for these cyclic groups may be rewritten by identifying C_n^{p-1} with g, and using a general expression for the characters, $\chi(j, g)$.

$$\psi(j) = \sum_g \chi(j, g) g \phi_1$$

$$= \sum_{p=1}^{n} \exp(2\pi i j(p-1)/n) C_n^{p-1} \phi_1 \qquad (7)$$

For the H_3^+ case it is easy to see that for $j=0$, $\chi(0, g) = 1$, and thus all the atomic orbitals are in phase in the $j=0$ (a_1) orbital in Figure 3.1. The reason for introducing what appears to be a cumbersome expression will become clear in the next section.

Let us now generate the molecular orbital energy levels of the BH_3 molecule. This is a transient species which has a trigonal planar geometry. Using the ideas of Lewis we would envisage this molecule as being held together by three electron pair bonds which Pauling would describe as the overlap of three sp^2 hybrid orbitals on boron with three hydrogen 1s orbitals. The six electrons present in the molecule then give rise to three B–H σ bonds. The symmetry properties of the hydrogen 1s orbitals are the same as in the H_3^+ molecule (i.e. they transform as $a_1 + e$) but now, since they are so far apart, their interaction is very small. We shall set it equal to zero. The strong interactions now occur between these orbitals and the central atom orbitals. Recall that only orbitals of the same symmetry may interact. The in-plane 2s and 2p orbitals of boron transform as $a_1 + e$, i.e. as the same symmetry species as the hydrogen 1s orbitals and so they may interact with each other. Figure 3.2 shows how a molecular orbital diagram for this situation is derived. The (empty) p_z orbital on boron has no interaction with the ligand orbitals and so remains unchanged in energy. Notice that the lower (occupied) orbitals are B–H bonding but that the higher orbitals are B–H anti-bonding. So just as before, climbing the stack of orbitals leads to an overall increase in anti-bonding character.

The problem with the scheme we have used here is that the boron-hydrogen interaction is described in terms of orbitals which are delocalized over the entire molecule, a feature of molecular-orbital theory which could be frightening for large molecules with a correspondingly large number of atomic, and thus molecular, orbitals. However, a little trick allows us to generate a localized

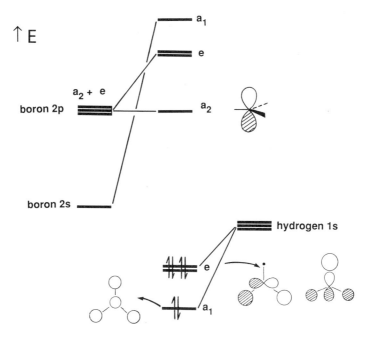

Figure 3.2 Generation of the orbitals of the molecule BH$_3$. For simplicity the form of the anti-bonding orbitals are not shown. They are, however, just the out-of-phase analogs of the bonding ones.

picture which has great similarities with the Pauling picture of three sp^2 hybrid orbitals involved in localized two-centre bonds between boron and hydrogen. If we adopt the viewpoint that it is the electron density that interests us when describing chemical bonding, then we may manipulate the molecular orbitals which contain electrons in almost any way we wish to derive a different picture. This is done in Figure 3.3 and the result of taking judiciously chosen linear combinations of these orbitals is a picture that has close similarities to the traditional localized one. There is one important qualification for this to work. There must be the same number of doubly occupied orbitals available for the localization process as there are bonds which we want to be able to make. With this comment in mind let us return to the case of H$_3^+$. Here there is only one doubly occupied orbital (a$_1$) but three 'bonds'. In this case we are not able to produce three two-centre–two-electron bonds. A similar restriction is true for the π orbitals of benzene. Here, there are only three doubly occupied orbitals but six close C–C contacts which are certainly 'bonds' of some type. In this case we have to resort to the use of resonance structures or use the delocalized model. Although the use of a circle inside the benzene hexagon is traditionally reserved for this molecule, the analogous picture for H$_3^+$ shown in Figure 3.1 is just as legitimate a description.

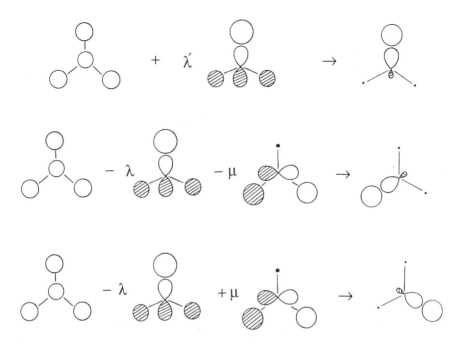

Figure 3.3 Localizaton of the delocalized orbitals of BH_3.

Thus, although a localized or delocalized picture for molecules is often a matter of choice, in many instances a delocalized viewpoint is the only one that may be used. Phrased in a different way, the delocalized picture is always applicable (as long as the molecular-orbital approach itself is valid) but the localized one only applies to specialized cases. It will be apparent to the enquiring reader that the localized wave functions of Figure 3.3 are not orthogonal and are thus not stationary solutions of the Schrödinger wave equation. If orbital results are thus to be used for comparison with electronic or photoelectron spectra, for example, the delocalized ones are those that must be used.

It is instructive to see how the orbital picture changes if there are a total of six atoms in the ring, of two different types as in Figure 3.4 instead of a homoatomic ring of three hydrogen atoms. The orbitals transform as $2(a_1+e)$, each A and B trio of orbitals giving rise to one a_1+e set. Using the formula of equation (1) with each of the two sets we can simply generate the two functions of a_1 symmetry as

$$\psi(a_1) = \phi_1 + \phi_3 + \phi_5 \tag{8}$$

and

$$\psi'(a_1) = \phi_2 + \phi_4 + \phi_6 \qquad (9)$$

corresponding to A and B type orbitals, respectively. Since these two functions are of the same symmetry and are not orthogonal they will interact with each other. Solution of a secular determinant (equation (10)) leads to the new energies:

$$\begin{vmatrix} \langle\psi|\mathcal{H}|\psi\rangle - E\langle\psi|\psi\rangle & \langle\psi|\mathcal{H}|\psi'\rangle \\ \langle\psi|\mathcal{H}|\psi'\rangle & \langle\psi'|\mathcal{H}|\psi'\rangle - E\langle\psi'|\psi'\rangle \end{vmatrix} = 0. \qquad (10)$$

The $\langle\phi_i|\mathcal{H}|\phi_j\rangle$ integrals linking the orbitals $\phi_{1,3,5}$ are zero since these orbitals are not bonded to each other. Thus $\langle\psi|\mathcal{H}|\psi\rangle - E\langle\psi|\psi\rangle$ is equal to $3\alpha_{AA} - 3E$. Analogously $\langle\psi'|\mathcal{H}|\psi'\rangle - E\langle\psi'|\psi'\rangle$ is equal to $3\alpha_{BB} - 3E$. The off-diagonal term $\langle\psi|\mathcal{H}|\psi'\rangle$ does contain interactions between adjacent orbitals and is equal to $6\beta_{AB}$. The energies of the two new orbitals are, then, from solution of equation (10) with these substitutions:

$$2E = (\alpha_{AA} + \alpha_{BB}) \pm [(\alpha_{AA} - \alpha_{BB})^2 + 4\beta_{AB}^2]^{\frac{1}{2}}. \qquad (11)$$

Expansion of the square root gives $E \simeq \alpha_{AA} + 4\beta_{AB}^2/(\alpha_{AA} - \alpha_{BB})$ and $\alpha_{BB} - 4\beta_{AB}^2/(\alpha_{AA} - \alpha_{BB})$. The energies of the two pairs of orbitals of e symmetry are obtained in the same manner and the result is shown in Figure 3.4. Notice that the form of the deeper-lying orbitals is such that they have a larger amount of A character than B character, commensurate with a larger electronegativity for A compared with B. The converse is true for the higher-lying orbitals. The discussion here is the one that we would use, for example, to study the π orbitals of the borazine molecule $B_3N_3H_6$. The nitrogen atom levels lie deeper than those of boron, in accord with the relative electronegativities of the two.

3.3 Energy bands of solids

The derivation of the energy bands of solids follows a route very similar to the discussion in Section 3.2. (for a review see Burdett, 1984). The only real difference is that for the molecule the number of levels is finite, but for the solid, the number is very large indeed. A simple little trick enables us to overcome this problem. For solids the system is described by a space group instead of a point group. One way to appreciate how the electronic structure unfolds, with nice connections with our molecular discussion, is to consider an infinite chain of atoms carrying a set of orbitals, $\{\phi_i\}$. We describe this by a translation group T with a collection of symmetry elements $t \in T$. The symmetry elements of this group are the translations which take a given atom to another in the

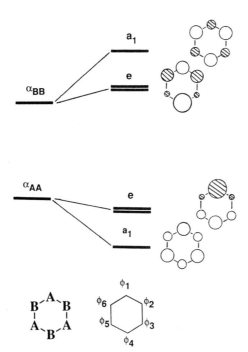

Figure 3.4 Generation of the orbitals of a cyclic A_3B_3 molecule. s orbitals are used on each of the atoms for simplicity but an identical picture would result if the orbitals are of π-type. Thus the diagram would be applicable to the π-levels of the borazine molecule, $B_3N_3H_3$, just as the diagram of Figure 3.1 is applicable to the π-manifold of the cyclopropenium cation $C_3H_3^+$.

chain. Recall in the cyclic group a C_3 operation took one H atom to an adjacent H atom in H_3^+. Symmetry adapted linear combinations of orbitals can be generated using this group in an exactly analogous way to that of equation (1). Of course the most obvious difference is that the group T is an infinite group with an infinite number of values of j. But, just as before in H_3^+, each representation occurs once in the collection of levels. We can write by analogy with the molecular case for a chain with n atoms

$$\psi(j) = \sum_g \chi(j,t) t \phi_1$$

$$= \sum_{p=1}^{n} \exp(2\pi i j(p-1)/n) T_n^{p-1} \phi_1 \quad (12)$$

T_n^{p-1} is the operator which moves $p-1$ atoms along this chain. This is clearly an unwieldy expression. All three labels j, p, and n are of the same order as the number of atoms in the crystal. However, with a little rearrangement the characters of the translation group take on a particularly simple form. If a is the repeat distance along the chain, then we can define r as the distance associated with a particular translation t. It will be a simple multiple of a. We shall set it equal to $(p-1)a$. We now define a new label k equal to $2j\pi/na$ so that

$$\chi(k, t) = \exp(ikr). \tag{13}$$

k is not only the label of the character of the group theoretical representation but is also the wave vector of the solid-state physicist. Thus while j is denumerably infinite for the solid-state case, k takes values from 0 to π/a and so avoids the problem associated with an infinite set of values. The space described by this spread of energy levels is called the first Brillouin zone. This solid-state equivalent of the molecular orbital method is usually called the tight-binding method. For a chain composed of NA atoms

$$\psi(k) = \cdots + \phi_1 + \exp(ik(a))\phi_2 + \exp(ik(2a))\phi_3 + \exp(ik(3a))\phi_4 + \cdots \tag{14}$$

with an energy therefore given by

$$\begin{aligned} E(k) &= [N\langle\phi_i|\mathcal{H}|\phi_i\rangle + 2N(\langle\phi_i|\mathcal{H}|\phi_j\rangle(\exp(ik(a)) + \exp(-ika)))]/N \\ &= \alpha_{AA} + 2\beta_{AA}\cos(ka) \end{aligned} \tag{15}$$

a result shown in Figure 3.5. This is an energy band of orbitals. Each one is labelled by a value of k, which now describes a continuum of levels, since the solid is of infinite extent. At the bottom of the band where $k=0$, (called the Brillouin zone centre) $E = \alpha_{AA} + 2\beta_{AA}$ and from equation (13) we see that the phase factor between adjacent orbitals is $+1$. That is, there are no nodes between the orbitals. It is the equivalent of the a_1 function for the H_3^+ molecule described earlier. Even its energy is the same for the cyclic case with $j=0$. It is bonding everywhere. At the top of the band where $k=\pi/a$ (the zone edge) the phase factor is -1 and there is a node between each adjacent orbital pair. It is anti-bonding and its energy is $\alpha_{AA} - 2\beta_{AA}$. k is then a node counter and the correspondence with the j of the molecular picture is thus a clear one. Since there is just one orbital per cell there is just one energy band, i.e. one energy value at each k point. Notice that the width of the energy band is proportional to the interaction energy (β_{AA}) between the adjacent orbitals.

The extension to the case of an alternating chain of A and B atoms follows from the discussion of the heteroatomic molecule above. A secular determinant (equation (16)) needs to be solved in just the same way as before, but now the basis orbitals are k-dependent, just as in equation (14).

$$\begin{vmatrix} \langle\psi(k)|\mathcal{H}|\psi(k)\rangle - E\langle\psi(k)|\psi(k)\rangle & \langle\psi(k)|\mathcal{H}|\psi'(k)\rangle \\ \langle\psi(k)|\mathcal{H}|\psi'(k)\rangle & \langle\psi'(k)|\mathcal{H}|\psi'(k)\rangle - E\langle\psi'(k)|\psi'(k)\rangle \end{vmatrix} = 0. \tag{16}$$

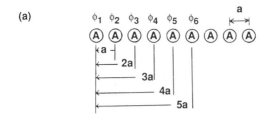

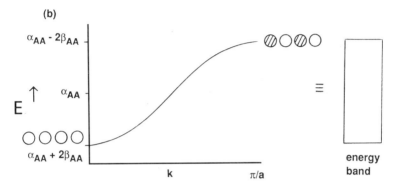

Figure 3.5 (a) Definition of the atom locations of the infinite homoatomic chain. (b) Generation of the energy band for the one atomic orbital per site of the chain. s orbitals are used on each of the atoms for simplicity but an identical picture would result if the orbitals are of π-type. Thus the diagram is just as applicable to the π-levels of polyacetylene, $(CH)_\infty$.

With reference to Figure 3.6 the basis orbital which describes the A orbitals is simply

$$\psi_A(k) = \cdots + \exp(jk(a))\phi_2 + \exp(ik(3a))\phi_4 + \exp(ik(5a))\phi_6 + \cdots \quad (17)$$

Since there is no direct overlap between the orbitals ϕ_i and ϕ_{i+2} there is no k-dependence of the energy of this function. If there are N orbitals of this type in the solid (N is very large of course) then the energy of $\psi_A(k)$ is just $N\alpha_{AA}$. For the B orbitals analogously

$$\psi_B(k) = \cdots + \phi_1 + \exp(ik(2a))\phi_3 + \exp(ik(4a))\phi_5 + \cdots \quad (18)$$

with an energy of $N\alpha_{BB}$. The off-diagonal terms do contain k however and are easily evaluated as $2N\beta_{AB}\cos(ka'/2)$ where a', the new unit cell measure is twice the old. By analogy with equation (10):

$$\begin{vmatrix} N\alpha_{AA} - NE & 2N\beta_{AB}\cos(ka'/2) \\ 2N\beta_{AB}\cos(ka'/2) & N\alpha_{BB} - NE \end{vmatrix} = 0. \quad (19)$$

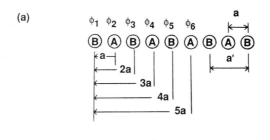

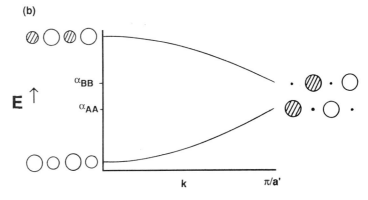

Figure 3.6 (a) Definition of the atom locations of the infinite heteroatomic chain. (b) Generation of the energy bands for the two atomic orbitals per cell of the chain. s orbitals are used on each of the atoms A,B for simplicity but an identical picture would result if the orbitals are of π-type. Thus the diagram is just as applicable to the π-levels of the hypothetical polymer, $(BNH_2)_\infty$.

The result is two energy bands

$$2E = (\alpha_{AA} + \alpha_{BB}) \pm [(\alpha_{AA} - \alpha_{BB})^2 + 16\beta_{AB}^2 \cos^2(ka'/2)]^{\frac{1}{2}}. \tag{20}$$

Their dispersion behaviour (i.e. k-dependence) is shown in Figure 3.6 obtained by expanding the square root of equation (20). This quickly gives $E \cong \alpha_{AA} + 4\beta_{AB}^2 \cos^2(ka/2)/(\alpha_{AA} - \alpha_{BB})$ and $\alpha_{BB} - 4\beta_{AB}^2 \cos^2(ka/2)/(\alpha_{AA} - \alpha_{BB})$. Compare this with equation (11) and Figure 3.4, the corresponding molecular case. An energy band of orbitals has replaced the finite collection of orbitals. At the bottom of the lower band at $k = 0$ the orbitals are in-phase from cell to cell from equation (13), as they are at the very top of the upper band. These correspond to maximum bonding and anti-bonding situations, respectively. At the zone edge the orbitals have to be out-of-phase and this gives rise to two non-bonding orbitals with energies of α_{AA} and α_{BB}.

We may derive this result in another fashion. Although it may be done mathematically a qualitative route will suffice. It leans on the form of the energy diagram generated by choosing a unit cell for the one-dimensional chain which contains two atoms rather than one. Now with two orbitals per cell there will be two energy levels at each k-value. But we know that the physical properties of the system should be independent of our choice of unit cell. The general result of doubling the unit cell is shown in Figure 3.7. It shows a 'folding back' of the dispersion curve associated with the one atom cell. $k = \pi/a'$ (where a' is the length of the new unit cell) corresponds to $k = \pi/(2a)$ (where a is the length of the old unit cell) since $a' = 2a$. The derivation of the result shown in Figure 3.6 for the AB chain case is thus easily understood from Figure 3.8, where an electronegativity perturbation is applied to the energy bands. The important result is the opening up of a gap at $k = \pi/a'$. (This result is of some importance. It is behind, for example, the result that graphite is metallic but isoelectronic, and isostructural BN is a white insulator.)

3.4 Localized or delocalized bonds in silicates and metallic oxides?

In this section we will first discuss the band structure for a part of the structure of a vertex-sharing chain of tetrahedra (Figure 3.9). Such tetrahedra occur in

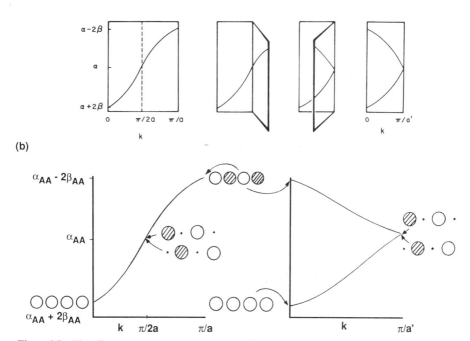

Figure 3.7 The effect on the band structure of doubling the unit cell. (a) Pictorial representation of the band folding process. (b) Correlation of the orbital character at various k-points.

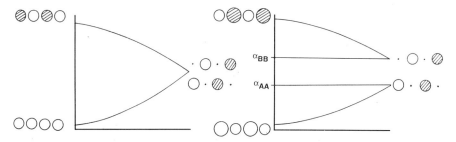

Figure 3.8 Effect of introducing an electronegativity difference between the two orbitals of the doubled cell of Figure 3.7. This might represent the Gedanken process $(CH)_\infty \to (BNH_2)_\infty$.

silicates such as the pyroxenes and phosphates such as Maddrell's salt, $NaPO_3$. The stoichiometry of the chain is TO_3^{-n} ($n=2$ for $T=Si$, and $n=1$ for $T=P$). We have chosen this particular example as it is the simplest infinite material of this type. Its structure is understood quite well by using the traditional ideas of two-centre, two-electron bonds. For the silicate case each silicon with four electrons forms two bonds to the oxygen atoms of the chain backbone, and two bonds to two pendant O^- units. Here, though, we examine the electronic structure using the band model. We will develop a qualitative picture but there is no reason why it could not be derived in a more quantitative way using the mathematics we have already described. We will study the band structure of a linear chain of alternating silicon and oxygen atoms, each containing one s and one p orbital. This will model the backbone of the tetrahedral chain. Slightly more complex results are found, which only differ in detail, not substance, if the real geometry of the chain is used.

First, we imagine a homoatomic linear chain which carries both s and p orbitals as in Figure 3.10(a). Simple symmetry arguments allow construction of the energy bands of this system. We know that the maximum bonding associated with the s band occurs at the zone centre, since from equation (13) this is where the phase factor is $+1$. However for the p band, lying at higher

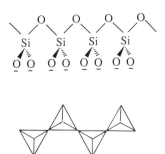

Figure 3.9 The vertex-sharing silicate (or phosphate) chain found in the pyroxenes, for example.

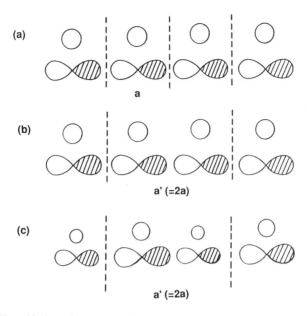

Figure 3.10 The orbitals used to generate the energy bands of the ... Si–O–Si–O ... backbone of Figure 3.9. (a) The s and p orbitals of the single atom cell of the homoatomic chain. (b) The s and p orbitals of the two atom cell of the homoatomic chain. (c) The s and p orbitals of the two atom cell of the heteroatomic chain.

energy, maximum antibonding occurs at the zone centre because the phase factor of $+1$ forces the arrangement of the anti-symmetrical p orbitals with adjacent lobes of opposite sign. Figure 3.11(a) shows the result we expect. (It is actually a little more complex since the s and p bands can mix together everywhere in the zone except at $k=0, \pi/a$. We look at this last.) Figure 3.11(b) shows what happens when this cell is doubled in size as in Figure 3.10(b). In an analogous way to that described above for the simple alternating chain, the bands are folded around $k=(1/2)\pi/a$. Now what happens when the orbital energies of the s and p orbitals are made different on the two centres? Again by analogy with our previous discussion a splitting of both bands occurs at the zone edge of the doubled cell. This is shown in Figure 3.11(c). We have labeled the bands I–IV in Figures 3.11(b),(c) to make the comparison easier. Notice that all four orbitals are orthogonal at the zone centre but that at the zone edge they are not. The result will be a mixing of two pairs of orbitals at the zone edge as shown in Figure 3.11(d) to give the final orbital picture of Figure 3.11(e). The result is a largely oxygen 2 s band (I), lying lowest in energy followed by an oxygen 2 p band (II). Each of these bands contain two electrons. To higher energy are two bands described as being composed largely of silicon 3 s (III) and silicon 3 p (IV). The bonding propensities of each are understandable using the simple rule we have described already.

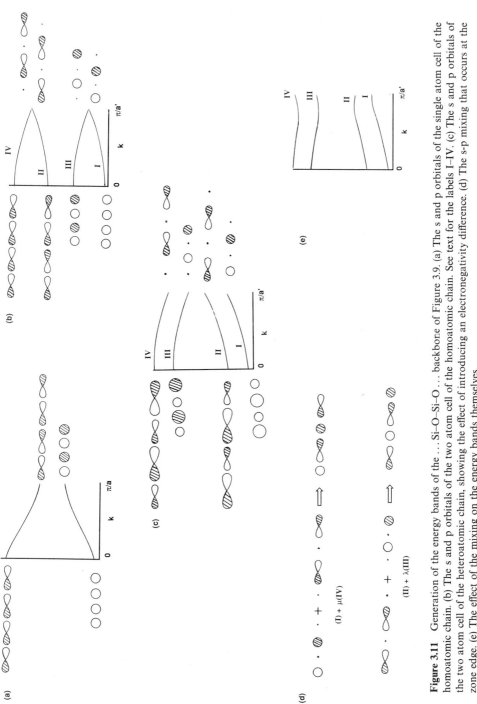

Figure 3.11 Generation of the energy bands of the ...Si–O–Si–O... backbone of Figure 3.9. (a) The s and p orbitals of the single atom cell of the homoatomic chain. (b) The s and p orbitals of the two atom cell of the homoatomic chain. See text for the labels I–IV. (c) The s and p orbitals of the two atom cell of the heteroatomic chain, showing the effect of introducing an electronegativity difference. (d) The s-p mixing that occurs at the zone edge. (e) The effect of the mixing on the energy bands themselves.

Notice the form of the bands of Figure 3.11. They are delocalized throughout the solid in much the same way as the molecular orbitals of BH_3 were delocalized over the molecule. However, we may use a similar technique to that used for the molecule in Figure 3.11 to localize these extended solid-state orbitals. To do this most accurately we should average the localization process over the whole zone, but the essence of the result may be gleaned from just doing the localization at the zone centre and the zone edge. Notice that the result shown in Figure 3.12 gives us two, two-centre–two-electron bonds. This is an important result and highlights the relationship between two very different ways to regard the bonding in these fundamental solids.

The concept of the metallic oxide follows naturally from the discussion above, and develops further the ideas of energy bands in solids. Sometimes it is perfectly acceptable to use a localized model to describe their chemical bonding as we have shown, but an energy-band model needs to be used to even begin to appreciate the observation of metallic oxides. As mentioned earlier the 'ionic' billiard-ball model of oxide structures has proven not only to be popular but also useful. However, a more detailed understanding of the electronic structure of the material is necessary in order to appreciate the observations such as the following. Anatase, one of the forms of TiO_2 is a white, insulating

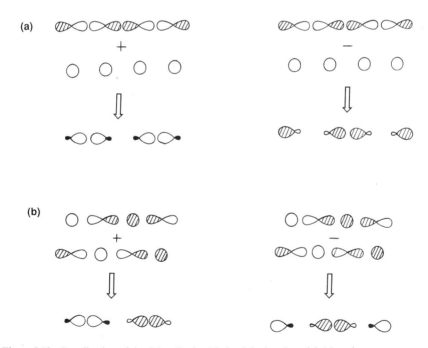

Figure 3.12 Localization of the delocalized orbitals of the band model, (a) at the zone centre and (b) at the zone edge.

material. It is in fact a 'typically ionic' oxide. However, reaction with the molecule Li^{t}Bu leads to generation of octanes and the incorporation of lithium into the lattice. For small amounts of lithium the anatase framework survives as the lithium intercalates into the material. Strikingly, the material is now black and a conductor of electricity.

Figure 3.13 shows a density-of-states diagram (Burdett *et al.*, 1987) computed for the rutile and anatase structures. Such diagrams show the number of energy levels per unit energy. There are three regions of interest here. The one which lies lowest is the band is largely oxygen p in character, and in order of increasing energy come the metal t_{2g} and e_g bands. The picture, then, is just a little bit more complicated than the AB chain of Figure 3.6 where there were just two bands, one largely A in character and the other largely B. For TiO_2 itself there are exactly the right number of electrons to fill the oxygen 2p band. The material is colourless (in its pure state) since there are no excitations to low-lying levels which would give rise to a band in the visible. The lowest energy excitation is across the band gap separating the full 'O 2p' band from the 'Tit_{2g}' bands. The material is an insulator since there are no accessible energy levels for the electrons to occupy to move through the crystal. The Pauli principle has ensured that each level is doubly occupied in this band so that there is no room for any more, even temporarily. However, addition of lithium to give Li_xTiO_2 leads to the transfer of electrons from the alkali metal into the metal t_{2g} band. Now there are accessible low-energy states and the metal

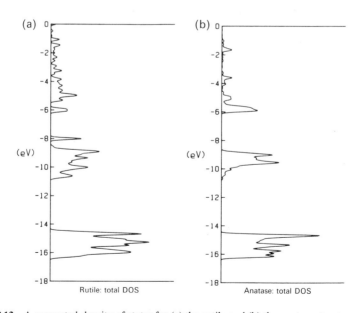

Figure 3.13 A computed density of states for (a) the rutile and (b) the anatase structure of TiO_2.

becomes a metal and black in colour. Later in this chapter we will find that use of a band model of this type is essential to understand in general the geometries of oxides with the rutile structure.

Similar observations, perhaps somewhat more dramatic, are seen in the series of high-T_c superconductors based on ternary, quaternary, and quinternary copper oxides. The oxide La_2CuO_4 is an insulator, but a little doping with divalent metals such as Sr to give $La_{2-x}Sr_xCuO_4$, where x is as small as ~ 0.06, gives a superconducting metal. The observation of metallic behaviour in these materials is not going to be understood using billiard-ball type of models. We do, however, note that some of the structural effects observed in these systems have been modelled using such an approach.

There are some restrictions concerning the use of orbital ideas in solids and these should be briefly mentioned. The molecular orbital, or Mulliken–Hund, model for molecules and the tight-binding method for solids relies on the dominance of the electronic problem by the strong orbital interactions between adjacent centres to produce a wave function where the electron is delocalized throught the solid or molecule. It ignores the terms in the energy such as the Coulombic repulsion experienced when two electrons are allowed to reside in the same atomic orbital. If such terms are large then the Mulliken–Hund wave function is not accurate and a wave function which forcibly localizes the electrons on different centres, the Heitler–London wave function, becomes appropriate. We refer the reader elsewhere (Burdett, 1990) for a discussion of the importance of the ratio of the one-electron to two-electron (e.g. Coulombic) energy terms in the electronic description of solids. Such considerations are in fact important in the area of the high-T_c oxides.

3.5 Interatomic distances and overlap populations

For some time now, since the pioneering work of G. V. Gibbs, molecular-orbital methods have been used to study several rather detailed aspects of the structures of silicates (Gibbs et al., 1981). For example, there is an excellent agreement between the results predicted using the Si–O–Si bending potential curve calculated via high-quality computations on molecules such as disiloxane, and the observed bond-angle distribution in these solids. Another impressive result is the correlation between bond angles and bond lengths in the same materials. There is though, as yet no consensus as to whether the rather soft bending potential, which is one of the underlying reasons behind the rich structural chemistry in this field, is a result of Si–O π-bonding or a result of Si–Si non-bonded interactions. It is quite clear though that high-quality ab initio calculations on fragments of these solids are quite adequate to describe numerically many of the vital parts of the solid-state problem. Thus, Gibbs' results tell us that, although the solid is infinite in extent, we can use local molecular orbital language to describe the electronic state of affairs, in the silicate backbone of the solid at least. In the language of molecular-orbital theory there is a calculable parameter, the bond overlap population which describes the strength of the chemical bonds formed between atoms. This is defined as

$$P_{ab} = 2\sum_j \sum_{i,k} c_{ij} c_{jk} S_{ik} N_j. \tag{21}$$

Here the first sum runs over all the molecular orbitals, j, and the second over all pairs of orbitals i,k, which are located on the two atoms a and b between which we wish to evaluate the overlap population. S_{ik} is the overlap integral between such an orbital pair and N_j is the number of electrons in the jth orbital. To see how this works we take two cases H_2 and H_3^+. Here there is only one occupied orbital, and only one orbital located at each atomic centre, so that there is only one term in the sum. The normalized bonding orbital in H_2 may be written (ignoring overlap in the normalization) as

$$\psi(H_2) = 2^{-\frac{1}{2}}(\phi_1 + \phi_2) \tag{22}$$

and the H–H bond overlap population in H_2 is thus $2 \times (2^{-\frac{1}{2}}) \times (2^{-\frac{1}{2}}) \times S \times 2 = 2S$. In H_2^+ it is just S. For H_3^+ the bonding orbital is written as

$$\psi(H_2) = 3^{-\frac{1}{2}}(\phi_1 + \phi_2 + \phi_3) \tag{23}$$

and, assuming the same H–H distance (implying the same overlap integral) the bond overlap population between a pair of hydrogen atoms is $2 \times (3^{-\frac{1}{2}}) \times (3^{-\frac{1}{2}}) \times S \times 2 = (4/3)S$. The considerably smaller number in H_3^+ represents the fact that the two electrons are delocalized over three 'bonds' in this molecule, but the two electrons are localized in one bond in H_2. In practice we find a longer (by about 0.1Å) H–H distance in H_3^+ than in H_2 and a longer (by about 0.2Å) H–H distance in H_2^+ than in H_3^+. Thus the bond overlap population provides a numerical recipe for the estimation of bond strengths, and as we have suggested here larger overlap populations imply shorter interatomic distances. Gibbs used this correlation to determine the electronic factors which determine the equilibrium Si–O distance as a function of geometry. We refer the reader to an excellent article (Gibbs et al., 1981) where the details of such studies have been elegantly described. There is a nice correlation between distance, bond overlap population, and the electrostatic bond strength sum, $p(O)$, to the oxide ions. The last parameter is a classical electrostatic parameter introduced by Pauling in 1929. Gibbs' papers show a striking isomorphism between the 'ionic' and molecular orbital approaches which will also surface below.

In this section we will use a rather simple example to highlight this relationship between the tenets of the electrostatic model and the predictions of molecular-orbital theory. Imagine a lead(II) atom lying in the eight-coordinate site shown in Figure 3.14(a). Such a state of affairs would apply to the

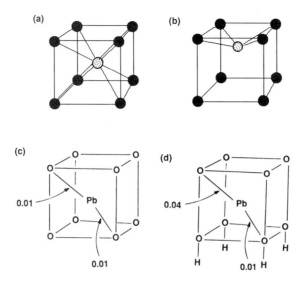

Figure 3.14 (i) Local geometries of CsCl and red PbO. (a) The CsCl structure type containing cubally coordinated atoms. (b) Movement of the atom at 1/2, 1/2, 1/2 towards one of the faces of the cube to generate the red PbO structure. (ii) Computed bond overlap populations for the Pb–O linkages for a structure containing a Pb(II) atom located centrally in a cube of oxygen atoms. (c) represents the case where all the oxygen atoms are identical. In (d) one set of oxygen atoms is coordinated to a set of hydrogen atoms.

hypothetical CsCl structure of PbO. We know that this structure type is not found in practice. The lead atom moves off-centre (Figure 3.14(b)) to generate the well-known tetragonal pyramidal lead(II) geometry in the tetragonal (red) polymorph of this material and elsewhere we have shown how this may be viewed in orbital terms (Burdett, 1981) *via* a second-order Jahn–Teller effect. The question we would like to answer is the following. If the environment of the lead atom is not cubal, for example, if the four oxygen atoms above the lead atom are coordinated to the rest of the structure in a different manner to the four below, how will the lead move? Such an arrangement is found in several minerals. The electrostatic model gives a ready answer. In order to maintain the largest number of correctly saturated oxygen atoms, the lead should move closer to the four oxygen atoms which are undercoordinated and away from the four oxygen atoms which are overcoordinated. The result will be values of $p(O)$ for the two types of oxygen which are as close as possible. The molecular-orbital approach makes similar predictions. Figures 3.14(c), and 3.14(d) show the calculated bond-overlap populations for this problem, where we have simulated the coordination asymmetry of the two sets of oxygen atoms by the attachment of hydrogen atoms to one set. The prediction is that the lead atom will move in the direction of the larger bond-overlap population obtained from the calculation where the two sets of internuclear distances are equal. This is a result quite equivalent to that obtained by use of the electrostatic bond sum.

It is important not to stop here. In order to really appreciate what is happening a study of the molecular-orbital diagram is important. Figure 3.15a shows the generation of the a_{1g} and t_{1u}, molecular orbitals derived from the 6s and 6p lead atomic orbitals. These orbitals, largely lead located, are anti-bonding between the lead and cube oxygen atoms. They represent the highest occupied molecular orbital (HOMO) and lowest unoccupied molecular orbital (LUMO), respectively, and thus play a vital part in determining the structural energetics. For the regular cube the contribution to the total overlap population from the

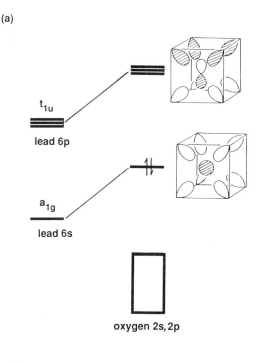

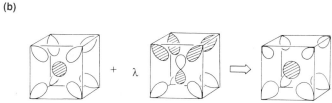

Figure 3.15 (a) Generation of a part of the molecular orbital diagram for the PbO_8 unit, showing the HOMO and LUMO. (b) Mixing of the HOMO and LUMO of (a) to give the orbitals of the PbO_8 unit to which hydrogen atoms have been attached to one side. The picture in (b) has been exaggerated for effect. There is, of course, admixture of a small amount of p orbital character into the central atom s orbital.

HOMO is the same (by symmetry) for both sets of oxygen atoms. For the asymmetric cube however, the a_{1g} and t_{1u} labels are inappropriate. The a_{1g} orbital (now a_1) mixes with one component of the t_{1u} set (now $a_1 + e$). The ideas or perturbation theory tell us how these orbitals will mix. The lower member will be stabilized by mixing in of the upper member in a bonding way. This is shown in Figure 3.15(b). As we can see the result is to decrease the anti-bonding contribution to the 'free' oxygen atoms, and increase the anti-bonding contribution to those oxygen atoms to which hydrogen atoms are attached. This shows up numerically in the calculated overlap populations.

3.6 Crystal-field and molecular-orbital stabilization energy

The crystal field theory has long been used in the study of minerals to understand many different aspects of the site preferences and geometries of transition metal ions in oxide minerals. In molecular chemistry, however, it has generally been discarded when it comes to understanding the nuances of transition metal complexes for both the Werner type (perhaps most akin to the situation in minerals) and for organometallic systems. In some areas of chemistry it is now often used only as a teaching tool. In this section we show how a molecular-orbital approach gives qualitatively indentical results, and also indicate the greater flexibility of such a scheme. The particular approach that we use has been termed the angular overlap model (Burdett, 1980). It received its initial development (as did the crystal field theory) in the area of electronic spectroscopy of transition metal complexes.

The general idea of the approach utilizes a perturbation expansion of the interaction energy between two orbitals on different atoms. In our cases here we shall identify these as a transition metal d orbital and a ligand orbital of some type. Atomic electronegativity has a strong correlation with atomic ionization potentials, and indeed on several scales is defined in these terms. Thus in general an oxygen 2s or 2p orbital will lie much deeper in energy than a metal d orbital, although as the metal becomes more electronegative the two can become of comparable energies as is apparent in recent studies of copper oxide chemistry. We start (Figure 3.16) with two orbitals i, j separated in energy by ΔE. The energy shift $\Delta \varepsilon$ on interaction is given to second order by $H_{ij}^2/\Delta E$ and to fourth order by $H_{ij}^2/\Delta E + H_{ij}^4/(\Delta E)^3$. An exactly analogous result was obtained, but only quoted to second order by the expansion of equation (11) described earlier. H_{ij}, the interaction integral between the two is usually set proportional to the overlap integral, S_{ij} between the two orbitals, an approximation introduced initially by Mulliken. This integral depends upon the distance (r) between the two centres holding the orbitals i, j, and also their angular orientation (θ, ϕ). In general we can write $S_{ij} = \sum_\lambda S_\lambda(r) f(\theta, \phi, \lambda)$ where λ represents contributions from σ, π and δ type overlap. The functional dependence of $S_\lambda(r)$ depends upon the details of the atomic wave functions $|i\rangle, |j\rangle$, but the $f(\theta, \phi, \lambda)$ are simply set by the form of the spherical harmonics which

ELECTRONIC PARADOXES IN THE STRUCTURES OF MINERALS

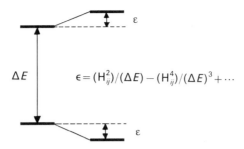

Figure 3.16 Perturbation theoretical description of the interaction of two orbitals, i and j separated in energy by ΔE. H_{ij} is the integral $<\phi_i|\mathcal{H}|\phi_j>$.

determine the angular behaviour of $|i>, |j>$. These are tabulated in several places (Burdett, 1980). If, because of the geometry of the problem, σ and π bonding are separable and δ interactions precluded by symmetry, as they are in the cases we shall discuss, then introduction of two energy parameters, e_λ and f_λ to describe the second-order and fourth-order interactions simplifies the state of affairs greatly. The energy shift expression contains terms such $S_\lambda^2(r)f^2(\theta', \phi', \lambda)$ in second order, and terms such as $S_\lambda^4(r)f^4(\theta', \phi', \lambda)$ in fourth order. In terms of these new parameters $\Delta\varepsilon_\lambda = e_\lambda(r)f^2(\theta', \phi', \lambda) + f_\lambda(r)f^4(\theta', \phi, \lambda)$ for a specific orientation (θ', ϕ') of a pair of metal and ligand orbitals. e_λ and f_λ of course depend on the metal-ligand distance and the identity of the orbitals $|i>$ and $|j>$. For the transition metal systems we shall study there are just two types of interaction, $\lambda = \sigma, \pi$. Because of its strong reliance on the ready identification of the form of the angular part of H_{ij}, this approach has been termed the angular overlap method.

Perturbation theory thus allows us to evaluate the total interaction energy between, for example, a central transition metal orbital and a set of ligand orbitals in terms of units of e_λ and f_λ. The general expression for the *total* interaction of (say) a metal z^2 orbital with a set of ligand orbitals for a metal ion surrounded by six oxygen atoms becomes $e_\lambda(r)[\sum f^2(\theta', \phi', \lambda)] + f_\lambda(r)[\sum f^2(\theta', \phi', \lambda)]^2$ where the sum runs over all of the interactions between the metal z^2 orbital and the ligand orbitals. Notice that the coefficient of the fourth-order term is just the square of the second-order one. If we know the $f(\theta', \phi', \lambda)$, then we can generate the interaction energies and hence the molecular orbital diagram. For our purposes here the leading (and thus larger) term in the perturbation expansion, e_λ, will be sufficient for our needs. In a later section we will find it necessary to study the form of the fourth-order term. In a strict sense just adopting the leading term in a perturbation expression is not going to be particularly accurate, but we shall find that the *qualitative* use of such an approach is very useful in understanding the electronic structure of the complex.

110

What do the $f(\theta', \phi', \lambda)$ look like? If we are just concerned with geometries of metal complexes where all the angles are either 90° or 180° then they take on a particularly simple form. Figure 3.17(a) shows them pictorially for this geometrical situation. The generation of the levels for the octahedral complex is simple, and is shown in Figure 3.17(b). For z^2 there are two interactions with

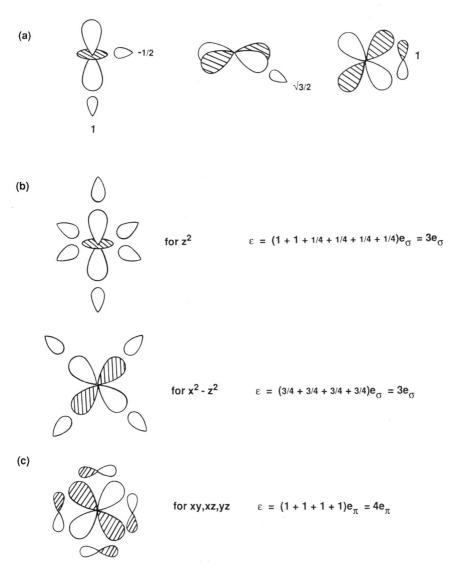

Figure 3.17 (a) The values of $f(\theta', \phi', \lambda)$ for the special cases of the octahedral geometry and for $\lambda = \sigma, \pi$. (b) Evaluation of the σ interaction energy associated with z^2 and $x^2 - y^2$ orbitals. (c) Evaluation of the π interaction energy associated with the xy, xz, and yz orbitals.

the large lobes of σ type (each contributing $1^2 e_\sigma$) and four interactions with the collar of z^2 (each contributing $(-1/2)^2 e_\sigma$). Thus the total interaction energy is $3e_\sigma$. One set of ligand orbitals is thus stabilized by $3e_\sigma$ and the z^2 orbital is destabilized by the same amount. We can use the same method to obtain the energy of $x^2 - y^2$. Here there are just four interactions with the lobes of $x^2 - y^2$ each contributing $(\sqrt{3}/2)^2 e_\sigma$. The total interaction is just $3e_\sigma$ the same figure we found for the z^2 orbital. This is just as it should be; these two orbitals form a degenerate e_g pair in this point group. The t_{2g} orbitals are of the wrong symmetry to interact with the ligands in a σ sense but can do so in a π sense. Figure 3.17(c) shows how the total π interaction with each member is $4e_\pi$. The energy scale of the complete molecular-orbital diagram of Figure 18 is in units of e_λ, and since σ bonding is generally more energetically important than π bonding $e_\sigma > e_\pi$. (Parenthetically we note that these parameters depend upon the internuclear separation. In general terms, over the region associated with chemical bonding distances, they increase in magnitude as the distance becomes shorter.) On this model, then, the $e_g - t_{2g}$ splitting (conventionally called Δ_{oct}) in the octahedral complex is just $3e_\sigma - 4e_\pi$. Spectroscopic results on a variety of systems have sought to generate numerical values for these parameters. Notice that the sum destabilization energy of the e_g pair is $2 \times 3e_\sigma = 6e_\sigma$. The coefficient is just the total number of σ orbitals presented by the ligands. Similarly the total π contribution is $3 \times 4e_\pi = 12 e_\pi$, and the coefficient here represents the total number of π orbitals presented by the ligands. This is a very general result. In a tetrahedral geometry the total σ interaction will be $4e_\sigma$.

Notice that there is considerable flexibility in the model. If the ligands are π-acceptors (CO is an example) rather than the π-donors we have used here then the t_{2g} set of orbitals are depressed in energy relative to the free d-orbitals and Δ_{oct} now becomes $3e_\sigma + 4e_\pi$. We shall see an application of this result in a later section, but note for now that such concepts give a much broader understanding concerning the origin of Δ_{oct} than the crystal-field ideas. Thus, for example, the much larger Δ_{oct} found when CO is a ligand compared to O is ascribed to CO being a π-acceptor and O being a π-donor.

For the tetrahedral geometry similar ideas apply, although because of the angular geometry the $f(\theta', \phi', \lambda)$ are a bit more complex. Here (Figure 3.18) the d orbitals split into two sets t_2 (involved in σ and π bonding) and e (involved in π bonding only). One useful result is easy to derive. Assume that π-bonding is not important and just look at the σ interactions. Since there are a total of four ligand σ orbitals, the sum total σ interaction is $4e_\sigma$. Since these are shared between the three components of the t_2 set, each receives an interaction energy of $(4/3)e_\sigma$. Thus Δ_{tet} is $(4/3)e_\sigma$ and $\Delta_{tet} = (4/9)\Delta_{oct}$. This is a result derivable using the crystal field theory but requires more than just a couple of lines.

Figure 3.19 shows a comparison between the d-orbital level structure obtained using the crystal-field and molecular-orbital approaches. A major difference is that in the molecular orbital model we need to consider the ligand-located orbitals in addition to the d-orbitals. We can readily evaluate

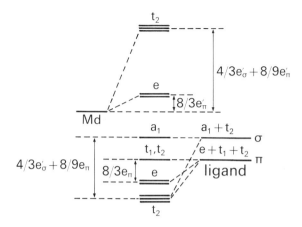

Figure 3.18 The d-levels of the ML_4 tetrahedron.

the molecular orbital stabilization energy (MOSE), the molecular orbital equivalent of the crystal field stabilization energy (CSFE) as a function of d-count. For the d^0 electron count the total MOSE is simply achieved by filling all the bonding orbitals located on the ligands with electrons. The total MOSE is thus $12e_\sigma + 24e_\pi$. As the d levels are filled, population of these metal–ligand and anti-bonding orbitals will reduce this figure. At the d^{10} electron count, bonding and anti-bonding orbitals are equally occupied and the MOSE is zero. This is shown in Figure 3.20(a). (Here, for simplicity we have put $e_\pi = 0$, but the reader can readily show that a plot using a realistic ratio of e_π/e_σ of 0.25 is qualitatively similar.) The plot looks very different from the CFSE (Figure

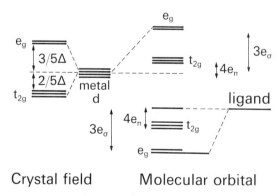

Figure 3.19 A comparison between the molecular-orbital and crystal-field approaches to the electronic structure of transition metal complexes.

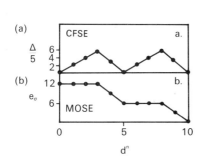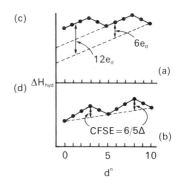

Figure 3.20 (a) The CFSE and (b) the MOSE for an octahedral transition metal complex. (c), (d) show the addition of a sloping background so as to mimic the variation in parameters such as the heat of formation of the oxides with d-count.

3.20(b)), but a little manipulation will show how they are related. A sloping background added to the MOSE plot will convert it into the well-known form found experimentally for the heats of formation of metal oxides, for example, This is shown in Figures 3.20(c) and 3.20(d) which compare the result with the traditional crystal-field one, where a sloping background is added to obtain agreement with experiment. The sloping background of the orbital model we claim comes from the strong interactions of the ligand orbitals with the central atom $(n+1)s$, p orbital which lies higher in energy than the metal nd ones. Of course past the d^{10} electron count these s and p orbitals dominate the chemical bonding. Notice that in quantitative terms since the two sloping backgrounds are different, the magnitude of the contribution from d-orbital effects is different on the two models. In general the d-orbital contribution is larger using the molecular orbital model.

Bond-length variations with electron configuration are interpreted in terms of the occupation of the d-orbitals of various types in the molecular orbital model. One particularly useful observation is the ability to include both π-acceptors and π-donors. Thus while a low-spin d^6 molecule with π-donor ligands results in occupation of t_{2g} orbitals which are metal ligand antibonding, a low-spin d^6 molecule with π-acceptor ligands results in occupation of t_{2g} orbitals which are metal-ligand bonding. Such concepts allow rationalization of the rather different behaviour found for the different types of ligand complex, impossible if the CFT is used. With π-acceptor ligands metal-ligand distances decrease as electrons are added to the metal-ligand π levels of t_{2g} symmetry. For the case of π-donor ligands the result is the opposite. In a comparison between ferro- and ferricyanide complexes, the ferrocyanide ion with a d^6 configuration has a shorter Fe-C distance by 0.026(8)Å than the d^5 ferricyanide ion. This is in spite of a difference in ionic radius of 0.06Å in the opposite direction. Another isoelectronic example concerns the V–C distances in V(CO)$_6$ and V(CO)$_6^-$ (2.001 and 1.931Å respectively).

3.7 Pauling's third rule: the instability of edge-sharing tetrahdra

Pauling's third rule is concerned with the sharing of polyhedral elements in solids. 'The presence of shared edges, and particularly of shared faces, in a coordinated structure decreases its stability; this effect is large for cations with large valence and small coordination number, and is especially large in case the radius ratio approaches the lower limit of stability of the polyhedron.' The idea for such a rule comes from the 'ionic' model of solid state structure. The absence of shared geometrical elements is attributed to the destabilizing effects of cation–cation Coulombic repulsions in structures in which they occur. In Figure 3.21 the cation–cation Coulombic repulsions in structures in which they occur. In Figure 3.21 the cation–cation distances in pairs of shared tetrahedra decrease in the order vertex-, edge-, face-sharing, resulting in an increase in electrostatic repulsion between these ions in the same order. In this section we will examine this rule of Pauling's using orbital arguments.

Many structures are known which violate this rule and they include that of β–BeO and the isostructural series BeCl$_2$, SiS$_2$, and SiO$_2$–w, all of which contain edge-sharing tetrahedra. In broad terms however the rule holds well. The way we will approach this problem (see Burdett and McLarnan, 1984) is to compare energetically a series of possible structures for a system of fixed stoichiometry which give rise to a spectrum of structural possibilities involving sharing of elements of more than one type. The system we choose is the comparison between the observed wurtzite-type structure of α-BeO, with alternative structures containing varying amounts of edge-sharing among the BeO$_4$ tetrahedra. Each of these structures is based on an hcp oxygen framework. Such structures can be generated by occupying some upward-pointing and some downward-pointing tetrahedra in the hcp framework, instead of filling only the upward-pointing sites as in wurtzite itself. In wurtzite (Figure 3.22(a)) all the tetrahedra point up, and no edges are shared as a result. In an alternative structure (Figure 3.22(b)) edge-sharing occurs since some tetrahedra

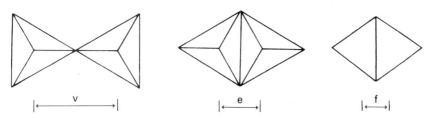

Figure 3.21 Variation in cation–cation distances with vertex, edge, and face-sharing of two tetrahedra.

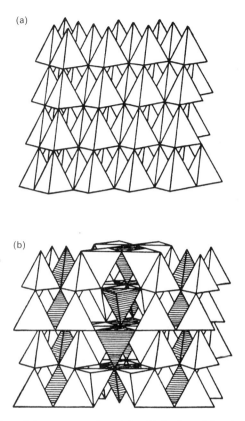

Figure 3.22 (a) A view of the structure of wurtzite. (b) A view of the structre of a wurtzite-like alternative.

point up, others point down. Here, we will consider only structures in which occupied tetrahedra share edges and vertices but not faces. Obviously, there are an infinite number of structures which may be generated using this algorithm, but only a finite number for a unit cell of a certain fixed size. There are, for example, 178,380 different ways to order the Be atoms over an hcp-anion array for a unit cell containing 24 oxygen atoms, but for our study it will be sufficient to study the 22 dipolar tetrahedral structures possible for a smaller cell. First, we need to rank them energetically by calculation, and then search for understanding of the results produced.

Both atoms lie in four-coordinate sites and, for the composition BeO, all the possibilities satisfy Pauling's electrostatic valence rule exactly. So, from the Pauling standpoint, they should differ in energy only because they differ in the extent of edge-sharing. We expect to see an increase in the energy with the extent of edge-sharing. The results are fascinating. Figure 3.23 shows the energy evaluated by using both band structure calculations and the most extreme and

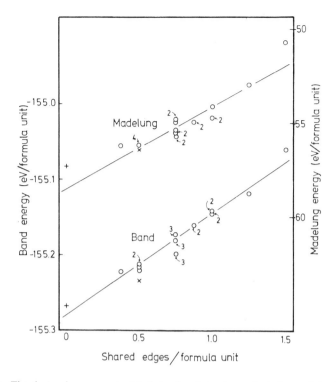

Figure 3.23 The electronic energy calculated *via* a band structure calculation and the electrostatic energy plotted against the average number of shared edges per cell for α-BeO (wurtzite structure), β-BeO, and 21 other systems with wurtzite-like structures.

simple ionic model, namely a Madelung sum [i.e. $E = \frac{1}{2} q_i q_j (r_{ij})^{-1}$, where q_i and q_j are the charges on atoms i and j separated by a distance r_{ij}]. The energy is clearly a linear function of the number of shared edges, in keeping with Pauling's third rule, but most importantly, either a purely electrostatic or a purely orbital model predicts that the number of shared edges controls the total energy. The variation in the electrostatic (Madelung) energy is understandable in terms of Pauling's initial ideas, but how about the variation in orbital energy?

The explanation is in fact an obvious one, but is one which turns on its head the traditional way of geometrically looking at structures of this type. Invariably, following the Pauling lead, structures such as these are envisaged as being composed of cation-centred polyhedra (tetrahedra in the present case) linked together into a three-dimensional structure. Such a viewpoint has proven to be a very useful one. However, in these structures most of the electron density lies on the oxide ions, and it is the coordination geometry of these which are important. In these 22 crystal structures, there are four possible anion coordination geometries, shown in Figure 3.24 and labelled 0, 1, 2, and 3, the number of shared edges in which an anion in that configuration participates. Notice that as the structure is increasingly distorted away from tetrahedral the number of shared edges mandated by the geometry increases. Obviously, by comparison with molecular chemistry it is the tetrahedral structure which has the lowest energy for the octet case. The tetrahedral structure of methane testifies to this. The relative energies of these four geometries (E_i) can be estimated by computing the energies of eight-electron OBe_4^{+4} molecules, where each of the 'ligands' is modelled by a Be atom which for simplicity contains a single 1s orbital. Approximate energies for the 22 crystal structures containing n_i anions per unit cell having coordination geometry i are computed by simply evaluating the sum $\Sigma n_i E_i$. The correlation between the energy calculated via these 'molecular' fragments and the crystal energy from the full band structure computation is very good, as shown in Figure 3.25. The reason behind the non-unit slope of this figure is not only because s orbitals alone have been used on the ligands but also because the ligand environment is one-, not four-coordinate.

The structure of β–BeO, shown in Figure 3.26 shows a three-dimensional framework composed of edge-sharing Be_2O_2 dimers that then link by vertex-sharing. (One way to describe the structure is to imagine the structure of rutile

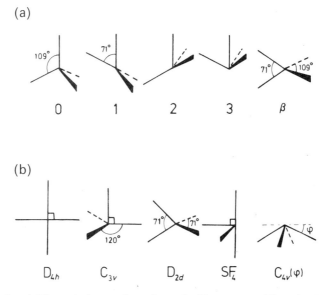

Figure 3.24 Local OBe_4 units in wurtzite and wurtzite-like structures. The numerical label gives the number of shared edges mandated by such a local oxygen geometry. The geometry labelled β is the local oxygen geometry in β–BeO. The other labels refer to geometries which will be used later.

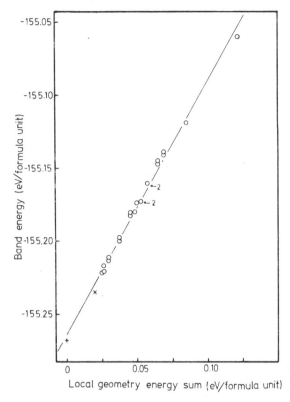

Figure 3.25 Estimation of the relative energy of the wurtzite and wurtzite-like structures *via* the energies (E_i) of local OBe_4^{+4} units. These energies for the 22 crystal structures containing n_i anions per unit cell having coordination geometry i are computed by evaluating the sum $\sum n_i E_i$.

where each TiO_6 octahedron has been replaced by this edge-sharing pair of BeO_4 tetrahedra.) If the BeO_4 tetrahedra in this arrangement are assumed to be regular, as in the 22 dipolar tetrahedral structures above, then the anion coordination is that shown in Figure 3.24 (labelled β). Calculations on an OBe_4^{+4} molecule with this geometry shows that it lies below any of the conformations 1, 2, and 3 in energy. Interestingly, as shown in Figure 3.23, the three-dimensional β-BeO crystal structure lies lower in energy on both models than any of the other structures except that of wurtzite (α-BeO). This occurs even though four of the 21 higher-energy types have the same number of shared edges as β-BeO, and one has fewer. Figure 3.25 shows that the energy of β-BeO relative to the α form is well approximated by considering the relative energies of 'molecular' OBe_4^{+4} units. The details of the anion coordination geometry are then vitally important in controlling the stability of solids of this type, and can enable an energetic ranking of similar structure types with different anion packings. Such considerations also help rationalize the occurrence of structures

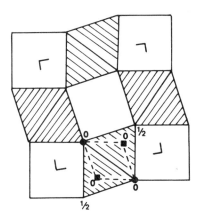

Figure 3.26 The structure of β–BeO.

like β–BeO with more shared edges (but also more stable anion environments) than some unobserved alternatives.

The 'isomorphism' between orbital and 'ionic' models is a fascinating one and encourages further thought. Be–Be interactions are not important for the orbital model. The same correlation is also found without their inclusion in the calculation. The energy differences are due almost entirely to the ability of the central oxygen atom to form stronger bonds when its coordinated Be atoms are arranged tetrahedrally, as indicated by (amongst other things) correlation between the Be–O overlap populations and $\Sigma n_i E_i$. Thus if we can understand why methane is a tetrahedral molecule then we can understand these results in a qualitative way. But the crucial issue is why for the 22 possible BeO structures do the energies associated with the orbital model correlate so strongly with those of the Madelung calculations. We can gain some insight by asking why these structures have different Madelung energies. The fixed hcp oxygen framework in these structures ensures that every anion sees the same site potential due to neighbouring anions in all 22 structures. Also each anion in any one of these structures has the same number of cations at any given distance as any other anion, so the contribution of cation–anion interactions to the Madelung energy will also be identical throughout these types. Thus the differences in energy can only occur via cation–cation repulsions. This part of the Madelung energy for the four molecular geometries is just $(\frac{1}{2})\Sigma(r_{ij})^{-1}$ where r_{ij} is the non-bonded Be–Be distance. It turns out that the total one-electron energy of these molecules too is proportional to $\Sigma(r_{ij})^{-1}$ as shown in Figure 3.27, a result which clearly locates the reason for the agreement between the covalent and ionic models in their energetic predictions for this family of structures. It is not though immediately clear in quantitative terms where such a result comes from on the orbital model. The result is, however,

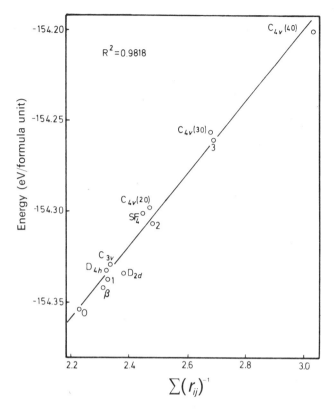

Figure 3.27 The calculated molecular orbital energies of the OBe_4^{+4} units shown in Figure 3.25 plotted against the sum of the inverse interatomic Be–Be separations, $\Sigma(r_{ij})^{-1}$.

in accord with the predictions of the valence shell-electron pair repulsion (VSEPR) or Nyholm–Gillespie model. Here the equilibrium molecular geometry is fixed by the so-called Pauli repulsions between localized pairs of electrons about a central atom. If these pairs interact by Coulombic forces, then the angular variation is likewise proportional to $\Sigma(r_{ij})^{-1}$. However calculations such as ours do not formally contain such repulsions, and we should certainly look for other explanations.

A purely one-electron result from the angular overlap model shows a qualitative relationship too. It may be argued that the 2s orbital on the oxygen atom really lies too deep in energy to strongly influence the geometry (the Rundle–Pimentel model) and just serves as a storage location for a pair of electrons. We may, then, use the angular overlap model to investigate the energetic changes with angle, by stressing the interactions of the central atom 2p orbitals with the ligand 1s orbitals of the OBe_4^{+4} unit. These are shown in Figure 3.28 for the case of a C_{3v} distortion of the tetrahedron. The overlap

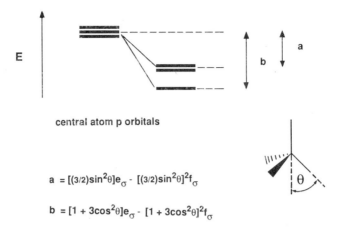

Figure 3.28 Molecular orbital diagram for a $C_{3v}OBe_4^{+4}$ unit using the angular overlap model.

integral of a ligand s orbital with a central atom p_z orbital is simply proportional to $\cos\theta$, where θ is the angle the internuclear axis makes with the z axis. As can be seen the quadratic terms sum to an angle independent figure (as they must do from Unsöld's theorem) but the quartic terms do have a strong angle dependency. Figure 3.29 plots $\Sigma(r_{ij})^{-1}$ against the sum of the quartic terms as a function of the displacement away from the tetrahedral geometry for $\theta = 45\text{–}80°$. There seems to be a linear relationship between the two for this type of distortion. (Not shown though is the somewhat poor agreement for $\theta = 90°$.)

Although we found that the molecular-orbital energy varies as $\Sigma(r_{ij})^{-1}$ for the case of BeO it is interesting to ask how the exponent changes as the

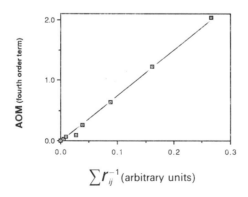

Figure 3.29 Plot of the sum of the inverse interatomic Be–Be separations, $\Sigma(r_{ij})^{-1}$ and the angular variation in the quartic part of the angular overlap energy against the angle θ.

electronegativity difference between the constituents varies. Molecular-orbital calculations for hypothetical octet AX_4 systems show that it hardens quite dramatically as the electronegative difference between A and X decreases. This tells us that it is probably quite difficult to really prove the reasons underlying the isomorphism noted above between 'ionic' and orbital models. However, using the result shown for the perturbation-theory calculation, one can imagine that higher terms in the expansion become important as the overlap increases. Very importantly, though, behind such a result may well be the reason for the failure of the ionic model in the area of 'covalent' materials.

3.8 The Jahn–Teller effect

The Jahn–Teller theorem is a well-used structural tool. Based on the result that orbitally degenerate electronic states in molecules are geometrically unstable, the theorem is quite a powerful one. Many of the textbook examples are from the solid state, one particularly popular series being the oxides and fluorides that adopt the rutile structure. The idea is that these species are 'ionic' so that the geometry around the metal ion adopts a geometry controlled by very local effects. This section though will show how this viewpoint is unable to rationalize some of the details associated with the Jahn–Teller distortion. We have known for some time that the local model has its limitations. There are, for example, cooperative interactions between ions on adjacent sites which lead to 'ferrodistortive' and 'antiferrodistortive' three-dimensional patterns (Reinen and Friebel, 1979). Although we will see that the electronic requirements of the local MX_6 geometry are important, so, too, are the extended interactions within the solid. In other words the orbital interactions between the atoms are of sufficient strength that the properties of the energy bands need to be examined. As we have mentioned earlier doping the typically insulating material TiO_2 with lithium gives rise to a metal where the electronic levels must be described in terms of bands.

The rutile structure itself is a tetragonal variant of the orthorhombic $CaCl_2$ type, generated by filling half of the octahedral holes of a hexagonal close-packing. Figure 3.30 shows the structure, with one of the edge-sharing chains of octahedra which run in the c direction highlighted. The Ti-O distances within the tetragonal structure fall into two symmetry-distinct sets, one containing four ('equatorial') and one containing two ('axial') linkages. In Figure 3.30(b) these two types of linkage are easy to see. In TiO_2 itself there are two long (L) and four short (S) Ti-O distances (Type 2 distortion). Since this corresponds to a d^0 configuration this distortion cannot be due to a Jahn–Teller effect. Elsewhere (Burdett et al., 1987) we have shown how this geometry is controlled by non-bonded interactions between the oxide ions. Such a distortion is not universal, however. The converse (where there are two short and four long (S) distances) is also found (Type 1 distortion). Table 3.1 gives details of the metal coordination geometry in all known rutile-type

ELECTRONIC PARADOXES IN THE STRUCTURES OF MINERALS

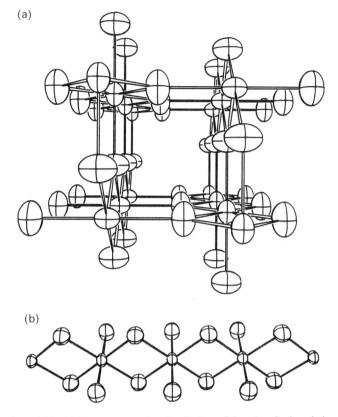

Figure 3.30 (a) The structure of rutile. (b) One of the edge-sharing chains.

Table 3.1 Local coordination in transition metal oxides and fluorides (Distances in Å)

System	4 linkages @	2 linkages @	difference	distortion type
MgF_2	1.998	1.979	−0.02	1
CrF_2	2.01[a]	2.43	+0.43	2
	1.98			
MnF_2	2.131	2.104	−0.03	1
FeF_2	2.118	1.998	−0.13	1
CoF_2	2.049	2.027	−0.01	1
NiF_2	2.022	1.981	−0.03	1
CuF_2	1.93	2.27	+0.34	2
ZnF_2	2.046	2.012	−0.03	1
TiO_2	1.945	1.986	+0.03	2
CrO_2	1.92	1.88	−0.04	1
RuO_2	1.984	1.942	−0.04	1
OsO_2	2.006	1.962	−0.04	1

[a]Monoclinic structure. Three pairs of distances.

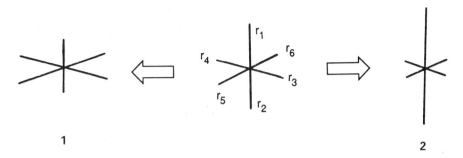

Figure 3.31 The two distortions of the regular octahedron.

transition metal oxides and fluorides. Examples of both types of distortion are found. For the cases of the fluorides of Cu^{II} and Cr^{II}, the tetragonal symmetry is lost and a monoclinic structure results. Although these are important structural observations, we shall restrict our discussion here to the series which remains tetragonal, which are those systems where the degeneracy occurs in the π-type orbitals on the metal.

One of the aspects of the Jahn–Teller theorem which continues to attract interest is the prediction of the distortion route away from octahedral (1 or 2 in Figure 3.31) as a function of the nature of the atomic composition. Recently, studies of the structures of some of the high-T_c copper oxide superconductors have drawn attention to the importance of this point. Here, as just noted, we shall concentrate on the energetic behaviour of the metal t_{2g} orbitals on distortion. Figure 3.32 shows these metal d orbitals of π type and how the levels

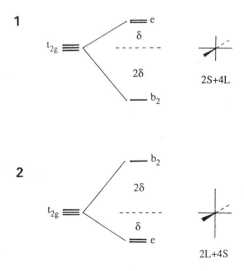

Figure 3.32 The change in the t_{2g} d-orbital energies on distortion. The orbital energy changes are expressed in terms of the change, δ, in the angular overlap parameter θ_π.

split apart in energy on distortion via the two modes for an isolated MX_6 unit. We may readily calculate the energy associated with the distortion as a function of electron count. This is done in Table 3.2 where we have used the angular-overlap model, although use of the crystal field theory gives exactly analogous results. Recall that the magnitude of the e_π parameters depends upon the metal-ligand distance, and so we can write the changes in energy of the d orbitals on distortion in terms of the parameters δ which labels the changes in magnitude of e_π, that is $e'_\pi = e_\pi + \delta$ on distortion. The energetics are then controlled by the variation in magnitude of π-bonding to the metal for cases involving t_{2g} degeneracies, respectively. The corresponding case for e_g degeneracies, of course, involves the varition in magnitude of σ-bonding to the metal. The form of one of the e_g distortion components which describes the motions of 1,2 is simply written in terms of the changes in the six metal-ligand distances, r_i, as

$$\Delta R = (2\sqrt{2})^{-1} \cdot (2\Delta r_1 + 2\Delta r_2 - \Delta r_3 - \Delta r_4 - \Delta r_5 - \Delta r_6) \quad (24)$$

Here, the Δr_i are the changes in the six bond lengths around the octahedron. The distortions 1,2 differ only in the sign of the Δr_i. Using this approach, at the regular octahedral geometry the t_{2g} energy levels are destabilized via M–O π^* interactions by $4e_\pi$ relative to the d orbital energy for the free atom, and the e_g levels are destabilized via M–O σ^* interactions by $3e_\sigma$. On distortion, for the t_{2g} set, the b_2 orbital moves by 2δ in energy while the e pair move by $-\delta$, where we have assumed that the changes in e_π are directly proportional to the change of distance in equation 1, shorter distances giving rise to larger

Table 3.2 Total d-orbital energies for the t_{2g} configurations of the regular, tetragonally compressed and tetragonally elongated octahedra[a]

d^n	Regular	2S+4L (1)	2L+4S (2)
d^0	0	0	0
d^1	$4e_\pi$	$4e_\pi - 2\delta$*	$4e_\pi - \delta$
d^2 (ls)[b]	$8e_\pi$	$8e_\pi - 4\delta$*	$8e_\pi - 2\delta$
d^2 (hs)	$8e_\pi$	$8e_\pi - \delta$	$8e_\pi - 2\delta$*
d^3 (ls)	$12e_\pi$	$12e_\pi - 3\delta$	$12e_\pi - 3\delta$
d^3 (hs)	$12e_\pi$	$12e_\pi$	$12e_\pi$
d^4 (ls)	$16e_\pi$	$16e_\pi - 2\delta$	$16e_\pi - 4\delta$*
d^4 (is)	$16e_\pi$	$16e_\pi - 2\delta$*	$16e_\pi - \delta$
d^5	$20e_\pi$	$20e_\pi - \delta$	$20e_\pi - 2\delta$*
d^6	$24e_\pi$	$24e_\pi$	$24e_\pi$

[a]The * indicates the lowest energy structure for π-donor ligands.
[b]ls = low-spin, hs = high-spin, is = intermediate spin.

e_π values. Since $e_\pi > 0$ this implies that $\delta < 0$ for 1 and $\delta > 0$ for 2. For case 1, the b_2 orbital has a smaller M–O π overlap than in the octahedral structure and now lies at $4e_\pi - 2|\delta|$. The increased M–O π overlap with the two *trans* ligands on distortion destabilizes the e set and they lie at $4e_\pi + |\delta|$. For case 2, the opposite effect ensues; the b_2 orbital rises to $4e_\pi + 2|\delta|$ and the e levels fall to $4e_\pi - |\delta|$. (Similar arguments apply to the orbitals of the octahedral e_g set, where we could use a parameter δ' to describe the variation in σ interaction. Importantly here the splitting pattern does not depend upon the distortion mode.)

Note, then, that a prediction of the distortion route is only possible for asymmetric t_{2g} configurations. For systems where there is partial occupancy of the e_g orbitals, the two pathways have equal distortion energies. The effect of d–s mixing via the second-order Jahn–Teller effect allows understanding of the virtually universal occurrence of 2 in the chemistry of Cu^{II} for example. However, our interest here will focus on those asymmetric t_{2g} configurations.

For the species involving degeneracies within the t_{2g} block there is a clear dependence of distortion energy on electronic configuration. The result is shown pictorially in Figure 3.33. We note, too, though that within the umbrella of the Jahn–Teller theorem, other types of distortions of T_{2g} and T_{1g} states are possible to give structures containing, for example, threefold or twofold rotation axes. Interactions between the ligands may also control the observed geometry.

One point which is particularly striking about Table 3.2 is that the structure predicted to be most stable depends on whether the ligands are π-donors or acceptors. In the table the most stable structure for π-donor ligands is indicated with an asterisk. The predictions are reversed if the ligands are π-acceptors. It is interesting to see that the distortion 1 found for low-spin d^5 $V(CO)_6$, a

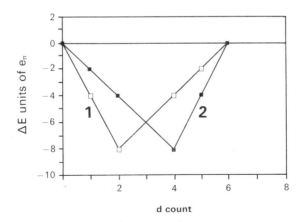

Figure 3.33 Results of Table 3.1 presented pictorially. The abscissa has units proportional to δ. Plotted is the energy difference between the octahedral and distorted structures.

molecular species, is correctly predicted by the model. Recall that the CO ligand is a π-acceptor. Coordinated oxygen is a π-donor. The crystal-field theory would have difficulty with this result, being inherently unable to distinguish between the two.

Our attention was drawn to the problem of the distortion modes in solid oxides and fluorides by the case of CrO_2 where the preferred distortion predicted from Table 3.2 is that which produces two long and four short Cr–O distances (2)—a prediction opposite to that observed *via* X-ray crystallography and from our own low-temperature, neutron diffraction studies. Similarly, the dioxides of Os and Ru (low spin d^4) have two short distances, whereas two long ones would have been predicted. CrO_2 is isostructural with rutile and is a high-spin d^2 system. There appears to be no structural evidence for the type of distortion which may be attributed to metal–metal bonding as found in the isoelectronic (but low-spin) systems MoO_2 and WO_2 as well as in d^1 VO_2 and d^3 α-ReO_2 and TcO_2. Here a pairing up of the metal atoms occurs along the chain axis. For high-spin d^2 systems such as CrO_2, neither the angular overlap model nor the crystal field theory give any clues as to why the form of the distortion is opposite to that predicted for the isolated molecular case.

Further inspection shows that this phenomenon, the failure of a local theory, is not restricted to the CrO_2 case. Thus, although FeF_2 has an orbital degeneracy associated with the t_{2g} orbital set, and the sense of the distortion is exactly as would be expected from Figure 3.33, why is no distortion of the opposite type found for CoF_2?

The problem with the theoretical approach described above, is that it is a local one. We know that in order to understand the metallic behaviour of oxides when they are doped with electron donors, we have to use a band model in the extended solid. Figure 3.34 shows a calculated (Burdett *et al.*, 1988) energy-difference curve as a function of d count (low-spin case) for the two types of distortion (1, 2) from the results, not of a local calculation, but of a band structure computation on the infinite solid dioxide. It predicts two regions of behavior in the 't_{2g}' block. For d^0 systems two long and four short distances are to be expected, a result of the demands of O–O non-bonded forces. For d^1–d^6 it predicts two short and four long distances. This is in quite good agreement with experiment. There is no stabilization of type 2 for any occupancy of the 't_{2g}' block. Compare the two curves shown in Figures 3.33 and 3.34. The difference is striking and tells us that we may not use the local ideas of the crystal field theory (or any local model) to probe the distortion mechanisms in these systems, but must use an orbital model which takes into account the translational periodicity of the material.

To uncover the electronic basis behind the form of this energy difference curve we need to investigate the form of the d-orbital levels of the infinite solid. The energetic behaviour with wave vector of the 't_{2g}' levels for a single one-dimensional chain of edge-linked octahedra is shown in Figure 3.35 using the angular overlap model. A strong 't_{2g}'/'e_g' mixing occurs in the chain, a

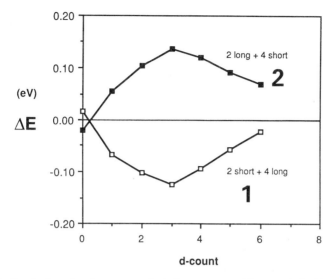

Figure 3.34 Results from band structure calculations on a transition metal dioxide in the rutile structure. Plotted is the energy difference between the octahedral and distorted structures. Notice the very different predictions concerning the preferred distortion route from that made by the local model of Figure 3.33. Here the two-short and four-long distortion is favoured for almost all electron counts.

process which is not allowed of course on the local model. One of the 't$_{2g}$' levels is not purely involved in π type interactions but has mixed with one of the 'e$_g$' orbitals of the same symmetry. Shown in Figure 3.35 is how such σ/π mixing can come about. This interaction (ε_{stab}) is quite large and strongly influences the energetics of the distortion. The details are given elsewhere (Burdett *et al.*,

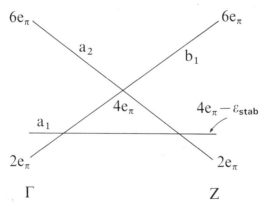

Figure 3.35 (a) Schematic showing the dispersion along the chain direction of the metal 't$_{2g}$' levels, showing the result of the 't$_{2g}$'-'e$_g$' mixing (ε_{stab}) demanded by the electronic structure of the extended array, (b).

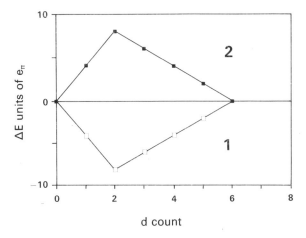

Figure 3.36 Calculated energetics of distortion of the octahedron using the extended model of Figure 3.35(b) and employing the angular overlap model.

1988). An approximation to the total energy of the system which will serve our purpose here is that evaluated at the mean value point ($\pi/2c$) of the Brillouin zone. Figure 3.36 shows the energy difference curve describing the distortion as a function of d count taking into acount this mixing. The result is a very similar plot to that computed using a full-band-structure calculation on the three-dimensional solid shown in Figure 3.34. Similar results probably apply to the two-dimensional sheet structures of the cadmium halide type. This is a striking illustration of a new explanation of an old structural problem, where the use of quite simple, molecular-orbital arguments has not only provided a much broader view but is in agreement with the experimental results.

3.9 Conclusion

It is clear from our discussion in this chapter that there is not a monopoly on ways to understand structural observations in solids. Any model, though, will be judged on its predictive powers and also, very importantly, on its breadth. Certainly breadth is one of the strong points of the orbital model we have used here. For the problems we have addressed it is quite apparent that such a viewpoint is a valuable one.

Acknowledgements

Our work at Chicago has been supported by The National Science Foundation, The Petroleum Research, The Office of Naval Research, and The Dow Chemical Company.

References

Albright T. A., Burdett, J. K., and Whangbo, M. H. (1985) *Orbital Interactions in Chemistry*, John Wiley & Sons, New York.

Burdett, J. K. (1980) *Molecular Shapes*, John Wiley & Sons, New York.

Burdett, J. K. (1981) Molecules within infinite solids, in *Structure and Bonding in Crystals*, vol. 1, (eds M. O'Keeffe and A. Navrotsky), Academic Press, New York, pp. 255–77.

Burdett, J. K. (1984) From bonds to bands and molecules to solids. *Progress in Solid State Chemistry*, **15**, 173–255.

Burdett, J. K. (1990) in *Chemistry of Superconducting Materials*, (ed. T. A. Vanderah).

Burdett, J.K. and T. J. McLarnan (1984) An orbital interpretation of Pauling's rules. *American Mineralogist*, **69**, 601–21.

Burdett, J. K., Hughbanks, T., Miller, G. J., *et al.* (1987) Structural-electronic relationships in inorganic solids: powder neutron diffraction studies of the rutile and anatase polymorphs of titanium dioxide at 15 and 295 K. *Journal of the American Chemical Society*, **109**, 3639–46.

Burdett, J. K., Miller, G. J., Richardson, J. W. and Smith, J. V. (1988) Low-temperature neutron powder diffraction study of CrO_2 and the validity of the Jahn–Teller viewpoint. *Journal of the American Chemical Society*, **110**, 8064–71.

Gibbs, G. V., Meagher, E. P., Newton, M. D., *et al.* (1981) A comparison of experimental and theoretical bond length and angle variations for minerals, inorganic solids, and molecules, in *Structure and Bonding in Crystals*, vol. 1, (eds M. O'Keeffe and A. Navrotsky), Academic Press, New York, pp. 255–77.

Pauling, L. (1929) The principles determining the structure of complex ionic crystals. *Journal of the American Chemical Society*, **51**, 1010–26.

Reinen, D. and Friebel C. (1979) Local and cooperative Jahn–Teller interactions in model structures. Spectroscopic and structural evidence. *Structure and Bonding* (Berlin), **37**, 1–60.

CHAPTER FOUR
Lattice vibration and mineral stability
Nancy L. Ross

4.1 Introduction

Prediction of phase stabilities in geological systems represents one of the fundamental problems challenging earth scientists today. Phase equilibria can be studied directly under controlled pressure and temperature conditions in laboratory experiments. In some cases, however, it is not possible to determine phase stabilities from experiments, because, for example, the pressures and temperatures of interest may not be attainable in the laboratory, samples may be non-quenchable, or equilibrium may not be attainable. Under such circumstances, a different approach is needed to study mineral stability. The purpose of this chapter is to describe how mineral stabilities can be studied from direct consideration of the mineral's lattice vibrations.

Since the decisive criterion for mineral stability under equilibrium conditions is governed by which phase has the lowest free energy, the question arises as to how we can determine the Gibbs free energy as a function of temperature, pressure, and chemical potential. It is well known that the heat capacity of a crystal can be determined from knowledge of the frequencies of all lattice vibrations (e.g. Debye, 1912; Born and von Karman, 1912; 1913; Born and Huang, 1954; Maradudin *et al.*, 1963). The vibrations of atoms in a crystal, the *lattice dynamics* of a crystal, determine its thermal properties that can, in the harmonic approximation, be expressed as averages over the crystal's frequency distribution. Furthermore, knowledge of lattice vibrations enables determination of a mineral's thermodynamic functions, such as the Gibbs free energy, at a certain pressure and temperature from which it is possible to calculate and interpret phase relations between different minerals. Sections 4.2 and 4.3 describe the basic elements of lattice dynamical theory and the relationship between the phonon density of states and thermodynamic properties.

Kieffer (1979a,b,c; 1980) recognized that lattice dynamics provide great power to calculate mineral properties and to interrelate elastic constant, spectral, thermodynamic, and phase equilibria data, thus elucidating the crystal's macroscopic properties in terms of its microscopic interactions. In the past, the application of rigorous, lattice dynamical calculations to problems of

The Stability of Minerals. Edited by G. D. Price and N. L. Ross.
Published in 1992 by Chapman & Hall, London. ISBN 0 412 44150 0

mineral stability has been hampered by our lack of knowledge of atomic interactions in solids and from the inherent complexity of minerals. Because of these difficulties, various models have been used that simplify the frequency distribution of lattice vibrations. One of the most enduring analytic models was developed by Debye (1912). This approximation works fairly well for solids of high symmetry containing only one kind of atom, or atoms of similar masses and bonding characteristics, but is inadequate for more complex phases. Two approximations that provide a more detailed model of a mineral's phonon density of states and have been used in phase equilibrium studies were proposed by Salje and Viswanathan (1976) and Kieffer (1979c). Both models require acoustic, spectroscopic, and crystallographic data as input. The Debye and Kieffer models are described in Section 4.4, and the effect of various approximations to the mineral's vibrational density of states on heat capacity and entropy calculations is examined in Section 4.5. Applications of vibrational models to mineral stability are described in Section 4.6.

In recent years, there has been a significant increase in the development of theoretical, computer-based models that enable prediction of the physical and thermodynamic properties of complex silicate phases at the extremes of temperature and pressure. Computer simulations range from *ab initio*, total energy calculations (e.g. Cohen *et al.*, 1987) to those involving the use of empirical potential models (e.g. Catlow and Mackrodt, 1982). Whereas the former method explicitly describes interactions between each and every electron in the solid, the latter method attempts only to describe the interactions between individual atoms or ions in the structure. The simplicity of this method, however, allows prediction of a wide range of physical and defect properties of crystals, as well as predictions of phase stability (e.g. Price *et al.*, 1987; 1989). An example of an application of this atomistic approach to prediction of the phonon density of states and thermodynamic properties of the high-pressure polymorphs of $MnTiO_3$ is presented in Section 4.7.

4.2 Background

In this section, the link between a crystal's thermodynamic properties and its lattice vibrations is introduced. All of the models described in this chapter utilize the *harmonic approximation*. In the harmonic approximation, the energy of a crystal is the sum of the energies of the independent normal modes, each equivalent to a simple harmonic oscillator. The energy of a linear, simple harmonic oscillator can only assume the values,

$$E_n = \left(n + \frac{1}{2}\right)\hbar\omega, \quad (1)$$

where n is a positive integer. The probability, P_n, that the simple harmonic oscillator has energy E_n is,

$$P_n = \frac{\exp(-E_n/k_B T)}{\sum_{n=0}^{\infty} \exp(-E_n/k_B T)}, \quad (2)$$

where k_B is the Boltzmann constant and T is the absolute temperature. The mean energy, $\bar{E}$, associated with the oscillator is given by,

$$\bar{E} = \sum_{n=0}^{\infty} E_n P_n. \quad (3)$$

Substitution of (1) and (2) into (3) leads to

$$\bar{E} = \hbar\omega \left\{ \frac{1}{2} + \frac{1}{\exp(\hbar\omega/k_B T) - 1} \right\} = \hbar\omega \left(\eta + \frac{1}{2} \right) \quad (4)$$

where the function, η, is the Bose–Einstein distribution function. The total energy of a crystal, the sum of energies of the independent harmonic oscillators, is therefore,

$$U = \sum_{k,j} \hbar\omega_j(k) \left(\frac{1}{2} + \frac{1}{\exp(\hbar\omega_j(k)/k_B T) - 1} \right), \quad (5)$$

where k is the 'wave-vector' ($\lambda = \pi/k$) and j represents each branch of the dispersion curves, $j = 1, 2, 3, \ldots, 3N$, explained in more detail in Section 4.3. It is apparent from (5) that the energy depends only on the temperature and on the frequencies of the modes. Furthermore, at absolute zero the crystal still has 'zero-point energy',

$$U_0 = \frac{1}{2} \hbar \sum_{k,j} \omega_j(k). \quad (6)$$

In a macroscopic crystal, where surface effects can be neglected, the modes are practically continuously distributed over a frequency range from zero to some maximum frequency, ω_m, and the sum in (5) can be replaced by an integral. The frequency distribution, $G(\omega)$, is defined such that $G(\omega)\,d\omega$ is the number of modes having frequencies in the range ω to $\omega + d\omega$. Equation (5) thus becomes,

$$U = \int_0^{\omega_m} \hbar\omega \left(\frac{1}{2} + \frac{1}{\exp(\hbar\omega/k_B T) - 1} \right) G(\omega)\,d\omega. \quad (7)$$

Since the total number of normal modes in a crystal of s unit cells and N atoms per unit cell is $3sN$,

$$\int_0^{\omega_m} G(\omega)\, d\omega = 3sN. \tag{8}$$

If $G(\omega)$ is known, other thermodynamic parameters can be calculated. The specific heat, for example, is obtained by differentiating U with respect to temperature,

$$C_v = \frac{dU}{dT} = k_B \int_0^{\omega_m} \left(\frac{\hbar\omega}{k_B T}\right)^2 \frac{\exp(\hbar\omega/k_B T) G(\omega)\, d\omega}{[\exp(\hbar\omega/k_B T) - 1]^2}. \tag{9}$$

Expressions for other thermodynamic functions (in the harmonic approximation) are given in Table 4.1. The integral in equation (9) can be evaluated numerically when $G(\omega)$ is known, thus enabling calculation of the specific heat at any temperature. In addition,

$$G_p = C_v + TV\alpha^2/\beta, \tag{10}$$

where α is the volume coefficient of thermal expansion, β the compressibility, and V the molar volume. The heat capacity, C_p, can be used to calculate the entropy, S_p, and the enthalpy, H_p, which yield the Gibbs free energy,

$$G_p = H_p - TS_p. \tag{11}$$

Thus the entire thermodynamic behaviour of a mineral, as determined by phonon processes, can be obtained from knowledge of $G(\omega)$.

Table 4.1 Harmonic expressions for the thermodynamic functions, internal energy ($\bar{U}$), heat capacity (C_V), and entropy (S), in terms of the phonon density of states, $G(\omega)$

$$\bar{U} = \int_0^{\omega_m} \hbar\omega \left(\frac{1}{2} + \frac{1}{\exp(\hbar\omega/k_B T) - 1}\right) G(\omega)\, d\omega$$

$$C_V = k_B \int_0^{\omega_m} \left(\frac{\hbar\omega}{k_B T}\right)^2 \frac{\exp(\hbar\omega/k_B T)}{[\exp(\hbar\omega/k_B T) - 1]^2} G(\omega)\, d\omega$$

$$S = \frac{1}{T}\int_0^{\omega_m} \left(\frac{\hbar\omega}{\exp(\hbar\omega/k_B T) - 1}\right) G(\omega)\, d\omega - k_B \int_0^{\omega_m} \ln[1 - \exp(-\hbar\omega/k_B T)]\, G(\omega)\, d\omega$$

4.3 Vibrational density of states, $G(\omega)$

In order to gain insight into the frequency distribution of a three-dimensional solid, consider $G(\omega)$ of a simple, one-dimensional linear chain consisting of one atom of mass, m, in a unit cell dimension, a. We assume the forces between the atoms extend only to nearest neighbours and that the crystal is 'harmonic' in the sense that the force acting on an atom displaced from equilibrium is proportional to the first power of the displacement. Interactions between adjacent atoms can be represented by a spring of force constant, f. We consider the *longitudinal* vibrations or those in which the atom displacements are along the length of the chain. The equation of motion is,

$$m \frac{\partial^2 u_n}{\partial t^2} = f\{u_{n+1} - u_n + u_{n-1} - u_n\}, \tag{12}$$

where u_n is the displacement from equilibrium of the nth atom. The equations of motion of all the atoms are in this form, differing only in the value of n. Substitution of a wave solution of the form,

$$u_n = A \exp\{i(ku_n^0 - \omega t)\}, \tag{13}$$

into (12), where k is the wave vector and u_n^0 is the undisplaced position of the nth atom, yields,

$$m\omega^2 = 4f \sin^2 \frac{ka}{2}, \tag{14}$$

where ω is the angular frequency ($=2\pi v$ where v is the frequency). This relation between ω and k, known as the *dispersion relation*, is shown in Figure 4.1(a) for our example of a one-dimensional monatomic chain. The frequency rises to a maximum of $2(f/m)^{1/2}$ at $k_{max} = \pi/a$ and falls to zero at $k = 2\pi/a$. In fact, $\omega(k)$ is periodic with periodicity $2\pi/a$, the dimension of the one-dimensional reciprocal lattice. Equation (14) can therefore be rewritten as

$$\omega(k) = \omega_m \sin \tfrac{1}{2}ka. \tag{15}$$

Thus the lattice acts as a filter, only passing frequencies from 0 to ω_m and blocking higher ones.

To evaluate the frequency distribution function, $G(\omega)$, we must consider an interval dk in k-space corresponding to an interval $d\omega$ in the frequency space. For one example of a one-dimensional linear chain, $G(k)$, the density of allowed modes in the k-space, is

$$G(k) = \frac{N}{2\pi/a} \tag{16}$$

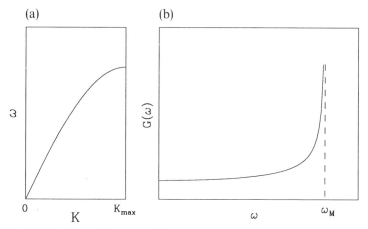

Figure 4.1 (a) Dispersion relation and (b) frequency distribution of the normal modes of vibration of a one-dimensional, monatomic crystal.

and, if we limit k to positive values,

$$2G(k)\,dk = G(\omega)\,d\omega. \tag{17}$$

Using $dk/d\omega$ from (15), the frequency distribution is,

$$G(\omega) = \frac{2N}{\pi(\omega_m^2 - \omega^2)^{1/2}} \tag{18}$$

as shown in Figure 4.1(b), The specific heat can be determined upon substitution of (18) in (9).

A simpler frequency distribution is obtained if dispersion in the $\omega(k)$ relation is ignored and the sine curve in equation (15) is replaced by a straight line,

$$\omega(k) = \tfrac{1}{2}\omega_m ka. \tag{19}$$

The corresponding frequency distribution is,

$$G(\omega) = \frac{2N}{\pi\omega_m} = \frac{N}{\omega_D} \quad (\omega < \omega_D), \tag{20}$$

where ω_D is the limiting frequency at the zone boundary. The Debye approximation for a three-dimensional crystal is discussed in more detail in the Section 4.4.

It can be seen from this simple example that determination of $G(\omega)$ requires knowledge of the phonon dispersion curves. In addition, for prediction of thermodynamic properties, it is necessary to integrate all phonons over the entire Brillouin zone. For a three-dimensional solid with N atoms in the unit cell, there are a total of $3N$ branches to the phonon dispersion curves. Three of these are acoustic modes and the remaining $3N-3$ modes are optic modes. The acoustic modes show the greatest variation in frequency with wave vector whereas the optic modes generally exhibit lower dispersion. See, for example, the dispersion relations measured by Elcombe (1967) for low quartz. For a three-dimensional lattice, it is not possible to obtain an analytical expression for the frequency distribution function and numerical determinations are required. The most direct method of calculating $G(\omega)$ is to determine the vibrational frequencies in a crystal by solving the Newtonian equations of motion, similar to (12), and resulting eigenvector–eigenvalue equations in which the eigenvalues are the squared frequencies of each of the normal modes of the crystal and the corresponding eigenvectors determine the pattern of atomic motion in the mode of vibration (e.g. Cochran, 1973). This approach will be considered further in Section 4.7. Another approach is to determine experimentally vibrational modes at the Brillouin-zone centre and to calculate dispersion curves using some appropriate theoretical model, like the shell model (e.g. Woods et al., 1963). An example of this procedure is the determination of $G(\omega)$ of low quartz by Striefler and Barsch (1975). This method, like the previous one, requires knowledge of interatomic forces present in the structure as well as considerable computational effort. Another method of interest, but limited application, is the method of moments (Montroll, 1942; 1943). In this approximation, the frequency distribution is expanded in terms of moments of the spectrum. This method has successfully accounted for thermodynamic properties of some cubic substances.

The phonon density of states can be obtained directly from experiment by using incoherent inelastic scattering of slow neutrons (e.g. Placzek and Van Hove, 1954). Unfortunately, the number of substances having a sufficiently large cross section for incoherent scattering is limited and the energy resolution of the scattering experiments is generally too poor to reveal the finer details of $G(\omega)$. The phonon density of states can be determined indirectly from coherent, inelastic neutron-scattering experiments on powder samples, as described by Ghose (1988). Coherent inelastic neutron spectroscopy can also be used to determine the phonon dispersion curves in a crystal. In order to calculate the phonon density of states from the dispersion curves, it is necessary to assume some force model. This is usually obtained by fitting a set of force constants to the dispersion curves and to the elastic constants. These force constants enable the evaluation of the frequencies for a large number of values of k in the first Brillouin zone of the appropriate reciprocal lattice. Once the frequencies, $\omega_j(k)$, have been determined, $G(\omega)$ can be obtained by numerical summation of $\omega_j(k)$ over all wave vectors, k (e.g. Gilat, 1972). Cowley et al. (1966), for example, determined the frequency distribution of potassium using this procedure. Although this method may be preferred, it requires substantial experimental effort and is limited by practical considerations to large single crystals of high symmetry.

Many authors have attempted to find simpler representations of $G(\omega)$ than that given by lattice dynamical theory. Lattice dynamical calculations require knowledge of interatomic forces among large numbers of atoms, the solution of equations of motion for large numbers of wave vectors, and the integration of the dispersion relations over all wave-vector space. In the next section two models, that have been utilized in a variety of problems in earth sciences, are described. Both models greatly simplify the vibrational density of states of a mineral. The first model describes the entire frequency distribution with a single parameter. The second model, which also simplifies $G(\omega)$, has its roots in lattice dynamical theory.

4.4 Approximations of $G(\omega)$

The motivating force for proceeding with a less rigorous model than the complete lattice dynamics formulation for lattice vibrations of minerals is that thermodynamic functions are expressed as *averages* over the frequency distribution and hence are relatively insensitive to the details of the spectrum. In this section, two models that have been used in many geophysical and geologic studies are described. The first is the Debye model, which has commonly been used in geologic studies to predict or extrapolate the heat capacity and entropy of minerals and to relate their thermal properties and acoustic velocities. In addition, Debye theory is so firmly established that it has become customary to present results of even more rigorous calculations in terms of a Debye temperature, θ_D. An alternative to the Debye approximation that was proposed by Kieffer (1979c) is also described. The Kieffer model has been used in numerous geologic applications, including phase-equilibria studies, isotopic fractionation studies, and prediction of thermodynamic properties of minerals at high pressure.

4.4.1 Debye model

One of the most enduring analytic models proposed for frequency distributions of lattice vibrations was developed by Debye (1912) who considered the solid as an isotropic elastic continuum. Quantization of vibrational energy implies that, at low temperatures, only the low-frequency modes will be appreciably excited. In a solid, the low-frequency modes are usually acoustic vibrations that have wavelengths much larger than atomic spacings at $k=0$ and can therefore be discussed in terms of an elastic continuum. Debye calculated the distribution of frequencies that result from propogation of acoustic waves of permitted wavelengths in a continuous isotropic solid and assumed the same distribution can be applied to crystals.

The Debye approximation assumes that all modes of vibration are acoustic and have the same wave velocity. The various branches of the $\omega_j(k)$ relation are replaced by a single acoustic branch,

$$\omega_j(k) = uk, \qquad (21)$$

that displays linear dispersion (u is a constant), as shown in Figure 4.2(a). The Brillouin zone is simplified by replacing the actual zone with a sphere of the same volume in reciprocal space. Since values of k are uniformly distributed throughout the Brillouin zone, it follows that

$$G(k)dk \propto 4\pi k^2 dk \qquad (22)$$

and from (21),

$$G_D(\omega) = B\omega^2 \qquad (23)$$

where the subscript 'D' indicates the 'Debye approximation' and B is a constant that depends on the elastic wave velocities. The Debye approximation to the frequency distribution in a crystal, $G_D(\omega)$, is shown in Figure 4.2(b).

Equation (23) will apply up to a maximum frequency corresponding to k at the Brillouin-zone boundary. This is the Debye frequency, ω_D, and can be determined from acoustic measurements,

$$\omega_D = \left(\frac{18\pi^2 N_A}{ZV}\right)^{1/3} \cdot (V_p^{-3} + 2V_s^{-3})^{-1/3}, \qquad (24)$$

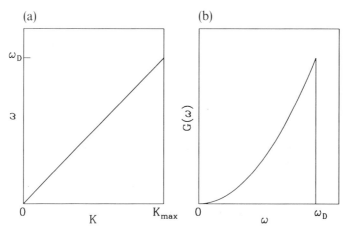

Figure 4.2 (a) Dispersion relation of the Debye approximation, and (b) the Debye approximation for the frequency distribution of the normal modes of vibration in a three-dimensional crystal.

where V_p and V_s are the compressional and shear wave velocities, respectively. The cut-off frequency, ω_D, is the quantity normally used to characterize the Debye spectrum of a crystal and is usually given in terms of the Debye temperature, θ_D, defined as,

$$\theta_D = \frac{\hbar \omega_D}{k_B}. \tag{25}$$

The Debye temperature calculated from acoustic velocities is called the elastic Debye temperature, θ_{el}.

The principal feature of the Debye approximation is the parabolic form of the frequency distribution, given by equation (23) and shown in Figure 4.2(b). If we are dealing with a monatomic crystal,

$$G_D(\omega) = 9N \left(\frac{\omega^2}{\omega_D^3} \right) \quad (\omega \leqslant \omega_D). \tag{26}$$

It should be noted that the total number of modes is 3N, and thus,

$$\int_0^{\omega_D} G_D(\omega) \, d\omega = 3N. \tag{27}$$

Therefore the dynamical properties of a crystal are expressed in terms of a single parameter, ω_D.

The thermal energy of a monatomic Debye solid, obtained from substituting equation (26) into (7), is given as

$$U = \frac{9N\hbar}{\omega_D^3} \int_0^{\omega_D} \frac{\omega^3 \, d\omega}{[\exp(\hbar\omega/k_B T) - 1]}. \tag{28}$$

It is convenient to introduce the dimensionless quantity,

$$x_D = \frac{\hbar \omega_D}{k_B T} = \frac{\theta_D}{T}, \tag{29}$$

and equation (28) thus becomes,

$$U = 9Nk_B T \left(\frac{T}{\theta_D} \right)^3 \int_0^{x_D} \frac{x^3 \, dx}{\exp(x) - 1}. \tag{30}$$

Differentiation of equation (30) with respect to temperature yields the heat capacity,

$$C_v = 9Nk_B \left(\frac{T}{\theta_D}\right)^3 \int_0^{x_D} \frac{x^4 \exp(x)\, dx}{[\exp(x)-1]^2}. \qquad (31)$$

From the definition of x_D Equation (29), it is apparent that $x_D \to \infty$ as $T \to 0$. Thus at low temperatures the limits of the integral can be taken to be 0 and ∞, and the thermal energy is,

$$U = \frac{3Nk_B \pi^4 T^4}{5\theta_D^3}, \qquad (32)$$

and

$$C_v = \frac{12}{5}\pi^4 Nk_B \left(\frac{T^3}{\theta_D}\right). \qquad (33)$$

This is the Debye T^3 law for the specific heat of a solid at low temperatures. At high temperatures, the molar heat capacity approaches a constant value of 3NR, the Dulong–Petit limit.

For polyatomic solids, the parabolic frequency distribution of the Debye model works reasonably well provided that the different atoms in the crystal play nearly equivalent mechanical roles in oscillations. In general terms, this means that the various atoms should have nearly equal masses, the coordination of different atoms should be nearly identical, the environments must be essentially isotropic and nearest-neighbour force constants should be approximately equal. Failure to meet one or more of these conditions results in changes in the vibrational spectrum such that the Debye model no longer provides an adequate approximation to the phonon density of states. The extent of departure from Debye-like behaviour can be shown by representing C_v at each temperature in terms of a 'calorimetric Debye temperature', $\theta_{cal}(T)$, which is the value of θ_D in equation (31) that will reproduce the observed $C_v(T)$, as distinguished from the 'elastic Debye temperature', θ_{el}, described earlier. If a compound was a perfect Debye solid, then the calorimetric Debye temperature, which would be equal to the elastic Debye temperature, would be constant with temperature. For simple solids such as the alkali halides, the general agreement of the $\theta_{cal}(T)$ curves with the predictions of the Debye model suggests that the parabolic representation of the vibrational spectrum is adequate for prediction of the thermodynamic properties. Silicates, however, generally show very different behaviour. Typically $\theta_{cal}(T)$ curves show a pronounced minimum at temperatures of a few degrees Kelvin and, with increasing temperature, generally approach a value greatly exceeding θ_{el}. See, for example, Figure 3 of Kieffer (1979a). These departures from Debye-like behaviour imply deviations of the vibrational spectrum from the parabolic approximation. The dip in the θ_{cal} curve at low temperature, for example, implies an excess heat capacity compared to that predicted from the Debye model and therefore implies excess modes of vibration occur at low frequencies. The rise of the $\theta_{cal}(T)$ curve to a value exceeding the Debye temperature implies a deficiency of oscillators at frequencies near ω_D and hence their presence at higher frequencies. Thus a more rigorous representation of the frequency distribution is needed to predict heat capacities and entropies of complex phases such as silicates.

4.4.2 Kieffer's approximation

Applicability of the Debye approximation to minerals is governed by the extent to which the actual lattice vibrational spectrum is approximated by a parabolic frequency distribution. Kieffer (1979a) postulated that the specific heat variations from a Debye model are caused by the following factors:

1. A mean sound velocity is inadequate for minerals with a high degree of acoustic anisotropy.
2. Lattice waves do not follow a linear dispersion relationship.
3. Excess optic modes may occur at low frequencies not accounted for in the Debye spectrum.
4. Optic modes may occur at higher frequencies than those predicted from a Debye model.

Kieffer (1979c) proposed a model that takes these four factors into account. The model is consistent with lattice dynamics in requiring the use of the primitive unit cell, in correctly enumerating acoustic modes, and in recognizing anisotropy and dispersion of acoustic branches. The success of the method results from the relative insensitivity of thermodynamic functions to details of the spectrum at most frequencies characteristic of the optic modes.

The main features of the Kieffer approximation are described below. The reader is referred to Kieffer (1979c) for a detailed description of the model. The fundamental vibrating unit of the crystal is taken as the primitive unit cell. Associated with this cell are 3N vibrational degrees of freedom where N is the number of atoms in the primitive unit cell. The Brillouin zone is replaced by a sphere of the same volume in reciprocal space with a radius of k_{max}. Three of the 3N degrees of freedom must be acoustic modes. Each acoustic branch may be characterized at long wavelengths ($k \to 0$) by a directionally averaged sound velocity estimated from acoustic velocities (Kieffer, 1979a). The frequency distribution for the acoustic part of the spectrum is obtained from an assumed sinusoidal dispersion relation for the three acoustic branches given by,

$$\omega(k) = \omega_{i,\max}(k) \sin\left(\frac{k}{k_{\max}} \cdot \frac{\pi}{2}\right), \tag{34}$$

where $\omega_{i,\max}$ is the maximum frequency reached at the zone boundary by each branch i (Fig. 4.3(a)). The slope of each branch is required to approach the acoustic speed, v_i at $k=0$. The maximum frequency of each branch is therefore,

$$\omega_{i,\max} = v_i k_{\max} \cdot (2/\pi). \tag{35}$$

Each acoustic branch i accounts for $1/3N$ of the total degrees of freedom per unit cell and the acoustic spectral distribution is give by,

$$G_A(\omega)\,d\omega = Z' \sum_{i=1}^{3} \frac{[\sin^{-1}(\omega/\omega_i)]^2}{(\omega_i - \omega^2)^{1/2}}\,d\omega, \tag{36}$$

where ω_i is the maximum frequency for each acoustic branch and Z' is a constant.

The remaining modes are optic modes. In Kieffer's approximation, the optic modes are represented in a way that satisfies spectroscopic observations of the maximum and minimum values of vibrational frequencies. In principle, $3N-3$ separate optic modes can be identified and enumerated, but, in practice, only one or two groups of modes can be enumerated. These are usually stretching

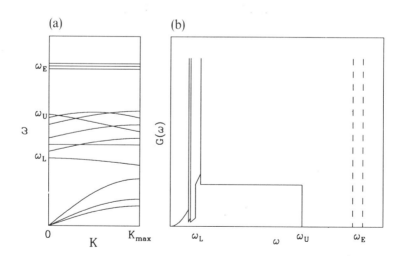

Figure 4.3 (a) Dispersion relations in the Kieffer approximation, and (b) the Kieffer approximation for the frequency distribution of the normal modes of vibration in a three-dimensional crystal.

modes, such as the internal Si–O stretching modes or Al–O stretching modes or O–H stretching modes, if present. These stretching modes can be represented in the model with weighted Einstein oscillators. A mode, for example, represented by a Einstein oscillator at frequency ω_E with a fraction, q, modes at this frequency has a density of states,

$$G_E = qN'\delta(\omega - \omega_E), \tag{37}$$

where δ is the Dirac delta function and N' is a constant.

The remaining $3N[1-(1/N)-q]$ modes per unit cell are assumed to be uniformly distributed over a band of frequencies spanning the spectrum between a lower limit, ω_l, and an upper limit, ω_u. The density of states of this band of frequencies known as the *optic continuum* is,

$$G_0 = S' \frac{[1-(1/N)-q]}{(\omega_u - \omega_l)}, \quad (\omega_l < \omega < \omega_u) \tag{38}$$

where S' is a constant.

The Kieffer approximation to the vibrational density of states of a mineral therefore consists of three acoustic branches, an optic continuum, and Einstein oscillators (optional) whose frequency distributions are given by equations (36), (37), and (38), respectively. The dispersion relations for each of these components and corresponding frequency distribution are shown in Figure 4.3.

The expression for the molar heat capacity using Kieffer's model, normalized to its monatomic equivalent, C_v^*, is:

$$C_v^* = \frac{3N_A k_B}{N} \sum_{i=1}^{3} \Upsilon(x_i) + 3N_A k_B \left(1 - \frac{1}{N} - q\right) \Omega\left(\frac{x_u}{x_1}\right) + 3N_A k_B q \Xi(x_E) \tag{39}$$

where Υ is the contribution from the acoustic branches,

$$\Upsilon(x_i) = \left(\frac{2}{\pi}\right)^3 \int_0^{x_i} \frac{[\sin^{-1}(x/x_i)]^2 x^2 \exp(x) \, dx}{(x_i^2 - x^2)^{1/2} [\exp(x) - 1]^2}, \tag{40}$$

$\Omega(X_l, X_u)$ is the optic continuum contribution,

$$\Omega\left(\frac{x_u}{x_l}\right) = \int_{x_l}^{x_u} \frac{x^2 \exp(x) \, dx}{(x_u - x_l)[\exp(x) - 1]^2}, \tag{41}$$

and Ξ is the heat capacity function of an Einstein oscillator,

$$\Xi(x_E) = \frac{x_E^2 \exp(x_E)}{[\exp(x_E) - 1]^2}. \tag{42}$$

Equations (39)–(42) have been simplified by introducing the non-dimensionalized frequencies, $x_j = \hbar\omega_j/k_B T$. At low temperatures the acoustic function has the same T^3 dependence as the Debye function. Similar expressions can be derived for the entropy, internal energy, and Helmholtz free energy and are given in Kieffer (1979c).

The Kieffer approximation therefore allows calculation of the thermodynamic functions given the following acoustic and spectroscopic data: three acoustic velocities (two shear and one compressional) from which the acoustic cut-off frequencies can be calculated, two spectral frequencies, ω_l and ω_u, that define the limits of the optic continuum, and frequencies and partitioning fractions for any modes that lie separate from the optic continuum. The Kieffer approximation is compared with the Debye approximation for the oxide, rutile, and the more complex framework silicate—low quartz—in Section 4.5.

4.5 Comparison of vibrational models

Vibrational spectra calculated from inelastic neutron scattering and lattice dynamical models are available for low quartz (Elcombe, 1967; Striefler and Barsch, 1975; Leadbetter, 1969; Galeener *et al.*, 1983) and rutile (Traylor *et al.*, 1971). In order to demonstrate how sensitive calculations of heat capacities and entropies are to approximations of a mineral's vibrational density of states, models closely approximating $G(\omega)$ for low quartz and rutile were constructed and compared with the Debye and Kieffer approximation. The two examples were chosen, in part, because of the large difference in their structures. Low quartz is a framework silicate in which all of the [SiO_4] tetrahedra share corners in a continuous three-dimensional network with an Si–O–Si angle of approximately 144°. The rutile structure, on the other hand, consists of 'ribbons' of edge-sharing [TiO_6] octahedra running parallel to c that are connected laterally by sharing corners. Thus with these two examples, we can not only explore differences in chemistry, but also note the effects that increasing coordination has on model predictions of heat capacities and entropies. The latter is relevant to studies of high-pressure phase transitions that involve increases in cation coordination (Navrotsky, 1980).

4.5.1 Example 1: low quartz

Low quartz occurs in two enantiomorphic crystal structures corresponding to the trigonal space groups $P3_121$ (D_3^4) and $P3_221$ (D_3^6). There are nine atoms in the primitive unit cell giving a total of 27 branches to the phonon dispersion curves. Elcombe (1967) determined the seven lowest branches of the dispersion curves along [001] using the method of inelastic neutron scattering with a $4 \times 4.6 \times 6.5 \text{ cm}^3$ single-crystal of low quartz. Phonon dispersion curves calculated from a modified rigid-ion model (Striefler and Barsch, 1975) show very good agreement for the optical branches measured by Elcombe (1967). The three acoustic branches exhibit the correct shape and position of the maximum of the highest branch, but lie up to 20% too low, reflecting the underestimation of the elastic constants, c_{33} and c_{44}. Striefler and Barsch (1975) also calculated the phonon density of states using this model by a root-sampling method from a total of 1000 wave vector points with a mesh width of 8 cm^{-1} for the entire Brillouin zone. The most noteworthy features of the resulting histogram, shown in Figure 4.4(a), are the relatively broad band of frequencies between 67 cm^{-1} and 800 cm^{-1}, a much narrower band of frequencies between 1050 and 1200 cm^{-1}, and the gap between the two. Within the lower-frequency band, there are pronounced peaks at 400, 470, and 770 cm^{-1}, with a marked decrease in the concentration of modes between the peaks at 470 and 770 cm^{-1}

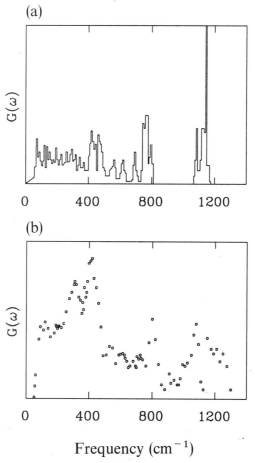

Figure 4.4 Vibrational density of states of low quartz: (a) calculated from a shell model (Striefler and Barsch, 1975), and (b) measured by inelastic neutron scattering (Galeener et al., 1983).

compared with the spectral region between 67 and 470 cm^{-1}. Striefler and Barsch (1975) found that the specific heat calculated from the model shows good agreement with experimental data between 100 and 500 K. Below 100 K, the calculations are up to 10% smaller than the experimental data due to the underestimation of the elastic constants. Above 500 K, the discrepancy arises from anharmonic effects and from the occurrence of the low- to high-quartz phase transformation at 850 K. The vibrational density of states of low quartz has also been determined by inelastic neutron scattering measurements on polycrystalline samples (Leadbetter, 1969; Galeener et al., 1983). The advantage of this method is that it provides a direct measure of the vibrational density of states; the disadvantage is that the energy resolution is not good enough to reveal the finer details of the frequency distribution. However, the measured spectrum by Galeener et al. (1983), shown in Figure 4.4(b), confirms many of the details noted in the calculated spectrum of Striefler and Barsch (1975) (Fig. 4.4(a)). There are two distinct bands of modes, one extending from 50 cm^{-1} to 800 cm^{-1} and a higher-frequency band extending from 1000 to 1300 cm^{-1}. Within the low-frequency band, there are pronounced peaks near 400, 475, and 800 cm^{-1}, with a smaller concentration of modes between 475 and 800 cm^{-1} as compared with the region between 50 and 475 cm^{-1} (Fig. 4.4(b)).

Four different approximations to the vibrational density of states of low quartz are shown in Figure 4.5. The first model, VDOS–1, closely approximates the experimentally determined G(ω) of Galeener et al. (1983) by using 27 'boxes' or optic continua to mimic the spectrum (Fig. 5.5(a)). The low-frequency band is approximated with 21 continua and 6 continua model the high-frequency region. The total number of modes is normalized to 1 with 22% of the modes assigned to the high-frequency region. The second approximation, VDOS–2, is less detailed than VDOS–1, and approximates G(ω) with seven optic continua (Fig. 5.5(b)). Six continua model the low-frequency band and a single continuum spans the high-frequency band. The third approximation, Kieffer's model, is a much more drastic approximation to G(ω) of low quartz than either VDOS–1 or VDOS–2. The band of frequencies below 850 cm^{-1} is represented by a single continuum extending from 90 to 810 cm^{-1} (ω_l was dispersed from 128 cm^{-1} at $k=0$ to 90 cm^{-1} at $k=k_{max}$) and the high-frequency band is represented by four Einstein oscillators at 1090, 1117, 1162, and 1200 cm^{-1} with 22% of the total number of modes. The lower and upper limits of the optic continuum (at $k=0$) and placement of oscillators are determined from far-infrared, mid-infrared, and Raman spectroscopic data. The acoustic modes comprise 11% (3/27) of the total number of vibrational modes and are characterized by directionally-averaged acoustic velocities of 3.76 km^{-1}, 4.46 km^{-1} and 6.05 km^{-1}. These acoustic modes are assumed to follow a sinusoidal dispersion relation, reaching 103 cm^{-1}, 122 cm^{-1}, and 165 cm^{-1}, respectively, at the Brillouin-zone boundary. When compared with Galeener and coworkers' (1983) experimental determination of G(ω), it is clear that the Kieffer model is a rather drastic simplification of G(ω), greatly underestimating the density of modes between 300 and 500 cm^{-1} and greatly overestimating the modes between 500 and 800 cm^{-1} (Fig. 4.5(c)).

The fourth and final model is the Debye approximation (Fig. 5.5(d)) in which the entire frequency distribution is described by a single parameter, ω_D (or θ_D). Both the 'elastic' Debye model and 'calorimetric' Debye model are shown in Figure 5.5(d). The elastic Debye model (solid line) is based on the averaged shear and compressional wave velocities of low quartz, 4.48 km^{-1}, that correspond to an elastic Debye temperature, θ_{el}, of 574 K and a cut-off frequency, ω_D, of 399 cm^{-1}. The calorimetric Debye model (dashed line) is based on an averaged thermal Debye temperature, θ_{th}, of 1062 K determined by Watanabe (1982) who fitted the measured heat capacity of low quartz between 350 and 700 K to Debye's theoretical expression. The corresponding cut-off frequency for this model is 737 cm^{-1}. It is clear that the single-parameter Debye model is an inadequate representation of the vibrational density of states of low quartz even when the Debye parameter is chosen as a free parameter to match the observed heat capacity. The elastic and calorimetric Debye models severely underestimate the density of modes at low frequencies and both cut-off frequencies are well below the second band, thus failing to account for the high-frequency Si–O stretching vibrations.

The effect that these various levels of approximation of $G(\omega)$, from the VDOS–1 with its 27 continua to the single-parameter Debye model, have on calculated heat capacities and entropies is shown in Figures 4.6 and 4.7. The model predictions are also compared with the experimental data of Lord and Morrow (1957) and Robie et al. (1975). Calculated heat capacities from the models, C_v, have been converted to C_p using equation (10). Heat capacities and entropies calculated from VDOS–1 show excellent agreement with experimental data. Between 150 and 800 K, heat capacities calculated from VDOS–1 lie within 1–2% of the experimental data (Fig. 4.6) and calculated entropies are within 2% of the observed values between 298 and 800 K (Fig. 4.7). The discrepancy between the observed heat capacities and entropies increases at lower temperatures, reflecting the poor resolution at low frequencies of the measured vibrational density of states using inelastic neutron scattering (Leadbetter, 1969; Galeener et al., 1983). Discrepancies from the experimental data above 800 K may be due to anharmonic effects and the occurrence of the low- to high-quartz phase transition at 850 K. Heat capacities and entropies calculated from the second approximation of $G(\omega)$, VDOS–2, that consists of 6 optic continua, also show excellent agreement with the experimental data between 150 to 800 K (Figs 4.6 and 4.7). Heat capacities and entropies calculated from VDOS–2 are almost identical to those calculated from VDOS–1. Thus, although VDOS–2 does not follow the measured frequency distribution as closely as VDOS–1, calculated heat capacities and entropies are not sensitive to details of the frequency spectrum.

Although the Kieffer approximation is a much more drastic approximation to the phonon density of states of low quartz than either VDOS–1 or VDOS–2

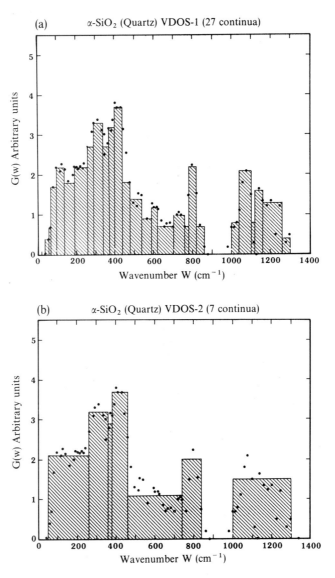

Figure 4.5 Approximations of G(ω) of low quartz: (a) VDOS-1, consisting of 27 optic continua, (b) VDOS-2, consisting of 7 optic continua, (c) Kieffer approximation, and (d) Debye approximation, including the elastic and calorimetric models. The phonon density of states measured by Galeener et al. (1983) is represented by solid circles.

(Fig. 4.5(c)), the model predicts heat capacities and entropies of low quartz remarkably well (Figs 4.6 and 4.7). Between 100 and 800 K, the approximation reproduces heat capacities and entropies within 3% of the experimental data. The major discrepancy between the model and the data is below 100 K, where

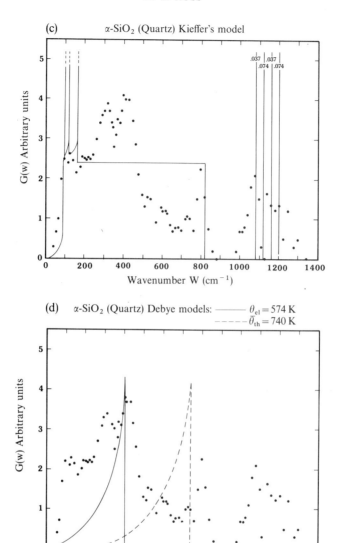

the lowest-frequency optic modes and acoustic modes make the dominant contributions to the heat capacity. The underestimation of C_p below 60 K is due to underestimation of dispersion of the acoustic branches in the model. Elcombe's (1967) experimental data for the normal modes along [001] and Leadbetter's (1969) determination of the spatially-averaged dispersion curve place the lowest shear branch in the range of 55 to 80 cm^{-1}, somewhat lower than the model value of 103 cm^{-1}. The overestimation of C_p between 60 and 100 K is most probably due to overestimation of dispersion of the lowest

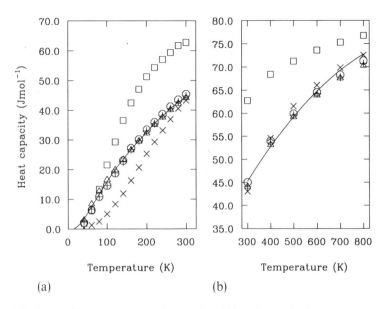

Figure 4.6 Comparison of heat capacities calculated from the models in Figure 4.5 with the experimental data (solid curve) at (a) low temperature, and (b) high temperature. The symbols represent heat capacity predictions from the various models: 'O': VDOS-1; '+': VDOS-2; triangle: Kieffer model; 'X': thermal Debye model; square: elastic Debye model.

frequency optic mode from 128 cm^{-1} at $k=0$ to 90 cm^{-1} at $k=k_{max}$. Both Elcombe's (1967) data and Striefler's and Barsch's (1975) calculations show that the lowest frequency optic mode shows little dispersion along some directions and actually shows a positive dispersion (i.e. an increase in frequency from $k=0$ to $k=k_{max}$) along other crystallographic directions.

The fourth approximation to G(ω), the single-parameter Debye model, is inadequate to predict the specific heat and entropy of low quartz. The elastic Debye model, which has a cut-off frequency at 399 cm^{-1}, greatly overestimates both the heat capacity and entropy of low quartz between 80 and 800 K (Figs 4.6 and 4.7). The best agreement between the model and experiment comes at very low temperatures where the heat capacity follows a T^3 dependence. The overestimation of heat capacity values by the model is due to the concentration of vibrational modes between 100 and 400 cm^{-1} and the failure to take into account any modes above 400 cm^{-1} (Fig. 4.5(d)). Thus the elastic Debye model is too drastic an approximation of G(ω) for low quartz and is useless for prediction of the thermal properties of low quartz. A similar conclusion is reached even when the Debye temperature has been chosen as a free parameter to match, in this case, the observed heat capacities between 350 and 700 K. Although the calorimetric Debye model reproduces observed heat capacities within 4% of the observed values in this region (Fig. 4.6(b)), it severely

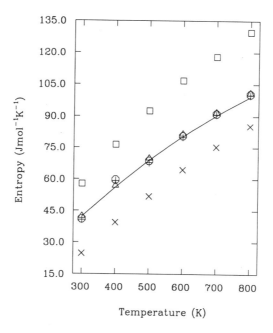

Figure 4.7 Comparison of entropies calculated from models in Figure 4.5 with experimental data (solid curve). The symbols represent predictions from the different models: 'O': VDOS-1; '+': VDOS-2; triangle: Kieffer model; 'X': thermal Debye model; square: elastic Debye model.

underestimates heat capacities below 300 K (Fig. 4.6(a)). Since the entropy is obtained by integrating the area under the C_p/T curve, the model also underestimates entropies. The entropy is underestimated by 40% at 300 K and 10% at 800 K (Fig. 4.7). Whereas the elastic Debye model concentrated too many modes between 100 and 400 cm^{-1}, the calorimetric Debye model underestimates modes below 400 cm^{-1} and concentrates the modes between 400 and 750 cm^{-1} (Fig. 4.5(d)).

In conclusion, model predictions of heat capacities and entropies are insensitive to details of the frequency distribution, as shown by the very similar values for the thermal properties calculated from VDOS-1, VDOS-2, and Kieffer's approximation. The single-parameter Debye model, however, is wholly inadequate for prediction of heat capacities and entropies of a phase such as low quartz. Kieffer (1979c; d) has shown that this conclusion is true, in general, for silicates with silicon in tetrahedral coordination. These minerals are characterized by high-frequency vibrational modes in the region of 1000 cm^{-1} that cannot be accounted for when describing the vibrational density of states with a simple parabolic distribution. The success of Kieffer's approximation is that, although it greatly simplifies $G(\omega)$, it does take into account acoustic anisotropy and dispersion and a distribution of optic modes that are consistent with spectroscopic data.

4.5.2 Example 2: rutile

Rutile crystallizes in the tetragonal space group $P4_2/mnm$ (D_{4h}^{14}) with six atoms per primitive cell giving a total of 18 branches to the phonon dispersion curves. Traylor *et al.* (1971) measured the phonon dispersion relations along the [100], [110], and [001] directions of the Brillouin zone using coherent inelastic scattering of neutrons. Theoretical models based on the rigid-ion and shell models were constructed and fit to the measured dispersion relations, with the shell model giving the best agreement with the experimental data. The phonon density of states calculated from this model is shown in Figure 4.8. The major feature of the frequency distribution is that the vibrational modes of rutile are

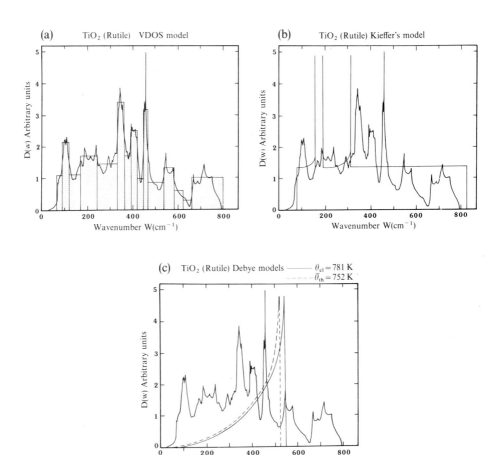

Figure 4.8 Approximations to $G(\omega)$ of rutile: (a) VDOS, consisting of 15 optic continua, (b) Kieffer approximation, and (c) Debye approximation, including the elastic and calorimetric models. The density of states calculated by Traylor *et al.* (1971) is represented by the solid line.

distributed almost continuously from 100 to 800 cm^{-1}. There is a small gap in the distribution near 630 cm^{-1} (<30 cm^{-1}) and there are pronounced peaks near 110, 335, 400, and 464 cm^{-1} (Fig. 4.8). The specific heat and Debye temperature associated with this model show reasonable agreement with the observed heat capacity data above 20 K. The room-temperature model, however, cannot reproduce low-temperature heat capacities because of the highly temperature-dependent, low-frequency A_{2u}(TO) mode. The frequency of this mode drops from 172 cm^{-1} at 298 K to 140 cm^{-1} at 4 K. The frequency distribution of Traylor et al. (1971), however, was used as a basis for comparison of thermodynamic properties calculated from various approximations to the vibrational density of states of rutile.

The first model, VDOS, closely approximates $G(\omega)$ determined from the shell model with a series of 15 contiguous rectangular boxes (Fig. 4.8(a)). Similar to the models for low quartz, the total area under the spectrum is normalized to 1.0, and $G(\omega)$ is used to calculate the heat capacity and entropy (Table 4.1). Whereas the VDOS model with 15 continua provides some structure to the phonon density of states, the Kieffer model consists of a single optic continuum extending from the lowest-frequency optic mode, 113 cm^{-1} at $k=0$ dispersed to 80 cm^{-1} at $k=k_{max}$, to the highest-frequency optic mode, 824 cm^{-1}. Unlike low quartz, there is no break in the spectral distribution between the low-frequency modes and the high-frequency Ti–O stretching modes occurring from 600 to 850 cm^{-1}. Thus the simplest Kieffer model that is consistent with the observed spectroscopic data is a single optic continuum. The three acoustic modes comprise 16.7% (3/18) of the total number of vibrational modes and are assumed to follow a sinusoidal dispersion to the Brillouin zone boundary. There is a large degree of acoustic anisotropy in rutile. In the Kieffer model of rutile, the acoustic branches reach 158, 191, and 309 cm^{-1} at the Brillouin-zone boundary. The acoustic branches measured by Traylor et al. (1971) are approximately sinusoidal along [100]. Along other high-symmetry directions, however, the branches deviate markedly and the slopes of the longitudinal acoustic branches along [100] and [001] differ by a factor of two, reflecting the high degree of acoustic anisotropy. The Kieffer approximation is compared with $G(\omega)$ obtained from the shell model of Traylor et al. (1971) in Figure 4.8(b). Although the two models agree at low frequencies in the representation of the optic continuum, the placement of the acoustic modes differs significantly between the two, with the acoustic peaks at considerably higher frequencies in the Kieffer model than those of the Traylor model. It should be noted that θ_{el} predicted from the Traylor model, 660 K, is substantially lower than the measured elastic Debye temperature of 781 K. The Kieffer model, on the other hand, has the correct acoustic limit.

The third approximation to $G(\omega)$ is the Debye approximation shown in Figure 4.8(c). Both the calorimetric Debye model and elastic Debye model are shown. The cut-off frequency for the elastic Debye model was calculated from averaged acoustic velocities of rutile, 5.72 km^{-1}, that gives θ_{el} of 781 K and ω_D

of 542 cm^{-1}. The calorimetric Debye model is based on an average thermal Debye temperature of 752 K calculated by Watanabe (1982) who fitted the measured heat capacity data of rutile between 350 and 700 K to Debye's theoretical expression. The corresponding ω_D for the latter model is 522 cm^{-1}. Thus unlike low quartz, the elastic and calorimetric Debye models are quite similar with only a 30 K difference in θ_D. Comparison of the Debye model with Traylor and coworkers' (1971) calculated frequency distribution (Fig. 4.8(c)) shows that the parabolic distribution of the Debye model underestimates the modes occurring at low frequencies and at high frequencies above ω_D. It is clear, however, that rutile is more 'Debye-like' than low quartz.

Heat capacities and entropies calculated from the models are compared with experimental data of rutile (Shomate, 1947; Dugdale *et al.*, 1954; Robie *et al.*, 1975) in Figures 4.9 and 4.10. The model values were corrected from C_v to C_p using equation (10). In general, all of the approximations to Traylor and coworkers' (1971) calculated density of states reproduce heat capacities within 4% of the experimental values between 300 and 1000 K (Fig. 4.9). All of the models, including Traylor and coworkers' (1971) calculated distribution, overestimate heat capacities at higher temperatures, which may be due to increased anharmonic effects. The Kieffer approximation is marginally better than the VDOS model, which is, in turn, better than either of the Debye models (Fig. 4.9(b)). Below 300 K, heat capacities calculated from the various models

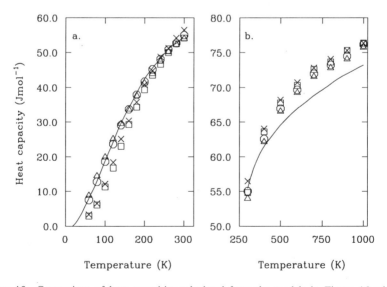

Figure 4.9 Comparison of heat capacities calculated from the models in Figure 4.8 with the experimental data (solid curve) at (a) low temperature and (b) high temperature. The symbols represent heat capacity predictions from the models as follows: 'O': VDOS; triangle: Kieffer model; 'X': thermal Debye model; square: elatic Debye model.

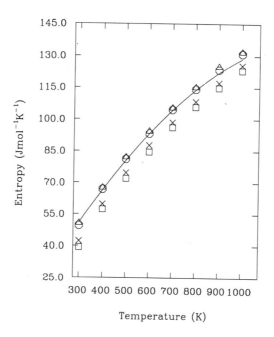

Figure 4.10 Comparison of entropies calculated from the models in Figure 4.8 with the experimental data (1975) (solid curve). The model predictions are represented by: 'O': VDOS; triangle: Kieffer model; 'X': thermal Debye model; square: elastic Debye model.

diverge. The 15-continua VDOS model and the Kieffer model predict heat capacities within 3% of the experimental data to 150 K whereas the Debye models underestimate the heat capacity below 250 K (Fig. 4.9(a)). The differences in the low-temperature heat capacities are reflected in calculations of entropies that provide a slightly more rigorous test of the adequacy of the models since entropy is obtained by integrating C_p/T over temperature. Unlike the heat capacities, only the VDOS model and the Kieffer model reproduce entropies within 2% of the experimental data between 300 and 1000 K (Fig. 4.10). Both of the Debye models underestimate the entropy by approximately 5 to 17% in this temperature interval, due to underestimating C_p below 300 K.

Thus, even though rutile has a simpler, more 'Debye-like' frequency distribution than low quartz, we find that the Kieffer approximation shows considerable improvement over the one-parameter Debye model for prediction of heat capacities and entropies. Both the Kieffer model and the VDOS model, that closely follow Traylor et al.'s (1971) calculated $G(\omega)$, predict similar values for C_p and S between 150 K and 1000 K. The effort involved in obtaining thermodynamic properties from the Kieffer model, however, is orders of magnitude less than the effort involved in the combined experimental and theoretical study of Traylor et al. (1971).

4.6 Application of vibrational models to mineral stability

Calculation of mineral stabilities and phase equilibria provides a more stringent test of approximations to the mineral's vibrational density of states than calculation of thermodynamic functions alone since small differences in the thermodynamic functions, rather than their absolute magnitude, determine the phase relations. Therefore errors resulting from averaging approximations become qualitatively important.

As the examples above and Kieffer (1979c; d) have shown, thermodynamic properties of minerals can be predicted quite accurately from approximating $G(\omega)$ with a model that integrates elastic, structural, and spectroscopic data. Kieffer (1982) further demonstrated that such a model could be applied to problems of mineral stability, calculating the phase equilibria of the SiO_2 polymorphs (quartz, coesite, and stishovite), the Al_2SiO_5 polymorphs (andalusite, kyanite, and sillimanite), and the phase boundary for the breakdown of albite to jadeite and quartz. Other workers found that the model was valuable in constraining thermodynamics properties of high-pressure, high-temperature phases which are seldom available in suitable quantity for complete thermodynamic characterization by conventional methods. Combination of the Kieffer model with high-temperature solution calorimetry, for example, provided independent cross checks and additional constraints on the phase equilibria of the olivine, modified spinel (wadsleyite) and spinel (ringwoodite) polymorphs of Mg_2SiO_4 (Akaogi et al., 1984), the olivine and spinel polymorphs of Mg_2GeO_4 (Ross and Navrotsky, 1987), the wollasonite, garnet and perovskite polymorphs of $CaGeO_3$ (Ross et al., 1986), and the orthopyroxene, clinopyroxene, and ilmenite polymorphs of $MgGeO_3$ (Ross and Navrotsky, 1988).

Another approach used to approximate $G(\omega)$ is worth mentioning because of its applications to problems of mineral stability. This model was developed by Salje and Viswanathan (1976) and described in detail by Salje and Werneke (1982a). Similar to the Kieffer model, the approximation relies on spectroscopically measured phonon frequencies at $k=0$ and elastic data to determine the slopes of the low-frequency, acoustic branches at $k=0$. Dispersion of the acoustic branches is assumed to be linear and therefore the phonon density of states in the low-frequency region increases as the square of the phonon frequency. This is equivalent to the Debye approximation with the exception that only the acoustic modes are considered in this way. The remaining $3N-3$ optic modes are determined at $k=0$ with infrared and Raman spectroscopy (and inelastic neutron scattering, if available). The model relies on knowledge of the position at $k=0$ of all of the optic modes. If some modes are inactive or unobserved, intelligent guesses must be made (usually with the aid of some theoretical model) as to where they occur. In addition, since it is necessary to integrate all phonons over the entire Brillouin zone for calculation of thermodynamic properties, some knowledge of dispersion of the optic modes is required. In most cases, however, such knowledge is unavailable and therefore the assumption is made that dispersion is the same for all optic modes. In the case of andalusite, all phonon lines are known and the phonon density of states approximated with this method is shown in Figure 4.11(a). A Kieffer model of anadalusite is shown for comparison in Figure 4.11(b). The major difference between the two approximations is that the phonon density of states of andalusite, like the previous examples of low quartz and rutile, is not a simple continuum. Salje and Werneke's (1982b) representation of $G(\omega)$ of anadalusite is probably a more realistic model of the mineral's 'true' vibrational density of states than Kieffer's model. However, a more complete knowledge of the symmetry and mode assignments is required for the Salje–Viswanathan model than the Kieffer model. Applications of the former model to problems of mineral stability include prediction of the phase equilibria of calcite and aragonite (Salje and Viswanathan, 1976), and the phase equilibrium boundary between andalusite and sillimanite (Salje and Werneke, 1982a; b).

More detailed approximations of $G(\omega)$ have also been made in applications

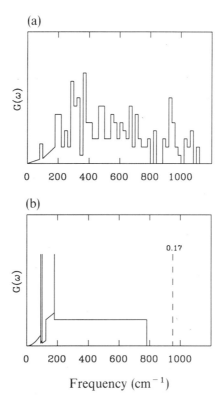

Figure 4.11 Vibrational density of states of andalusite based on (a) Salje and Werneke's (1982a; b) approximation and (b) Kieffer's model.

predicting a mineral's thermodynamic properties as a function of pressure and temperature. The approach, like that of Kieffer and Salje–Visnawathan, is to develop a vibrational density of states based on phonon modes measured by infrared and Raman spectroscopy. The band of modes, however, that would be treated with a single continuum using Kieffer's approach is represented instead by a series of continua with areas corresponding to the density of modes at $k = 0$, analogous to the VDOS models described for low quartz and rutile in the previous section. This method introduces more structure in the vibrational density of states.

Hofmeister (1987) and Chopelas (1990) used such an approach to study forsterite. Hofmeister (1987) noted that the zone-centre modes for fosterite are not uniformly distributed as a function of frequency and that a simple continuum, modelling all of the modes between 142 and 620 cm^{-1}, is inappropriate. She therefore divided the single continuum into six overlapping continua corresponding to the internal bending motions and stretching (v_2 and v_4) modes of the SiO_4 tetrahedra, translational modes of M1 and M2 cations, and translational and rotational modes of SiO_4. The high-frequency symmetric stretching (v_1) and antisymmetric stretching (v_3) modes of the SiO_4 tetrahedron are represented with four Einstein oscillators. Chopelas (1990) also proposed a frequency distribution for forsterite that is similar to Hofmeister's (1987) approximation, with the exception that the acoustic branches are represented as Debye oscillators. Chopelas divided the Kieffer single continuum into 7 continua with the high-frequency v_1 and v_3 SiO_4 modes represented with four Einstein oscillators. Both of these approximations of $G(\omega)$ for forsterite are consistent with the vibrational density of states proposed by Rao *et al.* (1988) determined from a combination of experimental inelastic neutron scattering and theory. Moreover, heat capacities and entropies calculated from Hofmeister's (1987) and Chopelas' (1990) representations of $G(\omega)$ for forsterite are within 2% of the experimental values between 50 and 1500 K.

The various approaches discussed above all rely on complete knowledge of the optic mode distribution at $k = 0$. In cases where optic modes are not infrared or Raman-active, inelastic neutron-scattering experiments are needed. In the absence of such data, theoretical models describing the interatomic forces present in the solid are needed. From such models, it is not only possible to predict the optic mode frequencies at $k = 0$, but also their dispersion relations and the phonon density of states. The development of interatomic potentials to completely describe the lattice dynamics of a mineral is a flourishing area of research in the earth sciences. It is not possible to describe all of the theoretical models that are being used to calculate lattice dynamics of different minerals, nor the advantages and disadvantages of each method. The reader is referred to Price *et al.* (1989) for a review of this subject. The remainder of this chapter describes in detail one of these theoretical approaches and an example of an application to mineral stability at high pressures and temperatures.

4.7 Lattice dynamics from atomistic simulations

The frequencies of atomic vibrations within a crystal are determined by the strength and nature of the bonding which holds the atoms in the crystal together. More generally, these bonding forces can be described in terms of interatomic potentials that not only determine the vibrational characteristics of a crystal, but also its structure and physical properties, such as its elastic and dielectric behaviour. In principle, the bonding and related physical properties of any silicate can be studied by quantum-mechanical methods, which directly describe the interactions of electrons and nuclei within a given system. However, because of the complexity of most minerals, such quantum-mechanical studies are limited to modelling simple structures or molecular fragments (e.g. Cohen et al., 1987; Gibbs, 1982). An alternative to quantum-mechanical calculations is the atomistic approach based on the classical Born model of solids. In this method, sets of interatomic potentials are developed, either from *ab initio* methods or empirically, to describe the net forces acting upon atoms within a structure. The most important contribution to the lattice energy comes from the electrostatic or Coulombic energy terms of the interatomic potential that result from the ionic charges of the atomic species. There is also a short-range repulsion term to model the effect of the overlap of the neighbouring charge clouds. Such short-range components can be represented by a Buckingham potential. For fully ionic, rigid-ion models, the Coulombic and short-range terms are generally the only components of the potential to be considered. In a semi-ionic structure, however, such as a silicate, a degree of directional, covalent bonding is expected between silicon and its coordinating oxygens. Bond directionality can be modelled by introducing an effective three-body interaction term, such as a bond-bending term, into the potential (e.g. Price et al., 1987). The interatomic potential can also be further developed to model electronic polarizability, which is essential for studies of a mineral's defect and high-frequency dielectric behaviour. A shell model (Dick and Overhauser 1958) provides a simple mechanical description of ionic polarizability. In this model, the atom or ion (it is frequently assumed that oxygen is the only polarizable atom in the structure) is described as having a core containing all the mass, surrounded by a massless, charged shell, representing the outer, valence electron cloud. The core and shell are coupled by a harmonic spring.

In the following sections, atomistic simulations involving interatomic potentials such as those described above are used to predict the phonon density of states, thermodynamic properties, and high-pressure behaviour of $MnTiO_3II$ and $MnTiO_3III$. $MnTiO_3$ is of interest to earth scientists because of its novel transformation behaviour at high pressure. Using $MnTiO_3$ ilmenite, the stable form under ambient conditions, as a starting product, Ko and Prewitt (1988) found that quench products from runs subjected to a pressure of 6 GPa and temperature of 1300°C possessed the $LiNbO_3$ structure. This was the first known occurrence of a II-IV compound with this structure type. In addition, Ross et al. (1989) found that $MnTiO_3II$, with the $LiNbO_3$ structure, undergoes a reversible transition to an orthorhombic perovskite structure at approximately 2.5 GPa and 298 K. This was the first transition of this type to be found, although Megaw (1969) postulated that such a transition, though unlikely, was possible given the close relationship between the two structures. The discovery of this new, high-pressure phase and reversible transformation to a perovskite structure raises the question as to whether such transformation behaviour occurs in other oxides or silicates of geological and geophysical interest. With atomistic simulations, we can predict thermodynamic properties of the $LiNbO_3$ and perovskite phases of $MnTiO_3$ and other compositions, thus providing insight into the factors controlling stabilities of these two phases and also a means of predicting whether such transformation behaviour is likely in other compounds.

4.7.1 Static and dynamic simulations

In the simulation of the high pressure polymorphs of $MnTiO_3$, a fully ionic pair-potential model was used (Table 4.2). The values for the short-range O–O interaction parameters were derived by Catlow (1977) using Hartree–Fock methods. The Mn–O short-range term was derived by Lewis and Catlow (1985) by fitting to the structural and elastic data of MnO while the Ti–O interaction term was obtained fitting the potential to the observed structural and elastic data for TiO_2. The polarizability of the oxygen ion was simulated by using a shell model. The shell charge and harmonic spring constant coupling the core and shell are also given in Table 4.2.

The computer code PARAPOCS (Parker and Price, 1989) was used to evaluate the interatomic potentials and to obtain the minimum energy configuration at zero Kelvin. The minimization of lattice energy to constant pressure involves removal of all mechanical strain acting on the unit cell, thus requiring calculation of the mechanical or static pressure as well as the elastic

Table 4.2 Interatomic potential parameters used in atomistic simulations of the $MnTiO_3$ polymorphs

	A(eV)	ρ(Å)	C(eV Å^6)	K_s(eV Å^{-2})
Mn–O	1007.4	0.3262	0.0	
Ti–O	754.0	0.3879	0.0	
O–O	22764.3	0.1490	27.88	70.00
	q_{Mn}	+2		
	q_{Ti}	+4		
	$q_{o,core}$	+0.80		
	$q_{o,shell}$	−2.80		

constants (e.g. Parker and Price, 1989). Calculations proceed, with adjustments of atomic coordinates and cell dimensions, until all residual strains are removed. This method yields the crystal structure at absolute zero.

In addition to predicting the minimum energy configuration at zero Kelvin and zero pressure, atomistic simulations can model the response of a structure to any required hydrostatic pressure by simply adding the hydrostatic pressure to the mechanical pressure and evaluating the resultant strain during the energy minimization procedure (e.g. Parker and Price, 1989). Moreover, simulation of temperature can be achieved by calculating the full, lattice dynamical behaviour of the crystal in the quasi-harmonic approximation. This procedure requires determination of the phonon frequencies by solving the Newtonian equations of motion, similar to equation (12), and resulting eigenvector–eigenvalue equations. The eigenvalues are the squared frequencies of each of the normal modes of the crystal and the eigenvectors describe the pattern of atomic displacement for each normal mode. Since the phonon frequencies are dependent on wave vector, k, determination of the phonon density of states involves calculation of the frequencies over all possible wave vectors. This is achieved by determining frequencies in a three-dimensional mesh of points within the Brillouin zone, using an appropriate weighting factor. Once the phonon frequencies have been determined for points within the irreducible Brillouin zone, the thermodynamic functions can be calculated.

4.7.2 Comparison of observed and simulated structures

A comparison of the observed structural and elastic properties of $MnTiO_3$ II and III at zero Kelvin and zero pressure with the predictions obtained from using the potential described above in computer code PARAPOCS is presented in Tables 4.3 and 4.4. The potential predicts the cell parameters of

Table 4.3 Comparison between observed and calculated structural data for $MnTiO_3$ II ($LiNbO_3$ structure). Experimental data from KO and Prewitt (1988) and Ross et al. (1989)

Parameter		Observed	Calculated
a (Å)		5.205	5.297
c (Å)		13.700	13.601
Vol. (Å^3)		321.4	330.5
Ti–O (Å)	[3x]	1.865	1.868
Ti–O (Å)	[3x]	2.114	2.104
⟨Ti–O⟩		1.990	1.986
Mn–O (Å)	[3x]	2.120	2.196
Mn–O (Å)	[3x]	2.280	2.360
⟨Mn–O⟩		2.200	2.278
K_T (GPa)		158	163

Table 4.4 Comparison between observed and calculated structural data for $MnTiO_3$ III (perovskite structure). Experimental data from Ross et al. (1989)

Parameter		Observed	Calculated
a (Å)		5.145	5.281
b (Å)		5.328	5.363
c (Å)		7.460	7.594
Vol. (Å^3)		204.6	215.1
Ti–O (Å)	[2x]	1.957	1.973
Ti–O (Å)	[2x]	1.946	1.946
Ti–O (Å)	[2x]	1.949	1.952
⟨Ti–O⟩		1.951	1.957
Mn–O (Å)	[1x]	2.107	2.189
Mn–O (Å)	[1x]	2.214	2.422
Mn–O (Å)	[2x]	2.113	2.218
Mn–O (Å)	[2x]	2.464	2.611
Mn–O (Å)	[2x]	2.635	2.711
⟨Mn–O⟩		2.343	2.461
K_T (GPa)		227	217

$MnTiO_3$ II (LiNbO$_3$ structure) within 2% of the observed values (Table 4.3). The a axis is slightly overestimated while c is sightly underestimated. The Ti–O bond lengths are reproduced within 0.5% of the observed values and the Mn–O bond lengths within 3%. The calculated bulk modulus, 163 GPa, is in excellent agreement with the observed bulk modulus, 158 GPa, determined by Ross et al. (1989).

The potential parameters were also used to simulate $MnTiO_3$ III (perovskite structure). Unlike the LiNbO$_3$ structure, where the Mn atoms are in octahedral coordination, the Mn atoms are coordinated to eight oxygen atoms in $MnTiO_3$ III. It is surprising how well the potential parameters model the perovskite which is one of the most disorted GdFeO$_3$–type perovskites known (Ross et al., 1989). The experimental results are compared with model predictions in Table 4.4. The cell parameters are overestimated by the model calculations. The differences between calculated and observed are 2.6%, 0.7%, and 1.8% for $a, b,$ and c, respectively (Table 4.4). The Ti–O bond lengths show excellent agreement with the observed bond lengths. The average ⟨Ti–O⟩ bond length, 1.957 Å, predicted from model calculations is slightly longer than the observed value, 1.951 Å. The Mn–O bond lengths, however, are less well predicted. The average ⟨Mn–O⟩ bond length predicted by the model calculations, 2.461 Å, is 5% greater than the experimental value. However, it should be remembered that we are comparing a static lattice simulation carried out at 0 K and 0 P with structural data determined at 298 K and 4.5 GPa, necessitated by the reversible nature of the transition from the LiNbO$_3$ to perovskite structure (Ross et al., 1989). Unlike the cell parameters which could be extrapolated back to 0.0001 GPa, the bond lengths given in Table 4.4 are those determined at 4.5 GPa. Thus the agreement between theory and experiment is better than that given. The calculated bulk modulus, 217 GPa, compares well with the bulk modulus determined from compressibility data, 227 GPa.

Thus, an interatomic potential incorporating short-range terms from empirical potentials derived to simulate simple oxides can successfully be transferred to study the more complex structures of $MnTiO_3$ II and $MnTiO_3$ III. In the next section, calculations of the thermodynamic properties and high-pressure behaviour of both polymorphs are presented.

4.7.3 Calculation of thermodynamic properties

The first step in determining the thermodynamic functions involves calculation of the phonon density of states, $G(\omega)$, which, in turn, requires information about the vibrational frequencies throughout the entire Brillouin zone. $MnTiO_3$ II has 10 atoms in its primitive unit cell and therefore 30 branches to its phonon dispersion curves. Vibrational frequencies were calculated at a set of points in a $4 \times 4 \times 4$ grid within the first Brillouin zone; hence the total number of modes evaluated was 1920 (30×64) which were then used to construct a histogram of the vibrational density of states of $MnTiO_3$ II. The phonon density of states of $MnTiO_3$ III, which has 20 atoms in its primitive unit cell and hence 60 branches to the phonon dispersion curves, was determined from a $3 \times 3 \times 3$ grid covering the first Brillouin zone. The total number of modes evaluated, 1620, was therefore comparable to the number evaluated for $MnTiO_3$ II. The resulting normalized density of states of $MnTiO_3$ II and $MnTiO_3$ III are shown in Figure 4.12. The similarity between these calculated phonon density of states is pronounced, which is not surprising given the similarity between the two structures (Megaw, 1969; Ross et al., 1989). The optic modes span a range of frequencies from 100 to 800 cm^{-1} in both $MnTiO_3$ II and $MnTiO_3$ III. The calculated frequencies of the vibrational modes of $MnTiO_3$ II are in good agreement with the Raman spectra of Ko et al. (1989). No spectroscopic data are available for $MnTiO_3$ III. The calculated phonon density of states for both pholymorphs show departures from a Kieffer-like distribution (Fig. 4.12). There are two gaps in the mode distribution, one near 440 cm^{-1} and the other around 700 cm^{-1}. The latter separates the high-frequency modes extending to 800 cm^{-1}, that involve predominantly Ti–O stretching motions, from the rest of the modes. The most notable differences between $G(\omega)$ for $MnTiO_3$ II and $MnTiO_3$ III are the greater concentration of modes in the pervoskite structure at frequencies below 200 cm^{-1} and the less pronounced gap between the high-frequency modes and the rest of the modes in the phonon density of states of the perovskite structure.

Heat capacities and entropies calculated from the atomistic simulations are presented in Table 4.5. The similarity in the calculated heat capacities and entropies reflects their similar vibrational density of states. $MnTiO_3$ II, with

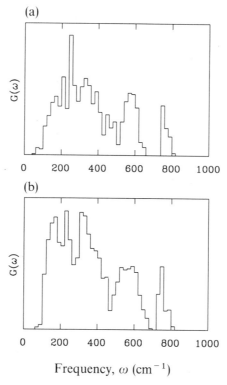

Figure 4.12 Vibrational density of states of MnTiO$_3$ II (LiNbO$_3$ structure) and MnTiO$_3$ III (orthorhombic perovskite structure) calculated from atomistic simulations of the two structures (see text for details).

the LiNbO$_3$ structure, has a slightly lower heat capacity and hence entropy than MnTiO$_3$ III, with the perovskite structure. The greater entropy of MnTiO$_3$ III is due to the greater concentration of modes below 200 cm^{-1}. The entropy of transition for MnTiO$_3$ II→MnTiO$_3$ III is predicted to be approximately +4 J mol^{-1}K^{-1} at 1000 K. In addition, since ΔV is negative and $dP/dT = \Delta S/\Delta V$, the equilibrium boundary between the two phases should have a negative gradient. These calculations are consistent with Ko et al.'s (1989) predictions of ΔS, based on combined thermochemical and phase equilibria data, in that both predict a small value for the entropy of transition. Ko et al. (1989), however, estimate that ΔS for MnTiO$_3$ II→MnTiO$_3$ III should be negative, -8 J mol^{-1}K^{-1}, rather than positive. It is possible that there are additional contributions to the entropy of MnTiO$_3$ II, not dependent on the lattice vibrations, that must be considered in the determination of phase equilibria.

The free energies of the LiNbO$_3$ and perovskite polymorphs of MnTiO$_3$ were calculated over a range of pressures from 0 to 10 GPa at 300 K. The free

Table 4.5 Heat capacities and entropies calculated from atomistic simulations of MnTiO$_3$ II (LiNbO$_3$ structure) and MnTiO$_3$ III (orthorhombic perovskite structure)

Temp. (K)	MnTiO$_3$ II: C_P (J/atom)	S^0 (J/atom·K)	MnTiO$_3$ III: C_P (J/atom)	S^0 (J/atom·K)
100	7.20	1.80	7.63	1.91
150	11.82	5.57	12.17	5.85
200	15.23	9.45	15.52	9.81
250	17.65	13.12	17.91	13.55
300	19.37	16.50	19.61	16.97
350	20.60	19.58	20.84	20.09
400	21.51	22.40	21.75	22.94
450	22.19	24.97	22.44	25.55
500	22.71	27.34	22.97	27.94
550	23.13	29.53	23.40	30.15
600	23.46	31.36	23.75	32.21
650	23.73	33.45	24.04	34.12
700	23.96	35.21	24.29	35.91
750	24.15	36.88	24.50	37.60
800	24.31	38.44	24.69	39.19
850	24.45	39.92	24.85	40.69
900	24.58	41.32	25.00	42.11
950	24.69	42.65	25.13	43.47
1000	24.79	43.92	25.25	44.76
1100	24.97	46.30	25.48	47.18
1200	25.12	48.47	25.67	49.41
1300	25.25	50.49	25.85	51.47
1400	25.37	52.37	26.00	53.39
1500	25.47	54.12	26.16	55.19

energy vs. pressure curve follows a linear trend for both phases. Between 0 and 10 GPa the lines cross over, with the intersection giving the pressure at which the perovskite structure becomes stable relative to the LiNbO$_3$ structure. The ΔG curve for MnTiO$_3$ II→MnTiO$_3$ III at 0 K and a function of pressure is shown in Figure 4.13. The calculated transition pressure at 300 K is 3 GPa, which is in fortuitously good agreement with the experimental value of 2.5 GPa and 298 K (Ross et al., 1989).

The success of the atomistic approach in simulating the high-pressure polymorphs of MnTiO$_3$ is encouraging. Similar calculations completed on other compositions by Ross and Price (1989) suggest that FeTiO$_3$ will exhibit similar transformation behaviour as MnTiO$_3$ at high pressures and temperatures, unless it breaks down to its component oxides first. Moreover, if FeTiO$_3$ with a LiNbO$_3$ structure is recovered from high-pressure, high-temperature runs, Ross and Price (1989) predict that it will undergo a transformation to an orthorhombic perovskite structure at approximately 10 GPa and 0 K. Unlike

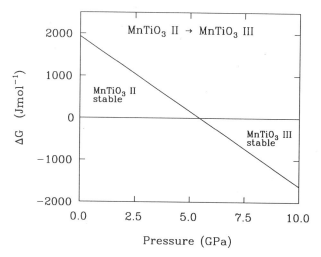

Figure 4.13 Calculated change in free energy, ΔG, for $MnTiO_3$ II→$MnTiO_3$ III at 0 K as a function of pressure, based on atomistic simulations.

$MnTiO_3$ and $FeTiO_3$, $CaTiO_3$ is not predicted to adopt a $LiNbO_3$ structure at high pressures and temperatures. At every pressure and temperature simulated, orthorhombic perovskite was found to be the stable phase for this composition (Ross and Price 1989). [Note added in proof: recently Leinenweber *et al.* (1991) have reported a $LiNbO_3$-perovskite transition in $FeTiO_3$ at 16 GPa and 300 K.]

4.8 Conclusion

In this chapter, I have attempted to show how knowledge of a crystal's lattice vibrations allows prediction of the phase's thermodynamic properties, and, hence, an ability to predict the equilibrium behaviour of the phase. The various approaches to model a crystal's phonon density of states have been reviewed, with emphasis on those methodologies that have been used in applications in earth science. The past decade has seen tremendous progress in our ability to simulate vibrational properties of minerals. We have followed, for example, how our approach to studying the lattice dynamics of a mineral has evolved from models that greatly simplify the mineral's vibrational density of states, such as the Kieffer model, to atomistic simulations which rely on knowledge of the interatomic potentials between the atoms of the structure. As our computational power increases and better interatomic potentials are developed, it is envisaged that we shall be able to simulate, with some reliability, even more complex minerals. In addition, interatomic potentials are being developed from *ab initio* methods, rather than empirical methods. One area that is especially challenging to theorists is the ability to model the microscopic interactions in a *disordered* mineral. This chapter has focused on the *vibrational* contribution to the thermodynamic properties of a phase and has neglected the *configurational* contributions to the thermodynamic properties of a phase. The latter are discussed in detail in the following chapters.

References

Akaogi M., Ross, N. L., McMillan, P., *et al.* (1984) The Mg_2SiO_4 polymorphs (olivine, modified spinel and spinel)—thermodynamic properties from oxide melt solution calorimetry, phase relations, and models of lattice vibrations. *American Mineralogist*, **69**, 499–512.

Born, M. and Huang, K. (1954) *Dynamical Theory of Crystal Lattices*, Oxford University Press, New York.

Born, M. and Von Karman, T. (1912) Über Schwingungen in Raumgittern. *Physikalische Zeitschrift*, **13**, 297–309.

Born, M. and Von Karman T. (1913) Theory of specific heat. *Physikalische Zeitschrift*, **14**, 15–71.

Catlow, C. R. A. (1977) Point defect and electronic properties of uranium dioxide. *Proceedings of the Royal Society of London*, **A353**, 533–61.

Catlow, C. R. A. and Mackrodt, W. C. (1982) Theory of simulation methods for lattice and defect energy calculations in crystals, (eds C. R. A. Catlow and W. C. Mackrodt), in *Lecture Notes in Physics 166: Computer Simulation of Solids*, Springer-Verlag, Berlin, pp. 3–20.

Chopelas, A. (1990) Thermal properties of forsterite at mantle pressures derived from vibrational spectroscopy. *Physics and Chemistry of Minerals*, **17**, 149–56.

Cochran, W. (1973) *The Structures and Properties of Solids 3: The Dynamics of Atoms in Crystals*, Edward Arnold Ltd, London.

Cohen, R. E., Boyer, L. L., and Mehl M. J. (1987) Lattice dynamics of the potential-induced breathing model: phonon dispersion in the alkaline-earth oxides. *Physical Review B*, **35**, 5749–60.

Cowley, R. A., Woods, A. D. B., and Dolling G. (1966) Crystal dynamics of potassium. I. Pseudopotential analysis of phonon dispersion curves at 9°K. *Physical Review*, **150**, 487–94.

Debye, P. (1912) Zur Theorie der spezifischen Wärmen. *Annalen der Physik* (Leipzig), **39**, 789–839.

Dick, B. G. and Overhauser, A. W. (1958) Theory of dielectric constants of alkali halide crystals. *Physical Review*, **112**, 90–103.

Dugdale, J. S., Morrison, J. A., and Patterson, D. (1954) The effect of particle size on the heat capacity of titanium dioxide. *Proceedings of the Royal Society of London*, **A224**, 228–35.

Elcombe, M. M. (1967) Some aspects of the lattice dynamics of quartz. *Proceedings of the Physical Society*, **91**, 947–58.

Galeener, F.L., Leadbetter, A. J., and Stringfellow, M. W. (1983) Comparison of neutron, Raman, and infrared vibrational spectra of vitreous SiO_2, GeO_2, and BeF_2. *Physical Review B*, **27**, 1052–78.

Ghose, S. (1988) Inelastic neutron scattering, in *Reviews in Mineralogy*, vol. 18: *Spectroscopic Methods in Mineralogy and Geology*, (ed. F. C. Hawthorne), Mineralogical Society of America, pp. 161–92.

Gibbs, G. V. (1982) Molecules as models for bonding in silicates. *American Mineralogist*, **67**, 421–50.

Gilat, G. (1972) Analysis of methods for calculating spectral properties in solids. *Journal of Computational Physics*, **10**, 432–65.

Hofmeister, A. M. (1987) Single-crystal absorption and reflection infrared spectroscopy of forsterite and fayalite. *Physics and Chemistry of Minerals*, **14**, 499–513.

Kieffer, S. W. (1979a) Thermodynamics and lattice vibrations of minerals: 1. Mineral heat capacities and their relationships to simple lattice vibrational models. *Reviews of Geophysics and Space Physics*, **17**, 1–19.

Kieffer, S. W. (1979b) Thermodynamics and lattice vibrations of minerals: 2. Vibrational characteristics of silicates. *Reviews of Geophysics and Space Physics*, **17**, 20–34.

Kieffer, S. W. (1979c) Thermodynamics and lattice vibrations of minerals: 3. Lattice dynamics and an approximation for minerals with application to simple substances and framework silicates. *Reviews of Geophysics and Space Physics*, **17**, 35–59.

Kieffer, S. W. (1980) Thermodynamics and lattice vibrations of minerals: 4. Application to chain and sheet silicates and orthosilicates. *Reviews of Geophysics and Space Physics*, **18**, 862–86.

Kieffer, S. W. (1982) Thermodynamics and lattice vibrations of minerals: 5. Applications to phase equilibria, isotope fractionation, and high-pressure thermodynamic properties. *Reviews of Geophysics and Space Physics*, **20**, 827–49.

Ko, J. and Prewitt C. T. (1988) High-pressure phase transition in $MnTiO_3$ from the ilmenite to the $LiNbO_3$ structure. *Physics and Chemistry of Minerals*, **15**, 355–62.

Ko, J., Brown, N. E., Navrotsky, A., et al. (1989) Phase equilibrium and calorimetric study of the transition of $MnTiO_3$ from the ilmenite to the lithium niobate structure and implications for the stability field of perovskite. *Physics and Chemistry of Minerals*, **16**, 727–33.

Leadbetter, A. J. (1969) Inelastic cold neutron scattering from different forms of silica. *Journal of Chemical Physics*, **51**, 779–86.

Leinenweber, K., Utsumi, W., Tsuchida, Y. et al. (1991) Unquenchable high-pressure perovskite polymorphs of $MnSnO_3$ and $FeTiO_3$. *Physics and Chemistry of Minerals*, **18**, 244–250.

Lewis, G. V. and Catlow, C. R. A. (1985) Potential models for ionic oxides. *Journal of Physics C*, **18**, 1149–61.

Lord, R. C. and Morrow, J. C. (1957) Calculation of heat capacity of α-quartz and vitreous silica from spectroscopic data. *Journal of Chemical Physics*, **26**, 230–2.

Maradudin, A. A., Montroll, E. W., and Weiss, G. H. (1963) *Lattice Dynamics in the Harmonic Approximation*, Academic Press, New York.

Megaw, H. (1969) A note on the structure of $LiNbO_3$. *Acta Crystallographica*, **A24**, 583–8.

Montroll, E. W. (1942) Frequency spectrum of crystalline solids. *Journal of Chemical Physics*, **10**, 218–29.

Montroll, E. W. (1943) Frequency spectrum of crystalline solids II: general theory and applications to simple cubic lattices. *Journal of Chemical Physics*, **11**, 481–95.

Navrotsky, A. (1980) Lower mantle phase transitions may generally have negative pressure–temperature slopes. *Geophysical Research Letters*, **7**, 709–11.

Parker, S. C. and Price, G. D. (1989) Computer modelling of phase transitions in minerals, in *Advances in Solid-State Chemistry*, vol. 1, JAI Press Inc., London, pp. 295–327.

Placzek, G. and Van Hove, L. (1954) Crystal dynamics and inelastic scattering of neutrons. *Physical Review*, **93**, 1207–14.

Price, G. D., Parker, S. C., and Leslie, M. (1987) The lattice dynamics of forsterite. *Mineralogical Magazine*, **51**, 157–70.

Price, G. D., Wall, A., and Parker, S. C. (1989) The properties and behaviour of mantle minerals: a computer simulation approach. *Philosophical Transactions of the Royal Society of London*, **A328**, 391–407.

Rao, K. R., Chaplot, S. L., Choudhury, N., et al. (1988) Lattice dynamics and inelastic neutron scattering from forsterite, Mg_2SiO_4: phonon dispersion relation, density of states and specific heat. *Physics and Chemistry of Minerals*, **16**, 83–97.

Robie, R. A., Hemingway, B. S., and Fischer, J. R. (1975) Thermodynamic properties of minerals and related substances at 298.15 K and 1 bar (10^5 pascals) pressure and at higher temperatures. *US Geological Survey Bulletin 1452*, US Government, Washington D.C.

Ross, N. L. and Navrotsky, A. (1987) The Mg_2GeO_4 olivine–spinel phase transition. *Physics and Chemistry of Minerals*, **14**, 473–81.

Ross, N. L. and Navrotsky, A. (1988) Study of the $MgGeO_3$ polymorphs (orthopyroxene, clinopyroxene, and ilmenite structures) by calorimetry, spectroscopy, and phase equilibria. *American Mineralogist*, **73**, 1355–65.

Ross, N. L.. and Price, G. D. (1989) Factors determining the stability of $LiNbO_3$ and ilmenite structures. *Transactions of the American Geophysical Union*, **70**, 350.

Ross N. L., Ko, J., and Prewitt C. T. (1989) A new phase transition in $MnTiO_3$:$LiNbO_3$ to perovskite structure. *Physics and Chemistry of Minerals*, **16**, 621–9.

Ross, N. L., Akaogi, M., Navrotsky, A., *et al*. Phase transitions among the $GaGeO_3$ polymorphs (wollastonite, garnet, and perovskite structures): studies by high-pressure synthesis, high-temperature calorimetry, and vibrational spectroscopy and calculation. *Journal of Geophysical Research*, **91**, 4685–96.

Salje, E. and Viswanathan, K. (1976) The phase diagram calcite–aragonite as derived from the crystallographic properties. *Contributions to Mineralogy and Petrology*, **55**, 55–67.

Salje, E. and Werneke, C. (1982a) How to determine phase stabilities from lattice vibrations, in *High-Pressure Researches in Geoscience*, (ed. W. Schreyer), E. Schwiezer. Verlag, Stuttgart, pp. 321–48.

Salje, E. and Werneke, C. (1982b) The phase equilibrium between sillimanite and andalusite as determined from lattice vibrations. *Contributions to Mineralogy and Petrology*, **79**, 56–67.

Shomate, C. H. (1947) Heat capacities at low temperatures of titanium dioxide (rutile and anatase). *Journal of the American Chemical Society*, **69**, 218–19.

Striefler, M. E. and Barsch, G. R. (1975) Lattice dynamics of α-quartz. *Physical Review B*, **12**, 4553–66.

Traylor, J.G., Smith, H. G., Nicklow, R. M., *et al*. (1971) Lattice dynamics of rutile. *Physical Review B*, **3**, 3457–72.

Watanabe, H. (1982) Thermochemical properties of synthetic high-pressure compounds relevant to the earth's mantle, in *High Pressure Research in Geophysics*, (eds S. Akimoto and M. H. Manghnani), D. Reidel, Boston, pp. 441–64.

Woods, A. D. B., Brockhouse, B. N., Cowley, R. A., *et al*. (1963) Lattice dynamics of alkali halide crystals II: experimental studies of KBr and NaI. *Physical Review*, **131**, 1025–39.

CHAPTER FIVE
Thermodynamics of phase transitions in minerals: a macroscopic approach

Michael A. Carpenter

5.1 Introduction

When calculating the stability fields of different mineral assemblages, petrologists have become used to dealing with two broad categories of reactions. In the first category are heterogeneous reactions between unrelated phases, such as jadeite $\rightleftharpoons$ albite + quartz, or between phases with the same composition but different structure, such as olivine $\rightleftharpoons$ spinel. The second category involves reactions which occur within individual phases and includes solid solution, cation ordering, displacive transitions, magnetic transitions, etc. Because the energy changes for the two types of processes are commonly on a similar scale, internal effects can have a profound influence on how the equilibrium boundaries for the heterogeneous reactions are distributed in PT space. For example, the stability field of cordierite with respect to enstatite + sillimanite + quartz is considerably reduced if the cordierite has a disordered, rather than an ordered distribution of Al and Si between tetrahedral sites (Figure 5.1(a), from Putnis and Holland, 1986). Similarly, the aragonite $\rightleftharpoons$ calcite phase boundary shows a marked curvature at least in part due to the orientational disordering of CO_3 groups in calcite (Figure 1(b), from Redfern *et al.*, 1989). At the same time, there has been an increasing awareness among physicists that structural phase transitions in minerals show characteristic properties worthy of investigation from a purely solid-state-physics point of view. There is a confluence of interests, therefore, which has resulted in a new approach to the thermodynamic analysis of phase transitions in minerals based on Landau theory. The main purpose of the present review is to outline some of these recent developments for the benefit of readers with an earth sciences background.

The principal element of Landau theory is a Gibbs free energy expansion which has the form of a Taylor series in a macroscopic order parameter, Q. In practice, variations of Q with temperature, pressure or any other external variable, are followed indirectly by measuring properties such as excess heat

The Stability of Minerals. Edited by G. D. Price and N. L. Ross.
Published in 1992 by Chapman & Hall, London. ISBN 0 412 44150 0

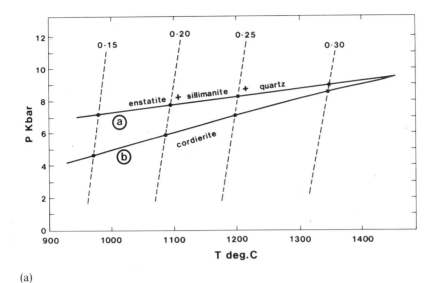

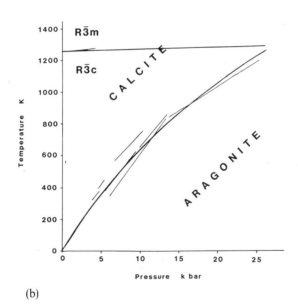

Figure 5.1 The effect of structural phase transitions on heterogeneous mineral equilibria. (a) PT relations for $En + Sill + Qz = Cord$ for ordered cordierite (labelled a) and for disordered cordierite (labelled b). (Dashed lines are isopleths of the fraction of Al in M1 sites in enstatite). (Putnis and Holland (1986). Reproduced by kind permission of A. Putnis.) (b) Calculated equilibrium boundary between calcite and aragonite. The marked curvature arises at least in part because of the $R\bar{3}m \rightleftharpoons R\bar{3}c$ transition in calcite. Short lines represent experimental data. (Redfern et al. (1989). Reproduced by kind permission of S. Redfern.)

capacity, excess enthalpy, and lattice strain, which vary explicitly wth Q and which lead to calibrations of the coefficients in the expansion. Real systems commonly do not conform to the behaviour predicted by the simplest Landau expansions, however, and it is necessary to develop more sophisticated expressions to include the effects of coupling between two or more order parameters. An entirely macroscopic approach can be adopted, to the extent that the microscopic origin of a given phase transition need not be known. Other contributions, from solid-solution effects, for example, can also be incorporated quite simply into the expansions. Many of these ideas have been introduced into the earth sciences literature by Salje and coworkers and this review draws heavily on the following papers: Salje (1985; 1987a; 1988a, b; 1990), Salje and Devarajan (1986), Salje et al. (1985), Carpenter (1988). Readers will also find books by Landau and Lifshitz (1980), Bruce and Cowley (1981), and Tolédano and Tolédano (1987) to be instructive. Lattice dynamical aspects of Landau theory, as reviewed by Bruce and Cowley (1981) and Bismayer (1988) will not be discussed here, however.

In the following six sections, the meaning of the order parameter is first discussed in some detail. The Landau free energy expansion is then introduced and modifications which may be incorporated to account for the effects of pressure, compositional changes, and lattice strains are also explained. A short section comparing Landau and Bragg–Williams expressions for describing the thermodynamics of order/disorder transitions is included to illuminate aspects of both approaches. Phase transitions involving only one order parameters are then discussed, with transitions in calcite, omphacite, and the alkali feldspars used as illustrative examples. It is more common to find that at least two order parameters are required to describe the transition behaviour of framework silicates and a section on order parameter coupling is included to demonstrate how these more complex systems may be treated. Finally, recent developments relating to the application of time-dependent Landau theory to minerals are briefly reviewed in order to show that the macroscopic approach can be applied to both equilibrium and kinetic aspects of phase transitions. Throughout, only thermodynamically continuous transitions are treated in any detail.

5.2 The order parameter, Q

The order parameter, usually designated as Q and scaled to vary between 0 in the high-symmetry form and 1 in the low-symmetry form at 0 K, is the macroscopic property of a crystal which is central to any thermodynamic approach to the treatment of structural phase transitions. Its behaviour is subject to rigorous symmetry constraints and, in effect, it controls the manner in which the thermodynamic properties change as a consequence of a transition. It often has a somewhat elusive quality, however, in that it is never measured directly. Values of Q are obtained only by reference to some other property of the crystal which reflects the extent of transformation; in a ferromagnetic transition this might be the magnetization, while in a ferroelastic transition it might be the lattice strain or the birefringence, for example.

In some cases the macroscopic order parameter can be understood in terms of structural changes on a microscopic scale. For example, the well-known cubic ⇌ tetragonal transition in $SrTiO_3$ occurs by displacements of oxygen atoms from their high-symmetry positions. Alternate TiO_6 octahedra, which are effectively rigid, rotate in opposite directions about a [001] axis in one layer of the structure while in the next layer the rotations have the reverse sense (Fig. 5.2a). The angle of rotation, ϕ, is a measure of the average displacement of the oxygen atoms and therefore corresponds to the order parameter. From the data in Figure 5.2(b), it is evident that ϕ and, hence, Q decrease smoothly to zero when crystals are heated up to the transition temperature of ~ 105 K (Müller et al., 1968; Müller and Berlinger, 1971; Müller and von Waldkirch, 1973).

Rather than measuring the angle ϕ directly in $SrTiO_3$ it might have been more convenient to measure changes in lattice parameters which accompany the phase transition. Rotations of the TiO_6 octahedra lead to a reduction of the unit cell edge that is proportional to $\cos \phi$. For small rotations this may be expanded as $[1-(\phi^2/2)]$ so that, whereas the rotation angle reflects Q directly, the associated lattice distortion (the spontaneous strain) varies with Q^2. Many properties will vary because of the phase transition and, if they can be followed as a function of temperature, will also provide an indication of the variation of the order parameter itself. In each case it is necessary to ensure that the dependence on Q of the property selected is expressed correctly, but the essential point is that the microscopic and macroscopic changes can be linked explicitly. In more complex structures the microscopic mechanisms may not be so clear though the macroscopic analysis can still be completed if the symmetry change is fully defined. Standard tables are useful for checking both the correct form of the order parameter and its relationship to properties such as the spontaneous strain for a given change in symmetry (see, for example, Tolédano and Tolédano, 1980; Stokes and Hatch, 1988; Salje, 1990).

In the case of order/disorder transitions the microscopic order parameter can usually be understood in terms of crystallographic site occupancies and can be defined quite simply. A general definition of Q for ordering in a phase consisting of equal proportions of A and B atoms would be:

$$Q = \frac{|A_s - A_{s'}|}{(A_s + A_{s'})} = \frac{|B_s - B_{s'}|}{(B_s + B_{s'})} \tag{1}$$

where A_s and B_s are the average proportions of A and B atoms occupying s site an $A_{s'}$ and $B_{s'}$ are the proportions occupying s' sites. The s and s' sites are related by symmetry in the high-symmetry disordered crystal but unrelated in the low-symmetry ordered crystal. For thermodynamic purposes the averaging must, of course, be over a large number of sites. This type of definition can also

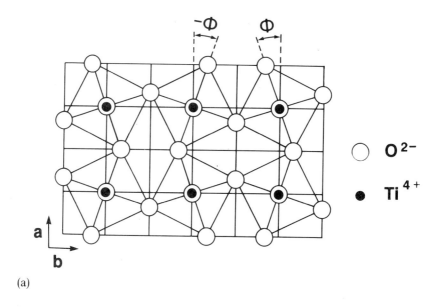

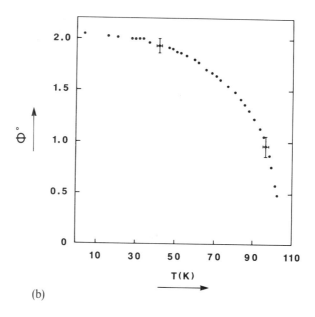

Figure 5.2 Displacive transition in $SrTiO_3$. (a) Alternate TiO_6 octahedra in a sheet of octahedra rotate in opposite directions through an angle ϕ about [001]. (b) Experimental data showing that the angle ϕ goes continuously to zero as the crystal is heated to the transition temperature of ~ 105 K (after Müller and von Waldkirch, 1973).

be used for orientational order/disorder, such as CO_3 group ordering in calcite, $CaCO_3$. A_s would be the proportion of CO_3 groups in one orientation at site s while B_s would be the proportion of CO_3 groups with opposite orientation at the same site; the order parameter, Q, is then $|A_s - B_s|/(A_s + B_s)$.

It has been recognized that the order parameter for a thermodynamically continuous phase transition shows three characteristic regimes between 0 K and the transition temperature, T_c (Salje, 1989a; 1990). Another perovskite example provides a convenient illustration of this. In $LaAlO_3$ the AlO_6 octahedra rotate about a [111] axis, and the symmetry change is from cubic to trigonal. The angle of rotation, ϕ, again corresponds to the order parameter and has been measured from paramagnetic resonance spectra of Fe^{3+} substituted for Al^{3+} (Müller et al., 1968). The variation of a spectroscopic parameter, D, where D is proportional to ϕ^2, is shown as a function of temperature in Figure 5.3 (from Müller et al., 1968 and Salje, 1989a). For most of the temperature range D and, hence, ϕ^2 and Q^2 vary linearly with T in accordance with the predictions of Landau theory for a second-order transition. The critical exponent, β, in $Q \propto (T_c - T)^\beta$, is 1/2. This is the Landau regime, denoted L in Figure 5.3. Close to T_c (~ 800 K), however, there is a small temperature interval over which β deviates significantly from 1/2 and is closer to $\sim 1/3$. This temperature range, denoted G in Figure 5.3, is usually referred to as the Ginzburg interval and is often associated with the presence of fluctuations in Q which become larger in amplitude than the mean value of Q in the crystal as a whole (see Landau and Lifshitz, 1980; Ginzburg et al., 1987). Finally, at the low temperature end Q levels off and becomes effectively temperature independent as 0 K is approached. This 'saturation' regime, labelled S in Figure 5.3, may be due to quantum effects (Salje, 1990).

For most practical purposes the Ginzburg interval, ΔT_G, may be ignored when phase transitions in framework silicates are being investigated. This is because order parameters operating in such structures are expected to have long correlation lengths. Ginzburg and coworkers (reviewed in Ginzburg et al., 1987, and see Salje, 1988b) have shown that the width of the temperature interval close to T_c in which long-wavelength fluctuations become important varies with the correlation length of the order parameter at 0 K, r_{co}, approximately as $\Delta T_G \propto 1/(r_{co}^6)$. The parameter r_{co} is related to the more familiar temperature-dependent correlation length of Q, ξ, by:

$$\xi \propto r_{co} \left| \frac{T_c - T}{T_c} \right|^{-v} \quad (v > 0) \qquad (2)$$

with $v = 1/2$ for a classical second-order transition. Both ξ and r_{co} can be understood physically as the length scale in a crystal over which a change in Q in one local region of crystal induces correlated changes in adjacent regions. Strain fields play an important role in this context (Ginzburg et al., 1987), and their large extent in framework silicates, i.e. at least 10–100 unit cells, will give

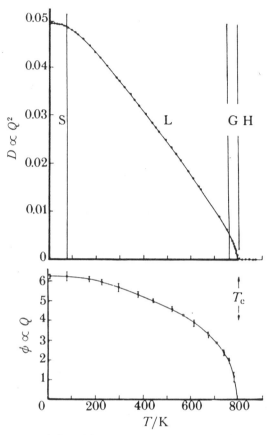

Figure 5.3 Temperature evolution of the order parameter for the displacive transition in LaAlO$_3$ showing four temperature regimes. H represents the stability field of the high-symmetry form. In the Ginzburg interval (G) fluctuations cause deviations from the classical critical exponent of $\beta = 1/2$. L indicates the range over which the Landau relationship $Q^2 \propto (T_c - T)$ operates and S indicates the temperature range of order parameter saturation. (Salje (1989a). Reproduced by kind permission of E. Salje.)

large effective correlation lengths. The long wavelength fluctuations will therefore be suppressed and Ginzburg intervals of less than ~ 1 K can reasonably be expected. This issue is discussed in more detail by Landau and Lifshitz (1980) and Ginzburg et al. (1987); see also Carpenter and Salje (1989) and references therein for discussion of silicate minerals.

Much less is known about the saturation regime. Landau theory is known to be inadequate at very low temperatures since it incorrectly predicts that $\partial Q/\partial T$ will remain finite down to 0 K. The third law of thermodynamics requires that the heat capacity of a crystal should asymptote to zero as 0 K is approached, and the order parameter must do likewise. A temperature interval over which Q levels off necessarily exists, therefore, and has indeed been observed for displacive phase transitions in alkali feldspars (Salje, 1986) and anorthite (Redfern et al., 1988). In both of these examples the saturation sets at well below room temperature, however, and may not be relevant for the characterization of natural systems evolving at geological temperatures. On the other hand, some cation-ordering transitions in minerals have very high transition temperatures and saturation effects might conceivably operate above room temperature when high degrees of order are reached. This important process awaits further theoretical analysis (see Salje, 1990).

In summary, there are sound reasons for expecting phase transitions in silicate minerals to conform to Landau theory over wide temperature intervals under geological conditions. The microscopic origin of the order parameter might not be understood in many cases but this need not hinder the thermodynamic analysis.

5.3 Landau free energy expansions

The Landau expansion for the excess free energy due to a phase transition is a power series in Q:

$$G = AQ + BQ^2 + CQ^3 + DQ^4 + \cdots \tag{3}$$

The stable state of a crystal, specified by Q_{eqm}, must be at a minimum in G with respect to Q, i.e.:

$$\left.\frac{\partial G}{\partial Q}\right|_{Q_{eqm}} = 0 \tag{4}$$

and:

$$\left.\frac{\partial^2 G}{\partial Q^2}\right|_{Q_{eqm}} > 0 \tag{5}$$

The high-symmetry form, for which $Q=0$, can only be stable if the linear term is absent and B is positive. If B is negative, the low-symmetry form is stable and the equilibrium value of Q is between 0 and 1. It is also easy to show that the criteria for stability (equations 4 and 5) require that all the odd-order terms must be strictly zero if the transition is to be continuous between states with $Q=0$ and states with $Q>0$ at the transition temperature, T_c. Landau demonstrated that as temperature changes there has to be a crossover in the sign of B at $T=T_c$, from positive ($Q_{eqm}=0$) to negative ($Q_{eqm} \neq 0$), and that this temperature dependence of the second-order coefficient can be expressed as a linear function of T (Landau and Lifshitz, 1980). In its familiar and most convenient form, the free-energy expansion thus becomes:

$$G = \tfrac{1}{2}a(T-T_c)Q^2 + \tfrac{1}{4}bQ^4 + \tfrac{1}{6}cQ^6 + \cdots \tag{6}$$

Three cases are usually considered. Firstly, if b is positive and the sixth-order term is negligibly small, the expansion describes a second-order transition. The equilibrium value of Q varies as:

$$Q = \left(\frac{T_c - T}{T_c}\right)^\beta \quad (T < T_c) \tag{7}$$

with $\beta = 1/2$, as observed in the example of $LaAlO_3$ (Fig. 5.3). Secondly, if $b = 0$, c is positive and the higher-order terms are negligible, the expansion describes a tricritical transition, with $\beta = 1/4$. If the fourth- and sixth-order terms are both positive and comparable in magnitude then values of β between $1/2$ and $1/4$ can be obtained over limited temperature intervals. Finally, if b is negative and c is positive, the transition becomes first order with a discontinuity in the equilibrium value of Q at the transition temperature, T_{tr}, where $T_{tr} > T_c$. These variations in Q are shown in Figure 5.4.

The free-energy expansions can be manipulated in the usual way to give the excess enthalpy, entropy, and heat capacity due to the transition. For a second-order transition:

$$S = -\frac{dG}{dT} = -\tfrac{1}{2}aQ^2, \tag{8}$$

$$H = G + TS = -\tfrac{1}{2}aT_cQ^2 + \tfrac{1}{4}bQ^4, \tag{9}$$

$$C_p = \frac{a^2}{2b} \cdot T \quad (T < T_c). \tag{10}$$

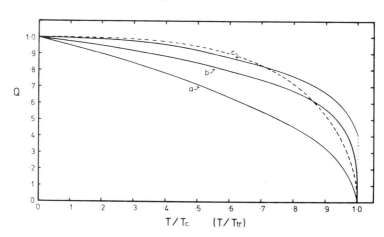

Figure 5.4 Variation of Q as a function of T/T_c for second-order (a) and tricritical expansions (b), and as a function of T/T_{tr} for a first-order transition (c) with a jump in Q of 0.4 at $T = T_{tr}$. The dashed line represents the variation of Q in the Bragg–Williams model.

Equivalent expressions for tricritical and first-order cases are listed by Carpenter (1988).

5.3.1 *The effect of pressure*

Implicit in the enthalpy part of the normal Landau free-energy expansion is the contribution of the excess volume, V, due to the transition and pressure, P. For systems in which a symmetry change is induced by varying pressure or in which the volume change is large, it may be instructive to include the PV effect explicitly. This can be achieved by expressing V as a series expansion in Q:

$$V = a_v Q + b_v Q^2 + c_v Q^3 + d_v Q^4 + \cdots \qquad (11)$$

where the coefficients have a subscript v to indicate that they are in units of volume. The contribution to G is PV but, once again, not all terms are allowed in the expansion. The linear term in Q must be strictly zero for $Q=0$ to be a stable solution and the higher odd-order terms must be zero if the transition is to be thermodynamically continuous. Thus, for a continuous transition.

$$V = b_v Q^2 + d_v Q^4 + \cdots \qquad (12)$$

Landau postulated that the excess entropy due to a phase transition is proportional to Q^2 and that higher-order terms are negligibly small (Landau and Lifshitz, 1980). In other words only the second-order term in the free energy is explicitly temperature-dependent. As a first approximation, exactly the same behaviour is expected when pressure is the principal variable and higher-order terms in the volume expansion may also be assumed to be small. The excess free energy for a continuous transition with the effect of pressure included explicitly therefore becomes (after relabelling the volume coefficient):

$$G = \tfrac{1}{2} a (T - T_c) Q^2 + \tfrac{1}{4} b Q^4 + \cdots + \tfrac{1}{2} a_v P Q^2. \qquad (13)$$

With rearrangement of terms this becomes:

$$G = \tfrac{1}{2} a \left[T - \left(T_c - \frac{a_v}{a} P \right) \right] Q^2 + \tfrac{1}{4} b Q^4 + \cdots \qquad (14)$$

from which it is clear that the temperature-independent terms in Q^2 give:

$$T_c^* = T_c - \frac{a_v}{a} P \qquad (15)$$

where T_c is the transition temperature at zero pressure and T_c^* is the transition temperature at non-zero pressure. The primary effect of changing P should therefore be to induce a linear change in the equilibrium transition temperature, with the slope dT_c^*/dP, given by the ratio of the Landau coefficients, $-a_v/a$. The sign of a_v may be positive or negative, or course, depending on whether the excess volume is positive or negative and T_c^* will increase or decrease with increasing P, accordingly. In the case of the tetragonal $\rightleftharpoons$ monoclinic transition in $BiVO_4$, for example, the excess volume is positive and dT_c^*/dP is negative (David and Wood, 1983; Hazen and Mariathasan, 1982).

An example of these simple relationships between V and Q^2 and between T_c^* and P is provided by the CO_3 group orientational order/disorder transition in calcite ($R\bar{3}m \rightleftharpoons R\bar{3}c$), which appears to be tricritical in character (Dove and Powell, 1989). Dove and Powell measured the intensity of a superlattice reflection at temperatures up to the transition temperature, ~ 1260 K. The excess volume can be calculated for different temperatures from their data, and a plot of $I_{11\bar{2}3}$ against V is linear within reasonable experimental error (Fig. 5.5). Using the temperature dependence of Q for a tricritical transition to calibrate $I_{11\bar{2}3}$ in terms of Q^2, a value of $a_v = -0.34$ J.mole^{-1}.K^{-1} is obtained. Taking $a = 24$ J.mole^{-1}.K^{-1} from Redfern et al. (1989) then leads to a calculated PT slope for the disordering transition of ~ 0.014 K.bar^{-1}. Ob-

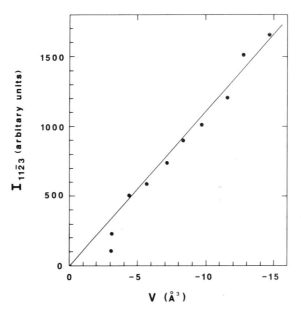

Figure 5.5 Relationship between the integrated intensity of a superlattice reflecton, $I_{11\bar{2}3}$, ($\propto Q^2$) and excess volume per unit cell, V, for the CO_3 group orientational order/disorder transition in calcite (data from Dove and Powell, 1989). Following Redfern et al. (1989), the excess volume was calculated using cell parameters for disordered calcite ($R\bar{3}m$) at 1260 K of $c = 17.823$ Å (Dove and Powell, 1989) and $a = 4.975$ Å (by extrapolation of the a values given by Dove and Powell), and by assuming that the volume of this phase does not vary significantly with temperature.

served values are much smaller, at $\lesssim 0.003$ K.bar^{-1} (Cohen and Klement, 1973; Mirwald, 1976), however. If this type of analysis is correct, coupling with a second-order parameter which has a different pressure dependence may be implicated but, in any case, a fuller investigation should be instructive.

A phase transition may be induced in some systems by changing pressure at constant temperature. If the experiment is carried out at an experimental temperature, T_{exp}, a transition pressure, P_c, will be identified from:

$$T_{exp} = T_c - \frac{a_v}{a} P_c \tag{16}$$

$$\Rightarrow \quad P_c = \frac{a}{a_v}(T_c - T_{exp}) \tag{17}$$

since T_{exp} is now also the transition temperature, T_c^*. Substituting:

$$T_{exp} - T_c = -\frac{a_v}{a} P \tag{18}$$

in equation (14) gives:

$$G = \tfrac{1}{2} a_v (P - P_c) Q^2 + \tfrac{1}{4} b Q^4 + \cdots \tag{19}$$

The equilibrium condition is, as usual, $\partial G/\partial Q = 0$, giving:

$$Q = \left[\frac{a_v}{b}(P_c - P)\right]^\beta \tag{20}$$

where $\beta = 1/2$ if the transition is second order and $\beta = 1/4$ if it is tricritical. The same approach can, of course, be used to generate equivalent expansions and relationships for first-order transitions with $b < 0$, $c > 0$.

It is worth noting that if the excess volume may be expressed as a function of Q^2 in this way, changing pressure alone cannot affect the thermodynamic character of the phase transition — only the transition temperature is renormalized. If a fourth-order term is significant, however, the free-energy expansion would become:

$$G = \tfrac{1}{2} a \left(T - T_c + \frac{a_v}{a} P\right) Q^2 + \tfrac{1}{4}(b + b_v P) Q^4 + \tfrac{1}{6} c Q^6 + \cdots \tag{21}$$

in which case a change in P could result in a change in sign of the fourth-order coefficient. For $(b + b_v P) > 0$, the transition would be second-order, for $(b + b_v P) = 0$, it would be tricritical, and for $(b + b_v P) < 0$, it would be first-order in the usual way. Similarly, if order parameter coupling occurred, the overall behaviour would be different, depending on the pressure dependence of the second-order parameter and the nature of the coupling.

5.3.2 Symmetry constraints

Because the primary order parameter for a phase transition must have properties which correctly reflect the symmetry change due to the transition, there are rigorous group-theoretical rules constraining the precise form that the Landau expansions can take (e.g. see Lyubarski, 1960, Chapter 7; Landau and Lifshitz, 1980, Chapter 14; or, for some specific examples of minerals, Hatch and Ghose, 1989; Hatch and Griffen, 1989; Hatch et al., 1990). In loose terms these rules stipulate that odd-order terms must be strictly zero when the high- and low-symmetry forms have a group/subgroup relationship and there is a change of point group, and/or a doubling of the unit cell volume, in orthorhombic, monoclinic, and triclinic systems. If the unit cell volume increased by a factor of three, however, or a change in point group allowed three equivalent orientational variants of the low-symmetry form with respect to the high-symmetry form, the odd-order terms would be allowed. An example of the latter situation is provided by the hexagonal $\rightleftharpoons$ orthorhombic transition in cordierite which is first order in character (Putnis, 1980; Salje, 1987b).

In high-symmetry systems (cubic, hexagonal, tetragonal, trigonal) the order parameter may also consist of more than one component, as set out in detail by Bruce and Cowley (1981). The physical origin of this is illustrated by the case of a phase transition in which the order parameter is some lattice distortion. Because of the high symmetry, two or more distortions can be identical in energy but different in orientation. The degeneracy is reflected in the form of the Landau expansion which, for a two-dimensional order parameter, becomes (ignoring odd order terms):

$$G = \tfrac{1}{2}a(T - T_c)(q_1^2 + q_2^2) + \tfrac{1}{4}b(q_1^2 + q_2^2)^2 + \tfrac{1}{6}c(q_1^2 + q_2^2)^3 + \cdots \qquad (22)$$

with q_1 and q_2 as the degenerate components (Salje, 1990). Component q_1 might represent a tetragonal distortion $(x_1 = x_2 \neq x_3)$ and q_2 an orthorhombic distortion $(x_1 = -x_2)$ in a crystal undergoing a cubic $\rightleftharpoons$ orthorhombic transition, where x_1, x_2 and x_3 represent linear strains of the lattice parallel to the crystallographic x, y, and z axes, respectively. Other causes of the degeneracy can occur and threefold or fourfold degenerate order parameters are also possible. Clearly, the underlying principle must be that the symmetry of the order parameter determines the form of the terms in the expansion. Tabulations of the correct Landau expansions for many different changes in symmetry are given by Tolédano and Tolédano (1976; 1977; 1980); Stokes and Hatch (1988) and Salje (1990).

Only systems with single-component order parameters are treated in this review.

5.4 Elaboration of single-order parameter expansions: coupling with strain and composition

During a phase transition there are several different contributions to the free energy which, when described in terms of a Landau expansion such as equation (6), are all in effect incorporated into the coefficients $a, b, c, \ldots$. As has already been seen for the excess volume, it is possible to separate out the different macroscopic origins of these energies by including them in the expansion explicitly. The spontaneous strain which accompanies a transition and variations in composition are also important and can be accounted for in a rather similar manner.

5.4.1 Coupling of Q with the spontaneous strain

The spontaneous strain, ε, is that part of the lattice distortion of a crystal which occurs as a consequence of a phase transition. It is an excess quantity which is always measured relative to an undistorted lattice at the same temperature and is therefore additional to the normal effects of thermal expansion. The experimental determination of ε involves measurement of lattice parameters at different temperatures and it is described formally by a second-rank tensor. The forms of this tensor for different changes in symmetry are described in detail by Aizu (1970), Janovec et al. (1975), Tolédano & Tolédano (1980), and Salje (1990).

With the strain energy included separately, the Landau expansion becomes (following Salje and Devarajan, 1986):

$$G = \tfrac{1}{2}a(T - T_c)Q^2 + \tfrac{1}{4}bQ^4 + \tfrac{1}{6}cQ^6 + \cdots + d\varepsilon Q + e\varepsilon Q^2 + \cdots + f\varepsilon^2 \quad (23)$$

The terms in εQ and εQ^2, often referred to as linear and quadratic coupling terms, and in $f\varepsilon^2$ describe the energy change which arises because a change in Q causes a change in ε. For completeness it is necessary to include the Hooke's law elastic energy, $f\varepsilon^2$, which can be given in full as $\tfrac{1}{2}\sum_{i,k} C_{ik} x_i x_k$; C_{ik} are elastic constants and x_i, x_k are elements of the strain tensor. The extra terms are again subject to symmetry constraints; in particular, linear coupling is disallowed for all zone boundary transitions and some zone centre transitions (see Salje, 1990).

At equilibrium the crystal must be stress free ($\partial G/\partial \varepsilon = 0$). In the case of transitions associated with the Brillouin zone centre for which linear coupling is allowed it is initially assumed that the linear term is adequate to describe the coupling, giving:

$$0 = dQ + 2f\varepsilon \quad (24)$$

$$\Rightarrow \quad \varepsilon = -\frac{d}{2f}Q. \quad (25)$$

If this is substituted back into the initial expansion:

$$G = \tfrac{1}{2}a\left(T - T_c - \frac{d^2}{2af}\right)Q^2 + \tfrac{1}{4}bQ^4 + \tfrac{1}{6}cQ^6 + \cdots \qquad (26)$$

In other words the strain energies are incorporated into the second-order term and might be manifest as a change in T_c relative to the same transition in a strain-free crystal. The renormalized transition temperature, T_c^*, is given by the temperature-independent part of the second-order term:

$$T_c^* = T_c + \frac{d^2}{2af}. \qquad (27)$$

For a transition associated with a special point on the Brillouin zone boundary the equivalent manipulations yield:

$$\varepsilon = -\frac{e}{2f}Q^2 \qquad (28)$$

and:

$$G = \tfrac{1}{2}a(T - T_c)Q^2 + \frac{1}{4}\left(b - \frac{e^2}{f}\right)Q^4 + \tfrac{1}{6}cQ^6 + \cdots \qquad (29)$$

In this case, therefore, the effect of the strain energy is to lower the value of the fourth-order coefficient. The contribution may be sufficient to change the thermodynamic character of the transition from second order in a strain-free crystal to first order in a crystal which experiences a significant strain.

An example of the strain associated with a zone-centre transition that has $\varepsilon \propto Q$ is provided by the monoclinic (C2/m) $\rightleftharpoons$ triclinic (C$\bar{1}$) displacive transition in crystals of albite with complete Al/Si disorder, as described by Salje (1985; 1988a; b) and Salje et al. (1985). For this symmetry change the only non-zero, symmetry-breaking strain-tensor components are x_4 and x_6 and the scalar spontaneous strain is defined as:

$$\varepsilon = \frac{1}{\sqrt{2}}(x_4^2 + x_6^2)^{1/2}. \qquad (30)$$

For small changes in the a, b, and c parameters, x_4 and x_6 may be given in terms of the lattice angles α^* and γ of the triclinic crystals as (after Salje et al., 1985):

$$x_4 \approx -\cos \alpha^* \qquad (31)$$

$$x_6 \approx \cos \gamma. \tag{32}$$

Salje and coworkers (1985) found experimentally that x_6 is negligibly small relative to x_4 so that, to a good approximation, the variation of ε is given directly by the variation of $\cos \alpha^*$. Since $\varepsilon \propto Q$ in this case the observation that $\cos^2 \alpha^*$ ($\propto Q^2$) is linear in T and goes continuously to zero at ~ 1238 K indicates that the transition is second order in character (Figure 5.6, from Carpenter, 1988). The experimental data do not reveal any obvious deviations from Landau behaviour in the vicinity of T_c or over a total range of temperature of nearly 1000 K.

These relationships refer to equilibrium conditions and should, strictly speaking, be determined for strain or order parameter data collected at the temperature of equilibration. They seem to be valid for data collected at room temperature if a high-temperature state of order can be preserved during rapid cooling, however. For example, ε determined from changes in room-temperature lattice parameters associated with the zone boundary ($C\bar{1} \rightleftharpoons I\bar{1}$) Al/Si ordering transition in anorthite varies linearly with Q^2 determined from tetrahedral site occupancies (Figure 5.7, from Carpenter et al., 1990a). The values of Q and ε were measured from crystals which had been quenched from their equilibration temperatures of 1300–1550°C. Al/Si ordering in anorthite is too slow for significant changes to occur during the quench.

Only relatively rarely are two or more of the strain-coupling terms needed to describe an observed spontaneous strain. An exceptional case is the cubic $\rightleftharpoons$ tetragonal transition in $K_2Cd_2(SO_4)_3$ for which linear, quadratic, and third-order terms were found to be significant (Devarajan and Salje, 1984). For the most part, the relationships $\varepsilon \propto Q$ and $\varepsilon \propto Q^2$ not only provide an

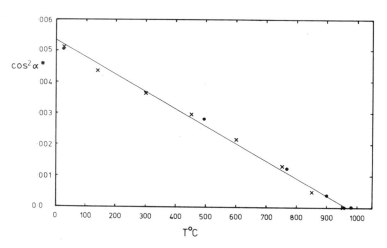

Figure 5.6 Variation of $\cos^2 \alpha^*$ ($\propto Q^2$) as a function of temperature for the C2/m$\rightleftharpoons$C$\bar{1}$ displacive transition in crystals of albite with complete Al/Si disorder (Carpenter, 1988).

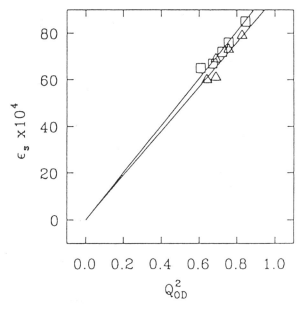

Figure 5.7 Variation of the spontaneous strain, ε_s (as measured at room temperature), with Q_{od}^2 (from X-ray structure refinement data), for the $C\bar{1} \rightleftharpoons I\bar{1}$, Al/Si order/disorder transition in anorthite. Squares = natural crystals with composition $\sim An_{100}$; triangles = natural crystals with composition $\sim An_{98}$. The range of states of order was produced by annealing crystals at temperatures of 1300–1550°C (Carpenter et al., 1990a).

adequate description of the strain behaviour at most phase transitions but also lead to a valuable experimental method for determining the temperature or pressure evolution of Q itself.

5.4.2 Coupling between Q and composition

Many mineral systems of geological interest form extensive solid solutions, and phase transitions which occur in end-member phases are often observed in crystals with varying compositions. The effect of solid solution can also be explored by adding composition terms to the Landau expansion. Before doing so, however, it is worth reflecting on the physical effect, at a microscopic level, of substituting impurity atoms into a pure crystal, as discussed by Salje (1990).

If one impurity atom was implanted into a pure crystal, any phase transition in most of the crystal would not be affected. On the other hand, in a region around the extraneous atom there would be a finite deformation cloud. In this small region the transition would be modified, with the distortion causing either an increase or a decrease in the local transition temperature. If further impurity atoms were added in a random manner, the bulk of the crystal would continue to transform at the transition temperature of the pure crystal but there would be a tendency for the transition to become smeared out because of the increasing total volume of the isolated distorted regions. Once the deformation clouds started to impinge on each other, however, the transition behaviour of the whole crystal would be modified. Two compositional regimes can therefore be distinguished. At low concentrations of impurities the transition temperature will not be affected, but, once some critical concentration has been reached, it may start to diverge from that of the pure crystal (Figure 5.8(a), after Salje, 1990).

A quick calculation is sufficient to reveal the likely extent of the constant T_c region since it is only necessary to estimate the concentration at which the deformation clouds begin to overlap. For example, the effect on the displacive transition in albite of substituting K^+ for Na^+ is to lower T_c. If the distortion around an individual K^+ ion can be represented by a sphere of radius r and the critical concentration limit is given by the composition at which the spheres just make contact in a close-packed arrangement, the range of constant T_c should extend to ~ 8 mole% Or for $r = 10$Å (~ 1 unit cell), ~ 1 mole% Or for $r = 20$Å (~ 2 unit cells), and ~ 0.3 mole% Or for $r = 30$Å (~ 3 unit cells). The critical concentration is expressed here in terms of the mole fraction of the orthoclase (Or) component, $KAlSi_3O_8$, substituting for $NaAlSi_3O_8$. Strain fields associated with twinning in alkali feldspars are known to extend at least 0.5 μm (see discussion in Carpenter, 1988), which suggests that it may not be unreasonable to postulate the existence of strained regions of greater than ~ 1–3 unit cells around individual K^+ ions replacing Na^+ ions in the feldspar structure. The implication is that the composition range of level T_c may be too small to be measurable in routine experiments.

From this type of semi-quantitative analysis it is apparent that, to a first approximation, it is the number of impurity atoms added which is important when considering the effect of solid solution on a phase transition. A compositional term, X, to represent the mole fraction of some second component added to a pure phase, might be included in the Landau expansion simply as another series expansion, giving:

$$G = \tfrac{1}{2}a(T - T_c)Q^2 + \tfrac{1}{4}bQ^4 + \tfrac{1}{6}cQ^6 + \cdots + mXQ + nXQ^2 + \cdots \tag{33}$$

where $X = 0$ in the pure phase. For the composition range of approximately level T_c (small X) the additional terms are effectively zero and, as usual, high-order terms in Q are assumed to be small when X is large. If the linear term, mXQ, becomes significant the phase transition will be fundamentally altered. This term in effect represents the consequences of changing composition as being equivalent to applying an external field (Salje, 1990). If the addition of the second component stabilizes the high-symmetry phase (m is positive), the transition will ultimately be suppressed, but, if it stabilizes the

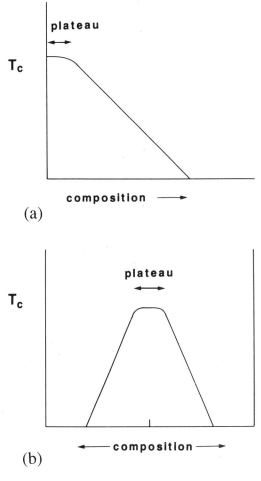

Figure 5.8 Variation of T_c with composition for phase transitions in solid solutions where the free-energy contribution due to coupling between composition, X, and the order parameter is described by nXQ^2. (a) Transition extends into solid solution from a pure end-member phase. Note that for a small composition range, T_c is nearly constant before it varies linearly with composition at higher concentrations of the second component in solid solution (Salje, 1990). (b) Phase transition centred on mid-composition of a binary solid solution. T_c now varies linearly away from the plateau region which represents small compositional deviations from the ideal 50:50 stoichiometry.

low-symmetry phase (m is negative), the transition will be pulled to higher temperatures (Figure 5.9, from Salje, 1990). In both cases the behaviour near T_c will not give classical critical exponents. Conversely, however, if classical behaviour is observed in crystals with different compositions, the linear term in Q must be small and the quadratic term will be dominant. Assuming $m=0$

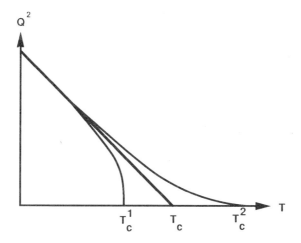

Figure 5.9 Effect of an applied field on a second-order transition as described by the contribution of a linear coupling term mXQ. In the present case X might be composition. In the pure phase (no effective field) the transition would occur at T_c. If the effect of solid solution was to stabilize the high-symmetry form (m is positive) the transition would occur at a lower temperature, T_c1. If the low-symmetry form was stabilized (m is negative), the transition temperature would occur at a higher temperature, T_c2. Note that significant deviations from classical Landau behaviour occur in the presence of an effective field. (Salje (1990). Reproduced by kind permission of E. Salje.)

in equation (33), then:

$$G = \tfrac{1}{2}a\left(T - T_c + \frac{2n}{a}X\right)Q^2 + \tfrac{1}{4}bQ^4 + \tfrac{1}{6}cQ^6 + \cdots \qquad (34)$$

One consequence of varying composition, therefore, can be to induce a change in the transition temperature that is linear in X:

$$T_c^* = T_c - \frac{2n}{a}X \qquad (35)$$

where T_c is the transition temperature in the pure phase and T_c^* is the transition temperature in the solid solution (for thermodynamically continuous-phase transitions).

Phase diagrams which contain a phase transition that extends from one end member into a binary solid solution will be expected to have the form shown in Figure 5.8(a) while those in which the phase transition is centered on the 50:50 composition might be expected to have the form shown in Figure 5.8(b). In both cases the composition range of approximately level T_c would be expected to depend on the extent of microscopic strain fields in the particular material of interest. Two examples of transitions in mineral systems for which

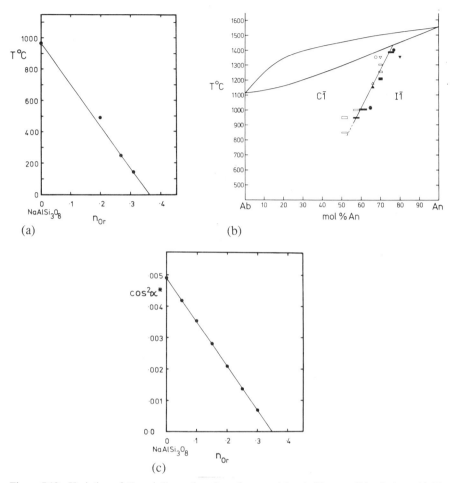

Figure 5.10 Variation of T_c and Q as a function of composition in binary solid solutions. (a) T_c for the $C2/m \rightleftharpoons C\bar{1}$ displacive transition in alkali feldspars with complete Al/Si disorder (Carpenter, 1988). (b) T_c for the $C\bar{1} \rightleftharpoons I\bar{1}$ Al/Si ordering transition in plagioclase feldspars (Carpenter and McConnell, 1984). (c) Variation of $\cos^2 \alpha^*$ ($\propto Q^2$) for the $C2/m \rightleftharpoons C\bar{1}$ displacive transition in alkali feldspars with complete Al/Si disorder (Carpenter, 1988).

T_c appears to vary linearly with composition are the $C2/m \rightleftharpoons C\bar{1}$ displacive transition in alkali feldspars with complete Al/Si disorder (Figure 5.10(a), from Carpenter, 1988) and the $C\bar{1} \rightleftharpoons I\bar{1}$, Al/Si ordering transition in plagioclase feldspars (Figure 5.10(b), from Carpenter and McConnell, 1984).

Applying the equilibrium condition, $\partial G/\partial Q = 0$, to equation (34) gives, as solutions at $T < T_c^*$ for second-order and tricritical transitions:

$$Q^2 = -\frac{a}{b}\left(T - T_c + \frac{2n}{a}X\right) \quad \text{(second order)} \tag{36}$$

and:

$$Q^4 = -\frac{a}{c}\left(T - T_c + \frac{2n}{a}X\right) \quad \text{(tricritical)}. \tag{37}$$

These may be rewritten as:

$$Q = \left(\frac{T_c^* - T}{T_c}\right)^\beta \tag{38}$$

where $\beta = 1/2$, $T_c = b/a$ for a second-order transition and $\beta = 1/4$, $T_c = c/a$ for a tricritical transition. At constant temperature and for coefficients which do not vary with composition equations (36) and (37) simplify to:

$$Q^2 \propto X \quad \text{(second order)} \tag{39}$$

$$Q^4 \propto X \quad \text{(tricritical)} \tag{40}$$

The $C2/m \rightleftharpoons C\bar{1}$ transition in alkali feldspars provides an example of the second-order relationship (Figure 5.10(c), from Carpenter, 1988) while the compositional dependence of Q for the $C2/c \rightleftharpoons P2/n$ transition (Na, Ca, Mg, Al ordering) in jadeite–augite pyroxenes is consistent with tricritical behaviour (Figure 5.11, from Carpenter et al., 1990b).

As in the analysis of strain, the use of macroscopic parameters provides a means of describing observed transition behaviour in a self-consistent and quantitative manner. In each case the coefficients have some real physical meaning but to investigate their origin further, it may be necessary to turn to microscopic models. Since these coefficients are being treated purely as phenomenological parameters here, the predicted relationships between macroscopic properties must be tested experimentally to ensure that sufficient terms are being used in the expansions. For example, it is possible that higher-order terms than XQ^2, such as X^2Q^2, X^3Q^2, etc., may be required in equation (34) for some particular systems. Such higher-order terms would be expected to lead to alternative variations of T_c with composition.

5.5 Comparison of Landau and Bragg–Williams treatments of order/disorder transitions

Over many years the most straightforward method of analysing the thermodynamics of order/disorder processes in minerals has been through the use of the Bragg–Williams model. The enthalpy of ordering is treated as arising from nearest-neighbour-type atomic interactions and the entropy is treated as being purely configurational. Both contributions to the free energy are

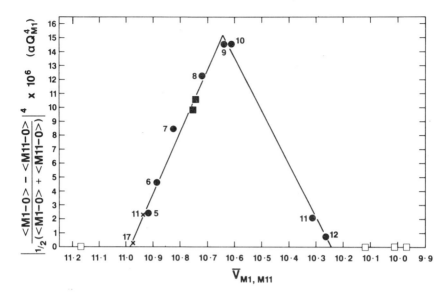

Figure 5.11 Variation of Q^4, as specified in terms of mean bond lengths for M1 and M11 sites, with composition, as specified in terms of the mean volume of M1 and M11 sites, for a suite of jadeite–augite pyroxenes from Nybø, Norway. The linear variation away from the mid-composition is consistent with tricritical behaviour for the C2/c$\rightleftharpoons$P2/n cation-ordering transition in omphacite (Carpenter et al., 1990b).

expressed as functions of a long-range order parameter which is defined in terms of the mean occupancies of crystallographic sites by different atoms (equation (1)). Given its simplicity, the model is remarkably effective at generating reasonable energies and phase diagram topologies (e.g. see Greenwood, 1972; Navrotsky and Loucks, 1977). A comparison of the Bragg–Williams and Landau approaches is instructive in the present context because it highlights the way in which energies with different microscopic origins are assigned in the Landau free-energy expansion.

The well-known solution for Q as a function of temperature for the Bragg–Williams model is:

$$Q = \tanh\left(\frac{T_c}{T}Q\right) \qquad (41)$$

and this variation is shown in Figure 5.4 for comparison with the Landau solutions. The closest resemblance is with a Landau tricritical solution, though the critical exponent, β, is $\approx 1/2$ for the range $Q \approx 0.7$ to $Q=0$ (Figure 5.12(a)).

Defining the internal energy, U_{BW}, as being the excess due to the phase transition, i.e. $U_{BW}=0$ at $Q=0$, and approximating U_{BW} for the enthalpy of ordering, H_{BW}, gives the Bragg–Williams enthalpy in a slightly different form from usual as:

$$H_{BW} = -\frac{n}{2}RT_cQ^2 \tag{42}$$

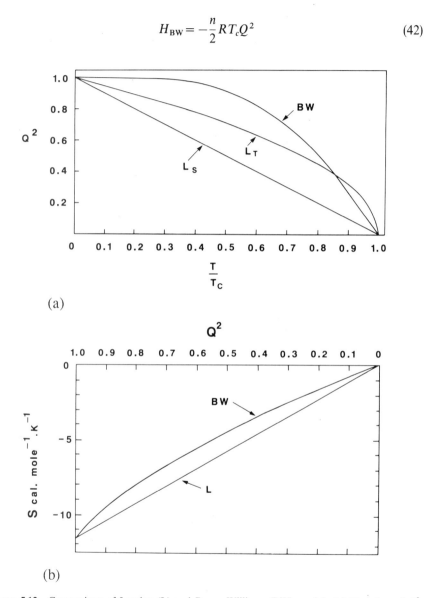

Figure 5.12 Comparison of Landau (L) and Bragg–Williams (BW) models. (a) Variation of Q^2 with reduced temperature, T/T_c. The BW model gives $Q^2 \propto T$ for the range of $Q<0.7$ but gives a much steeper dependence on T than the Landau solution. L_S = second-order Landau; L_T = tricritical Landau. (b) BW and L excess entropies for ordering in an AB-type compound assuming that the Landau entropy at $Q=1$ can be treated as being purely configurational.

where n is the number of sites per formula unit involved in the ordering ($n=2$ for an alloy with composition AB, $n=4$ for Al/Si ordering in $CaAl_2Si_2O_8$, etc.). The Q^2 dependence arises simply from the assumption that the internal energy is due entirely to pairwise interactions between nearest neighbour atoms involved in the ordering process. This compares with the Landau enthalpy, H_L, given by:

$$H_L = -\tfrac{1}{2}aT_cQ^2 + \tfrac{1}{4}bQ^4 + \tfrac{1}{6}cQ^6 + \cdots \tag{43}$$

Both H_{BW} and H_L have a Q^2 term but the higher-order terms in H_L can allow for the additional energy contributions from interactions between second nearest neighbours, many-body interactions, elastic effects due to the relaxation of the structure with changing Q, changes in the vibrational states, etc. The Landau coefficients must, of course, be calibrated from experimental data relating to some measured enthalpies whereas only a value for T_c is needed to determine H_{BW}. A more accurate representation of the real excess enthalpy should result, however.

The Bragg–Williams excess entropy, S_{BW}, defined with respect to $Q=0$ has the form:

$$S_{BW} = -\frac{nR}{2}\{(1+Q)\ln(1+Q)+(1-Q)\ln(1-Q)\}. \tag{44}$$

This compares with the Landau entropy, S_L, given by:

$$S_L = -\tfrac{1}{2}aQ^2. \tag{45}$$

Since the Landau a coefficient must be calibrated from experimental measurements, of the excess heat capacity, for example, non-configurational contributions associated with structural relaxations and vibrational effects will be automatically included. However, for some order/disorder systems in which these extra effects are small, a value for the a coefficient can be estimated by assuming that S_L at $Q=1$ is the purely configurational entropy value. For a two-site, two-atom type of ordering scheme this would give $S_L = 2R\ln(1/2) = -11.526$ J. mole^{-1}. K^{-1} at $Q=1$ and, hence, $a=23.05$ J. mole^{-1}. K^{-1}. The linear variation in S_L with Q^2 based on this value actually differs only by a small amount from S_{BW}, as illustrated in Figure 5.12(b). Alternatively, the observed Landau entropy change extrapolated to $Q=1$ can be compared with the purely configurational value to estimate the importance of non-configurational effects. For example, in the case of the orientational order/disorder transition of CO_3 groups in calcite, Redfern and coworkers (1989) obtained a value of $a = 24$ J. mole^{-1}. K^{-1} by drop calorimetry, giving $S_L = -12$ J. mole^{-1}. K^{-1} at $Q=1$. As discussed by Redfern et al., this is a factor of 2 larger than could be accounted for by the configurational contribution of the CO_3 group ordering alone ($n=1$ for quantities expressed in terms of one mole of

CaCO$_3$). By way of comparison, the a coefficient associated with Al/Si ordering in albite, NaAlSi$_3$O$_8$, gives $S_L = -20.81$ J.mole^{-1}.K^{-1}, which is much closer to the purely configurational entropy change of -18.7 J.mole^{-1}.K^{-1} (Salje et al., 1985; Carpenter 1988). A notable feature of the transition in calcite is that it is accompanied by an unusually large volume change, indicating a substantial relaxation of the structure in response to changing Q (Dove and Powell, 1989; Redfern et al., 1989). The origin of non-configurational entropy effects, such as from lattice vibrations, and this possible correlation with lattice relaxations need to be investigated for other systems.

As more order/disorder transitions in minerals are analysed using Landau formalism, it is becoming clear that approximately tricritical behaviour is not uncommon (see Carpenter, 1988; Carpenter et al., 1990b). There may be some fundamental reason for this but one notable implication is that the Bragg–Williams solution for Q (Figures 5.4 and 5.12(a)) is relatively close to the real behaviour of many systems. Perhaps the correct assessment is that if relaxational effects were added to the Bragg–Williams model, a Landau-type solution would emerge. The Bragg–Williams model gives a more realistic description of the asymptotic approach of Q to $Q = 1$ at temperatures approaching 0 K, however.

5.6 Phase transitions involving only one order parameter

A standard strategy for characterizing the thermodynamics of phase transitions requires, as a first step, the determination of how the order parameter varies both at and below the transition point. Since Q is a parameter which cannot be determined directly this means that some other property which depends on Q must be identified and then followed as a function of temperature (or pressure, composition, field strength, stress, etc). Suitable properties include mean site occupancies from structure refinements or spectroscopic measurements, birefringence, spontaneous strain, intensities of superlattice reflections, heat capacity; and the details of frequency and line widths from vibrational spectra. Many of these techniques are well illustrated in recent studies of the ferroelastic transition in As$_2$O$_5$ by Bismayer et al. (1986a), Salje et al. (1987), Redfern and Salje (1988), and Schmahl and Redfern (1988). If an observed temperature dependence of Q conforms to the prediction of a simple Landau expansion in one order parameter and the transition temperature is known, it may be necessary to perform only one further measurement to calibrate the coefficients. Three examples may help to demonstrate the overall approach as applied to minerals.

The displacive transition in albite with complete Al/Si disorder is known to be second order from the observed variation of the spontaneous strain with temperature (Fig. 5.6). T_c for the transition, 1238 K, is obtained from the linear extrapolation of $\cos^2 \alpha^*$ to zero and the appropriate Landau expansion has only second- and fourth-order terms. Since $T_c = b/a$ in this case, one thermochemical measurement is sufficient to give unique values for a and b. Salje et al. (1985) measured the variation of C_p through the same transition in crystals with composition $Ab_{69}Or_{31}$ and observed a small step in C_p at T_c, as expected for a second-order transition (Fig. 5.13). Rather than assuming that the Landau coefficients are independent of composition, they assumed that this step in C_p at the transition temperature is the same in $Ab_{69}Or_{31}$ as it is in pure albite, giving values of $a = 5.479$ J. mole^{-1}. K^{-1} and $b = 6854$ J. mole^{-1} for the transition in the end-member phase. The alternative assumption, that the Landau coefficients are constant, gives values of $a = 16.3$ J. mole^{-1}. K^{-1} and $b = 20180$ J. mole^{-1} which would imply a jump in C_p of 8.15 J. mole^{-1}. K^{-1} at $T = T_c$ for pure albite. Further experiments are required to resolve this detail.

From the variation of superlattice reflection intensities, Dove and Powell (1989) found that a value of $\beta = \frac{1}{4}$ describes the temperature dependence of the order parameter over a range of at least 550 K up to the transition at 1260 K for the $R\bar{3}m \rightleftharpoons R\bar{3}c$ transition in calcite (Fig. 5.14). Redfern et al. (1989) measured the excess enthalpy due to the transition by drop calorimetry and so were able to calibrate the coefficients, $a = 24$ J. mole^{-1}. K^{-1} and $c = 30$ kJ. mole^{-1}, in the tricritical expansion:

$$G = \tfrac{1}{2}a(T - T_c)Q^2 + \tfrac{1}{6}cQ^6. \tag{46}$$

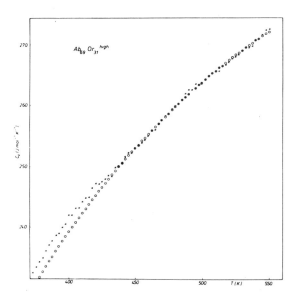

Figure 5.13 C_p of an alkali feldspar (composition $Or_{31}Ab_{69}$) with complete Al/Si disorder showing an excess heat capacity below the $C2/m \rightleftharpoons C\bar{1}$ transition ($T_c = 416$ K) consistent with a second-order transition. Crosses = observed values; circles = values for the high symmetry form extrapolated to low temperatures (Salje et al. (1985). Reproduced by kind permission of E. Salje.)

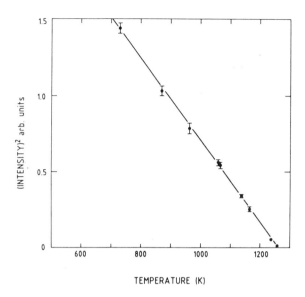

Figure 5.14 Square of the integrated intensity of the $11\bar{2}3$ reflection from calcite as a function of temperature. The data are consistent with a tricritical transition with $T_c = 1260$ K. (Dove and Powell (1989). Reproduced by kind permission of M. Dove.)

Redfern and coworkers then used the resulting excess entropy to test the effect of CO_3 group disordering in displacing the calcite $\rightleftharpoons$ aragonite equilibrium in PT space. A pronounced curvature of the phase boundary results from the continuous character of the phase transition in calcite (Fig. 5.1(b)).

Comparable experiments to follow the variation of Q as a function of T are not possible for the $C2/c \rightleftharpoons P2/n$ cation-ordering transition in omphacite because of the slow diffusion rates of Na^+, Ca^{2+}, Mg^{2+}, Fe^{2+}, and Al^{3+} in the pyroxene structure. The variation of Q as a function of composition in a suite of crystals from a single geological locality, and therefore with approximately the same equilibration temperature, provides some indication of the thermodynamic character of the transition, however. Carpenter and coworkers (1990b) showed on the basis of structure refinement data that $Q^4 \propto X$ describes the variation of Q across the binary jadeite–augite system, with a maximum near the stoichiometric composition for ordering of $Jd_{50}Aug_{50}$ (Fig. 5.11). If T_c varies linearly with X, the mole fraction of jadeite or augite components, and the Landau coefficients are independent of composition, this is consistent with tricritical behaviour with only one order parameter. The thermodynamics of ordering in crystals with compositions close to $Jd_{50}Aug_{50}$ can then be described using a Landau expansion with the coefficients calibrated on the basis of two further pieces of information. Wood and coworkers (1980) measured an excess enthalpy of ordering of -8.28 ± 1.46 kJ.mole^{-1} for some natural P2/n crystals with compositions near this ideal stoichiometry and

Carpenter (1981) obtained an estimate of $T_c = 1138$ K from disordering experiments on the same material. These natural crystals had $Q = 0.91$ so that values of $a = 22.8 \pm 4.0$ J.mole^{-1}.K^{-1} and $c = 25.9 \pm 0.5$ kJ.mole^{-1} are obtained (Carpenter et al. 1990b). If the variation of T_c with composition was known this description of the transition could be extended to the rest of the solid solution (see, for example, Holland, 1990).

It would be surprising if the apparently tricritical behaviour in calcite and omphacite corresponded with the condition that the fourth-order Landau coefficient is exactly zero. Rather, it is likely that the fourth-order term is only small. Phase transitions in other minerals might also be close to tricritical but with larger (positive or negative) fourth-order coefficients. Landau expansions in one order parameter should still provide satisfactory thermodynamic descriptions, however, though more experiments might be needed to calibrate the coefficients and determine the magnitude of any jump in Q at the transition temperature for first-order transitions. Deviations from this simple behaviour can be attributed to coupling effects with a second order parameter, as discussed in Section 5.7.

5.7 Systems with more than one phase transition: order parameter coupling

It is quite commonly found that the order parameter for a phase transition in a particular material does not conform to the temperature dependence predicted by a Landau expansion in one order parameter. Close to T_c and at very low temperatures the effects of fluctuations, impurities or saturation might be the cause, as discussed earlier. Experience of a number of such systems is beginning to show that apparently non-Landau behaviour at intermediate temperatures is often due to coupling between more than one order parameter (see, for example, Salje et al., 1985; Salje and Devarajan, 1986; Bismayer et al., 1986b; Redfern et al., 1988; Salje, 1988a; b; Palmer et al., 1989; 1990; Schmahl and Salje, 1989; Hatch et al., 1990). Direct evidence for the existence of two order parameters operating in a single system would be provided by the observation of two discrete phase transitions. In framework silicates these might be a displacive transition and an Al/Si ordering transition with quite different transition temperatures. It is also possible that the second order parameter is never actually responsible for a symmetry change but merely couples with the first order parameter over a limited temperature range, as observed in $NaNO_3$, for example (Schmahl and Salje, 1989). In both cases the interactions can be expressed explicitly using a free energy expansion incorporating the free energy due to changes in each order parameter and the contribution of a coupling term.

Salje and coworkers have identified strain coupling as an important mechanism by which two, discrete order parameters may interact. The physical picture is quite simple. A first phase transition will cause a spontaneous lattice distortion. A second transition will also cause a lattice distortion but this will be modified according to the manner in which the lattice is already distorted. In other words, the order parameters, Q_1 and Q_2, for the two transitions will influence each other via those distortions which are common to both. A simple way of expressing the interactions formally is to set out the coupling of Q_1 and Q_2 with the common strain, ε, separately, and then solve the resulting expansion for the equilibrium condition that the crystal must be stress free. Since Q_1 and Q_2 can vary over a very wide temperature range, the resulting coupling effects may be quite marked, even if the two transitions of interest have transition temperatures which are hundreds or thousands of degrees apart. Following Salje and Devarajan (1986) this is illustrated below for three cases: both order parameters associated with zone-centre transitions and having linear coupling between ε and Q, both associated with zone-boundary transitions, and Q_1 associated with a zone-centre transition, but Q_2 associated with a zone-boundary transition. Finally, when the interactions are between one order parameter and gradients in the second, incommensurate structures may become stable.

5.7.1 Bilinear coupling

For the case of a crystal with two phase transitions which have linear strain/order parameter coupling, the excess free energy may be expressed as:

$$G = \tfrac{1}{2} a_1 (T - T_{c1}) Q_1^2 + \tfrac{1}{4} b_1 Q_1^4 + \tfrac{1}{6} c_1 Q_1^6 + \cdots$$
$$+ \tfrac{1}{2} a_2 (T - T_{c2}) Q_2^2 + \tfrac{1}{4} b_2 Q_2^4 + \tfrac{1}{6} c_2 Q_2^6 + \cdots$$
$$+ d_1 \varepsilon Q_1 + d_2 \varepsilon Q_2 + f\varepsilon^2. \tag{47}$$

The subscripts 1 and 2 distinguish the contributions of the two order parameters and ε refers to the strain component common to both Q_1 and Q_2. At equilibrium the crystal must be stress free so that:

$$\frac{\partial G}{\partial \varepsilon} = 0 = d_1 Q_1 + d_2 Q_2 + 2f\varepsilon \tag{48}$$

$$\Rightarrow \quad \varepsilon = -\left(\frac{d_1 Q_1 + d_2 Q_2}{2f}\right). \tag{49}$$

Substituting for ε in equation (47) gives:

$$G = \tfrac{1}{2} a_1 \left(T - T_{c1} - \frac{d_1^2}{2 a_1 f} \right) Q_1^2 + \tfrac{1}{4} b_1 Q_1^4 + \tfrac{1}{6} c_1 Q_1^6 + \cdots$$
$$+ \tfrac{1}{2} a_2 \left(T - T_{c2} - \frac{d_2^2}{2 a_2 f} \right) Q_2^2 + \tfrac{1}{4} b_2 Q_2^4 + \tfrac{1}{6} c_2 Q_2^6 + \cdots$$
$$- \frac{d_1 d_2}{2f} Q_1 Q_2. \tag{50}$$

The coupling of each order parameter with ε results in renormalization of the second-order coefficients, as already discussed but, in addition, the two order parameters couple with each other through the bilinear coupling term, $-(d_1 d_2/2f)Q_1 Q_2$. This may also be written more simply as:

$$G = \tfrac{1}{2}a_1(T-T^*_{c1})Q_1^2 + \tfrac{1}{4}b_1 Q_1^4 + \tfrac{1}{6}c_1 Q_1^6 + \cdots$$
$$+ \tfrac{1}{2}a_2(T-T^*_{c2})Q_2^2 + \tfrac{1}{4}b_2 Q_2^4 + \tfrac{1}{6}c_2 Q_2^6 + \cdots$$
$$+ \lambda Q_1 Q_2 \tag{51}$$

where T^*_{c1} is the transition temperature that would be observed for a crystal which undergoes a transition governed by Q_1 alone and T^*_{c2} is the transition temperature that would be observed if the crystal experienced only the second phase transition. The general bilinear coupling term is indicated by the coefficient λ and can be temperature-dependent if, for example, the elastic constants which constitute the f coefficient vary with temperature.

At equilibrium, crystals must be at a free energy minimum with respect to both Q_1 and Q_2, i.e.:

$$\frac{\partial G}{\partial Q_1} = \frac{\partial G}{\partial Q_2} = 0 \tag{52}$$

giving:

$$a_1(T-T^*_{c1})Q_1 + b_1 Q_1^3 + c_1 Q_1^5 + \cdots + \lambda Q_2 = 0 \tag{53}$$

$$a_2(T-T^*_{c2})Q_2 + b_2 Q_2^3 + c_2 Q_2^5 + \cdots + \lambda Q_1 = 0 \tag{54}$$

Salje and coworkers (1985) used these expressions to describe the equilibrium behaviour of albite, in which displacive and Al/Si order/disorder processes could each cause the same symmetry reduction, $C2/m \rightleftharpoons C\bar{1}$. For the calibration of the Landau coefficients, Salje and coworkers used high-temperature lattice parameter data to follow the displacive order parameter (Q_1) as a function of temperature in crystals with different fixed values of the degree of Al/Si order (Q_2), together with thermodynamic data for the displacive transition in crystals with $Q_2 = 0$. Their determination of the equilibrium variation of the two order parameters is shown in Figure 5.15. Several features should be noted. Firstly, there is only one phase transition in which both order parameters are implicated. Secondly, neither of the order parameters conforms to the variation expected for a single-order parameter system. Thirdly, the equilibrium transition temperature is higher than the transition temperature for either of the pure displacive or pure order/disorder transitions. The

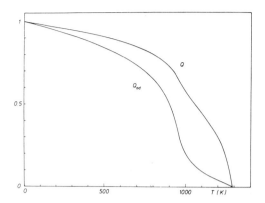

Figure 5.15 Temperature dependence of the displacive order parameter, Q, and the Al/Si order parameter, Q_{od}, for equilibrium transition behaviour of albite. Note the apparently non-Landau evolution of both order parameters which arises because they are coupled. (Salje et al. (1985). Reproduced by kind permission of E. Salje.)

coupling is favourable and provides additional stabilisation of the low-symmetry form.

5.7.2 Biquadratic coupling

The second type of coupling behaviour dealt with by Salje and Devarajan (1986) is for systems in which two zone-boundary transitions can occur. In this case, each of the order parameters couples with the common strain through terms in εQ_1^2 and εQ_2^2. The initial expansion describing the separate contributions to the excess free energy is:

$$G = \tfrac{1}{2}a_1(T - T_{c1})Q_1^2 + \tfrac{1}{4}b_1 Q_1^4 + \tfrac{1}{6}c_1 Q_1^6 + \cdots$$
$$+ \tfrac{1}{2}a_2(T - T_{c2})Q_2^2 + \tfrac{1}{4}b_2 Q_2^4 + \tfrac{1}{6}c_2 Q_2^6 + \cdots$$
$$+ e_1 \varepsilon Q_1^2 + e_2 \varepsilon Q_2^2 + f\varepsilon^2. \tag{55}$$

At equilibrium the crystal must be stress free, giving:

$$\frac{\partial G}{\partial \varepsilon} = 0 = e_1 Q_1^2 + e_2 Q_2^2 + 2f\varepsilon \tag{56}$$

$$\Rightarrow \quad \varepsilon = -\left(\frac{e_1 Q_1^2 + e_2 Q_2^2}{2f}\right) \tag{57}$$

Substituting this for ε in equation (55) and rearranging terms gives:

$$G = \tfrac{1}{2}a_1(T-T_{c1})Q_1^2 + \frac{1}{4}\left(b_1 - \frac{e_1^2}{f}\right)Q_1^4 + \tfrac{1}{6}c_1 Q_1^6 + \cdots$$

$$+ \tfrac{1}{2}a_2(T-T_{c2})Q_2^2 + \frac{1}{4}\left(b_2 - \frac{e_2^2}{f}\right)Q_2^4 + \tfrac{1}{6}c_2 Q_2^6 + \cdots$$

$$- \frac{e_1 e_2}{2f} Q_1^2 Q_2^2 \qquad (58)$$

The fourth-order terms are renormalized as in the single order parameter case discussed earlier, but, in addition, there is now a coupling between Q_1 and Q_2 that is described by the biquadratic coupling term. The coupling coefficient, $-(e_1 e_2)/2f$, can vary with temperature if, for example, the elastic constants are temperature-dependent.

Salje and Devarajan showed that there are a number of different solutions to this expansion depending on the values of the coefficients. Separate stability fields may exist for crystals with $Q_1 \neq 0$, $Q_2 = 0$, and $Q_1 = 0$, $Q_2 \neq 0$ or $Q_1 \neq 0$, $Q_2 \neq 0$, and a sequence of phase transitions can occur with falling temperature. A complete solution along these lines has not yet been produced for a mineral system but the $I\bar{1} \rightleftharpoons P\bar{1}$ displacive transition in anorthite has been used to illustrate some of the effects of biquadratic coupling (Salje, 1987a; 1988a; b; Redfern and Salje, 1987; Redfern et al., 1988; Hatch and Ghose, 1989).

The $I\bar{1} \rightleftharpoons P\bar{1}$ transition, to which the order parameter Q_1 may be assigned, occurs in anorthite near ~ 510 K and is largely displacive in character. It is strongly affected by the extent of Al/Si order present in the crystals which, as discussed above, can be described in terms of the order parameter for the $C\bar{1} \rightleftharpoons I\bar{1}$ transition (Q_2). Both transitions are associated with symmetry points on the Brillouin-zone boundary so that the interactions should be described by equation (58). The experimental observations are that in crystals with $Q_2 \approx 0.92$, the displacive transition occurs at $T_c = 510$ K and is tricritical (Wruck, 1986; Redfern and Salje, 1987; Salje, 1988b; calibration of Q_2 from Carpenter et al., 1990a). In crystals with a slightly lower degree of order ($Q_2 \approx 0.91$) and a small proportion of albite in solid solution, the transition is second order with $T_c = 530$ K (Redfern et al., 1988). The change from tricritical to second order behaviour in crystals with fixed but different degrees of Al/Si order implies a change in the fourth-order coefficient for the $I\bar{1} \rightleftharpoons P\bar{1}$ transition from $[b_1 - (d_1^2/f)] \approx 0$ to $[b_1 - d_1^2/f] > 0$ as a function of Q_2. Since the elastic constants of anorthite which constitute the f coefficient are likely to be sensitive to the state of Al/Si order and the value of the coupling coefficient d_1 may itself depend on the same elastic constants, it is quite reasonable to expect that changing the degree of Al/Si order should cause a renormalization of the fourth-order term in this way. In addition, since the second-order term is unaffected by biquadratic coupling, large variations in T_c should not be expected in response to changing the degree of Al/Si order. Compositional changes, on the other hand, should cause a renormalization of T_c.

Recently, Angel et al. (1989) have followed the displacive transition as a function of pressure at room temperature. The same transition as observed in well-ordered crystals at 510 K, 1 bar, appears to occur at 298 K, 25.3 kbar but is first order instead of tricritical at high pressure. From the earlier analysis, pressure alone would not be expected to cause a renormalization of the fourth-order coefficient and hence cause this change in thermodynamic character. As discussed by Angel et al. (1989) it is necessary to consider other factors and one obvious possibility might be coupling with Q for the $C2/m \rightleftharpoons C\bar{1}$ transition which, in the absence of melting, would occur at $T_c \gg 2000°C$ in pure anorthite. The complete solution may in fact depend on coupling between three order parameters, Q_1 for the $I\bar{1} \rightleftharpoons P\bar{1}$ transition, Q_2 for the $C\bar{1} \rightleftharpoons I\bar{1}$ transition, and Q_3 for the $C2/m \rightleftharpoons C\bar{1}$ transition. An important consideration in analysing the response to increasing pressure will be the fact that the excess volume associated with Q_1 is negligibly small (Redfern and Salje, 1987, Redfern et al., 1988) relative to the excess volume associated with Q_3.

5.7.3 Linear-quadratic coupling

An example of linear-quadratic coupling has not yet been definitely identified among mineral systems though it has been postulated for the cubic $\rightleftharpoons$ tetragonal transition in leucite (Palmer et al., 1990, Hatch et al., 1990). The appropriate equations are set out here for a general case. If the first order parameter, Q_1, is linearly coupled with strain and the second, Q_2, is associated with a zone-boundary transition, the total excess free energy is given by:

$$G = \tfrac{1}{2}a_1(T-T_{c1})Q_1^2 + \tfrac{1}{4}b_1 Q_1^4 + \tfrac{1}{6}c_1 Q_1^6 + \cdots$$
$$+ \tfrac{1}{2}a_2(T-T_{c2})Q_2^2 + \tfrac{1}{4}b_2 Q_2^4 + \tfrac{1}{6}c_2 Q_2^6 + \cdots$$
$$+ d_1 \varepsilon Q_1 + e_2 \varepsilon Q_2^2 + f\varepsilon^2. \tag{59}$$

At equilibrium the crystal is stress free:

$$\frac{\partial G}{\partial \varepsilon} = 0 = d_1 Q_1 + e_2 Q_2^2 + 2f\varepsilon \tag{60}$$

$$\Rightarrow \quad \varepsilon = -\left(\frac{d_1 Q_1 + e_2 Q_2^2}{2f}\right). \tag{61}$$

Substituting this back into equation (59) gives:

$$G = \tfrac{1}{2}a_1\left(T - T_{c1} - \frac{d_1^2}{2a_1 f}\right)Q_1^2 + \tfrac{1}{4}b_1 Q_1^4 + \tfrac{1}{6}c_1 Q_1^6 + \cdots$$

$$+ \tfrac{1}{2}a_2(T - T_{c2})Q_2^2 + \frac{1}{4}\left(b_2 - \frac{e_2^2}{f}\right)Q_2^4 + \tfrac{1}{6}c_2 Q_2^6 + \cdots$$

$$- \frac{d_1 e_2}{2f} Q_1 Q_2^2. \tag{62}$$

Several different sequences of phase transitions will probably exist, depending on the values of the coefficients, but two specific features can be anticipated for crystals which have one slow process (e.g. Al/Si ordering) and one rapid process (e.g. atomic displacements). The transition temperature of the transition governed by Q_1 might vary widely for crystals which are prepared with different values of Q_2 or, alternatively, the thermodynamic character of the transition governed by Q_2 might change between crystals prepared to have different values of Q_1.

5.7.4 Gradient coupling: incommensurate structures

It is generally recognized that an important mechanism for stabilizing incommensurate structures is by coupling of two ordering processes (see McConnell, 1988 for a review of mineral systems). In essence, one order parameter couples with gradients in the second to stabilize variations in the amplitudes of both as a function of distance in the crystal. To describe the excess free energy it is necessary to include the effects of the presence of gradients in Q_1 and Q_2 and of the coupling. For comparison with the coupling behaviour described above in which Q_1 and Q_2 remain homogeneous throughout a crystal, the simplest form of Landau expansion for an incommensurate structure is given here as:

$$G = \tfrac{1}{2}a_1(T - T_c)(Q_1^2 + Q_2^2) + \tfrac{1}{4}b(Q_1^4 + Q_2^4) + \cdots$$
$$+ d[Q_1(\nabla Q_2) - Q_2(\nabla Q_1)] + e[(\nabla Q_1)^2 + (\nabla Q_2)^2] \tag{63}$$

where the coupling terms in $Q_1(\nabla Q_2)$ and $Q_2(\nabla Q_1)$ provide much of the stabilization energy.

Two characteristic features of many phase transitions that lead to incommensurate structures are that they are thermodynamically continuous and that the vector describing the incommensurate repeat in reciprocal space varies with temperature. Apart from the incommensurate behaviour in quartz (Dolino et al., 1984a; b; Aslanyan et al., 1983) the macroscopic Landau approach has not yet been applied to this class of phase transitions in minerals. The microscopic mechanisms of coupling are to some extent understood in systems such as mullite, cordierite, nepheline, and intermediate plagioclase feldspars, however (McConnell, 1981; 1983; 1985; 1988; this volume; McConnell and Heine, 1985; Heine and McConnell, 1984).

5.8 Kinetics: the Ginzburg–Landau equation for time-dependent processes

Simple Landau expansions in effect define surfaces in G–T–Q space (Figure 5.16). Under equilibrium conditions the free energy of a crystal which is homogeneous with respect to Q follows a well-defined valley across such a surface as a function of temperature. Under non-equilibrium conditions the free energy of the crystal will still be represented by some point on the surface but it will be away from the valley. For thermodynamically continuous phase transitions the evolution with time can be represented in terms of a pathway across the free-energy surface towards the equilibrium point (Fig. 5.16). Intuitively one might expect that the rate of progress will depend on the steepness of the surface—a steep surface, implying a rapid change in G with respect to Q, should lead to a faster rate of reaction than a shallower surface. The Landau expansion in Q therefore provides a bridge between equilibrium and kinetic effects, the formal expression of which is the Ginzburg-Landau

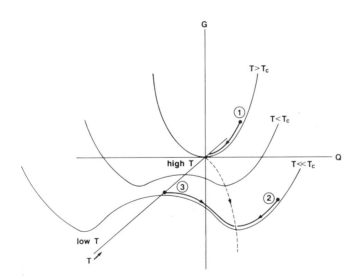

Figure 5.16 Free energy (G), order parameter (Q), temperature (T) surface for a second-order transition showing the variation of Q during equilibrium cooling (dashed line), and during isothermal kinetic experiments for crystals which remain homogeneous in Q (heavy lines). Pathway 1: disordering above T_c; pathway 2: disordering below T_c; pathway 3: ordering below T_c (Carpenter and Salje, 1989).

(GL) equation (Landau and Lifshitz, 1980; Salje, 1988c; 1989b; Carpenter and Salje, 1989):

$$\frac{\partial Q}{\partial t} = \frac{-\gamma\lambda \exp(-\Delta G^*/RT)}{2RT} \frac{\partial G}{\partial Q} \qquad (64)$$

where γ and λ are constants and ΔG^* is the free energy of activation.

As discussed by Salje (1988c; 1989b), Salje and Wruck (1988), and Carpenter and Salje (1989), this equation is valid for phase transitions in crystals where the correlation length of Q is large with respect to the length scale of conservation. It leads to a variety of specific rate laws and two examples, cation disordering in omphacite and Al/Si ordering in anorthite, can be used to illustrate the practical implications. The meaning and significance of the different length scales are reviewed by Carpenter and Salje (1989).

5.8.1 Q remaining homogeneous

If Q remains homogeneous within the crystals being studied, the structural states which develop with time under non-equilibrium conditions are identical to equilibrium structural states which occur in response to changing temperature. The Landau expansion can therefore be used directly to give $\partial G/\partial Q$ in the GL equation. Ordered crystals of omphacite which are held above T_c for the order/disorder ($C2/c \rightleftharpoons P2/n$) transition appear to conform to this behaviour and their evolution follows pathway 1 in Figure 5.16. Substituting the tricritical expression for the free-energy driving force into the GL equation leads to a rate law of the form (from Carpenter et al., 1990c):

$$\int_{t_o}^{t} dt = \int_{Q_o}^{Q} \frac{-2RT \exp(-\kappa_s Q^2/R)}{\gamma\lambda \exp(-\Delta G_0^*/RT)[a(T-T_c)Q + cQ^5]} dQ. \qquad (65)$$

In this case the free energy of activation is itself thought to be a function of Q:

$$\Delta G^* = \Delta G_0^* - T\kappa_s Q^2 \qquad (66)$$

where κ_s is a constant. This merely implies that the entropy of activation, which depends on the number of possible activated states that may lead to a change in Q, depends on the degree of order of the crystal.

Carpenter et al. (1990c) monitored the rate of disordering in natural omphacite crystals by measuring the intensity of superlattice reflections between successive annealing episodes at high temperatures. Reasonable agreement between the observed change in Q^2 with time, t, and a numerical solution to equation (65) is shown by the results in Figure 5.17.

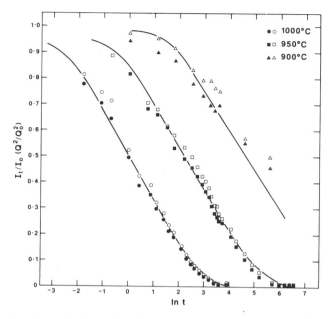

Figure 5.17 Variation of the intensities of superlattice reflections ($I_t/I_0 = Q^2/Q_0^2$) with annealing time, t, for cation disordering in omphacite at constant temperature ($T > T_c$). Solid lines represent a numerical solution to equation (65) while open and filled symbols represent observed values for two superlattice reflections (Carpenter et al., 1990c).

5.8.2 Fluctuations and inhomogeneities in Q

The equilibrium state of a crystal is within some well-defined free energy well (with respect to Q). Any fluctuations in Q away from the equilibrium value are therefore subject to an effective restoring force. Away from equilibrium, fluctuations in Q can lead to reductions in free energy and the effective restoring force no longer operates. As a consequence, the fluctuations will grow in amplitude and, since the rate of amplification of individual fluctuations is a function of wavelength (long wavelength fluctuations will grow or decay more slowly than those with short wavelengths), the crystals can become locally inhomogeneous in Q. A poorly defined modulated structure with a wavelength corresponding to the slowest changing fluctuation might even develop by this purely kinetic mechanism. In addition, however, the presence of gradients in Q in the inhomogeneous crystal might result in gradient coupling with a second-order parameter, thus allowing a rather well-defined modulated structure to develop. Inhomogeneous crystals (with respect to Q) in general and modulated crystals in particular are indeed commonly observed in non-equilibrium experiments. Appropriate free-energy expansions for these circumstances must include the gradient energy terms and these will not necessarily be present in the normal Landau expansion used to describe the equilibrium behaviour of a given system. It may not be so easy to produce a complete solution to the GL equation, therefore, and more empirical solutions must be developed. A practical approach is to follow the evolution of Q experimentally and then to rationalize an observed rate law in terms of various possible models.

Al/Si ordering in anorthite occurs by the exchange of Al and Si between tetrahedral sites such that the number of Al–O–Al linkages is progressively reduced. Carpenter (1991) has shown that ordering in highly disordered crystals produced by crystallization of glass of anorthite composition leads to a distinct but metastable incommensurate superstructure before the equilibrium $I\bar{1}$ state develops. From both strain and enthalpy measurements (e.g. see Figure 5.18), it is apparent that for a limited range of Q at a given temperature the ordering follows a rate law of the form (from Carpenter, 1991):

$$Q^2 \propto \ln t. \tag{67}$$

Salje (1988c) showed that this is a valid solution to the GL equation for systems in which Q becomes inhomogeneous. The full solution over a limited range of Q is:

$$Q^2 = AR(T - T_c)\left[\ln t - \frac{\Delta H^*}{RT} + \frac{\Delta S^*}{R} - \ln \tau_0\right] \tag{68}$$

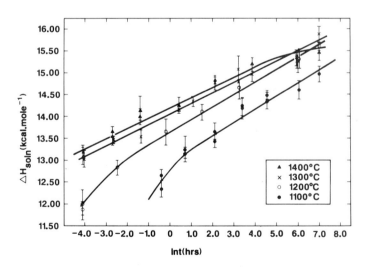

Figure 5.18 Variation of calorimetric heat of solution with annealing time for synthetic anorthite. The excess enthalpy due to the $C\bar{1}\rightleftharpoons I\bar{1}$ Al/Si ordering transition is linear with ΔH_{soln} and, from observations of the spontaneous strain, is also proportional to Q^2 for the transition. Thus, over a range of Q, the kinetic behaviour may be described by $Q^2 \propto \ln t$ (Carpenter, 1991).

where ΔH^* and ΔS^* are the enthalpy and entropy of activation and τ_0 is a characteristic jump frequency associated with the atoms that move during the transition. This describes the ordering kinetics adequately and yields values of T_c and ΔH^* which are consistent with other observations. A more analytical solution should be accessible once more is known about the gradient energy and coupling coefficients of the incommensurate anorthite structure.

5.9 Conclusion

The purpose of this review has been to demonstrate how a purely macroscopic approach, based on Landau theory, may be used to give quantitative descriptions of both the thermodynamics and kinetics of phase transitions in minerals. A common thread is provided by the use of a macroscopic order parameter, its relationship to other physical properties and its dependence on symmetry. From a purely practical point of view, the simple Landau free-energy expansions which have been developed show a number of attractive features. Apart from the fact that they provide reasonable descriptions of observed transition behaviour over many hundreds of degrees, the effects of pressure, strain and composition, or any other applied variable, can be incorporated without introducing excessive mathematical complexity. Moreover, experimental calibration of the coefficients frequently requires only familiar and widely available techniques and equipment. It is hoped that the overall methodology will lead to improvements in the accuracy of thermodynamic data available for real mineral systems and will also allow easy manipulation of that data in broader equilibrium and kinetic models of geological processes.

Acknowledgements

It is a pleasure to acknowledge the help of many colleagues over the last few years in highlighting the main themes of this review. Discerning readers will note, in particular, the pervasive influence of the ideas of Ekhard Salje who is thanked for his encouragement and patient tuition in all aspects of Landau theory. Ross Angel is thanked for his thorough review which led to a number of improvements to the manuscript. Cambridge Earth Sciences contribution number ES1764.

References

Aizu, K. (1970) Determination of the state parameters and formulation of spontaneous strain for ferroelastics. *Journal of the Physical Society of Japan*, **28**, 706–16.

Angel, R. J., Redfern, S. A. T., and Ross, N. L. (1989) Spontaneous strain below the $I\bar{1}\rightleftharpoons P\bar{1}$ transition in anorthite at pressure. *Physics and Chemistry of Minerals*, **16**, 539–44.

Aslanyan, T. A., Levanyuk, A. P., Vallade, M., *et al.* (1983) Various possibilities for formation of incommensurate superstructure near the α-β transition in quartz. *Journal of Physics C*, **16**, 6705–12.

Bismayer, U. (1988) New developments in Raman spectroscopy on structural phase transitions, in *Physical Properties and Thermodynamic Behaviour of Minerals*, (ed. E. K. H. Salje), NATO ASI series C 225, Reidel, Dordrecht, Boston, Lancaster, Tokyo, pp. 143–83.

Bismayer, U., Salje, E., Glazer, A. M., et al., (1986b) Effect of strain-induced order parameter coupling on the ferroelastic behaviour of lead phosphate-arsenate. *Phase Transitions* **6**, 129–51.

Bismayer, U., Salje, E., Jansen, M., et al. (1986a) Raman scattering near the structural phase transition of As_2O_5: order parameter treatment. *Journal of Physics C*, **19**, 4537–45.

Bruce, A. D. and Cowley, R. A. (1981) *Structural Phase Transitions*, Taylor & Francis, London.

Carpenter, M. A. (1981) Time–temperature–transformation (TTT) analysis of cation disordering in omphacite. *Contributions to Mineralogy and Petrology*, **78**, 433–40.

Carpenter, M. A. (1988) Thermochemistry of aluminium/silicon ordering in feldspar minerals, in *Physical Properties and Thermodynamic Behaviour of Minerals*, (ed. E. K. H. Salje), NATO ASI series C 225, Reidel, Dordrecht, Boston, Lancaster, Tokyo, pp. 265–323.

Carpenter, M. A. (1991) Mechanisms and kinetics of Al/Si ordering in anorthite, II: energetics and a Ginzburg–Landau rate law. *American Mineralogist*, **76**, 1120–33.

Carpenter, M. A. and McConnell, J. D. C. (1984) Experimental delineation of the $C\bar{1} \rightleftharpoons I\bar{1}$ transformation in intermediate plagioclase feldspars. *American Mineralogist*, **69**, 112–21.

Carpenter, M. A. and Salje, E. (1989) Time-dependent Landau theory for order/disorder processes in minerals. *Mineralogical Magazine*, **53**, 483–504.

Carpenter, M. A., Angel, R. J., and Finger, L. W. (1990a) Calibration of Al/Si order variations in anorthite. *Contributions to Mineralogy and Petrology*, **104**, 471–80.

Carpenter, M. A., Domeneghetti, M.-C., and Tazzoli, V. (1990b) Application of Landau theory to cation ordering in omphacite I: equilibrium behaviour. *European Journal of Mineralogy*, **2**, 7–18.

Carpenter, M. A., Domeneghetti, M.-C., and Tazzoli V. (1990c) Application of Landau theory to cation ordering in omphacite II: kinetic behaviour. *European Journal of Mineralogy*, **2**, 19–28.

Cohen, L. H. and Klement W., Jr. (1973) Determination of high-temperature transition in calcite to 5 kbar by differential thermal analysis in hydrostatic apparatus. *Journal of Geology*, **81**, 724–7.

David, W. I. F. and Wood, I. G. (1983) Ferroelastic phase transition in $BiVO_4$: V. Temperature dependence of Bi^{3+} displacement and spontaneous strains. *Journal of Physics C*, **16**, 5127–48.

Devarajan, V. and Salje, E. (1984) Phase transitions in $K_2Cd_2(SO_4)_3$: investigation of the non-linear dependence of spontaneous strain and morphic birefringence on order parameter as determined from excess entropy measurements. *Journal of Physics C*, **17**, 5525–37.

Dolino, G., Bachheimer, J. P., Berge, B., et al. (1984a) Incommensurate phase of quartz: I. Elastic neutron scattering. *Journal de Physique*, **45**, 361–71.

Dolino, G., Bachheimer, J. P., Berge, B., et al. (1984b) Incommensurate structure of quartz: III. Study of the coexistence state between the incommensurate structure and the α-phases by neutron scattering and electron microscopy. *Journal de Physique*, **45**, 901–12.

Dove, M. T. and Powell, B. M. (1989) Neutron diffraction study of the tricritical orientational order/disorder phase transition in calcite at 1260 K. *Physics and Chemistry of Minerals*, **16**, 503–7.

Ginzburg, V.L., Levanyuk, A. P., and Sobyanin, A. A. (1987) Comments on the region of applicability of the Landau theory for structural phase transitions. *Ferroelectrics*, **73**, 171–82.

Greenwood, H. J. (1972) Al^{IV}–Si^{IV} disorder in sillimanite and its effect on phase relations of the aluminium silicate minerals. *Geological Society of America Memoir*, **132**, 553–71.

Hatch, D. M. and Ghose, S. (1989) A dynamical model for the $I\bar{1}$–$P\bar{1}$ phase transition in anorthite, $CaAl_2Si_2O_8$ II. Order parameter treatment. *Physics and Chemistry of Minerals* **16**, 614–20.

Hatch, D. M. and Griffen, D. T. (1989) Phase transitions in the grandite garnets. *American Mineralogist*, **74**, 151–9.

Hatch, D. M., Ghose, S., and Stokes, H. T. (1990) Phase transitions in leucite, $KAlSi_2O_6$ I. Symmetry analysis with order parameter treatment and the resulting microscopic distortions. *Physics and Chemistry of Minerals*, **17**, 220–7.

Hazen, R. M. and Mariathasan, J. W. E. (1982) Bismuth vanadate: a high-pressure, high-temperature crystallographic study of the ferroelastic-paraelastic transition. *Science* **216**, 991–3.

Heine, V. and McConnell, J. D. C. (1984) The origin of incommensurate structures in insulators. *Journal of Physics C*, **17**, 1199–1220.

Holland, T. J. B. (1990) Activities of components in omphacitic solid solutions. An application of Landau theory to mixtures. *Contributions to Mineralogy and Petrology*, **105**, 446–53.

Janovec, V., Dvorak, V., and Petzelt, J. (1975) Symmetry classification and properties of equi-translation structural phase transitions. *Czechoslovak Journal of Physics B*, **25**, 1362–96.

Landau, L. D. and Lifshitz, E. M. (1980) *Statistical Physics*, 3rd edn, part 1. Pergamon Press, Oxford, New York, Toronto, Sydney, Paris, Frankfurt.

Lyubarskii, G. Y. (1960) *The Application of Group Theory in Physics*, Pergamon Press, New York, Oxford, London, Paris.

McConnell, J. D. C. (1981) Electron-optical study of modulated mineral solid solutions. *Bulletin de Minéralogie*, **104**, 231–5.

McConnell, J. D. C. (1983) A review of structural resonance and the nature of long-range interactions in modulated mineral structures. *American Mineralogist*, **68**, 1–10.

McConnell, J. D. C. (1985) Symmetry aspects of order–disorder and the application of Landau theory, in *Reviews in Mineralogy*, vol 14: *Microscopic to macroscopic*, (eds S. W. Kieffer and A. Navrotsky), Mineralogical Society of America, 165–86.

McConnell, J. D. C. (1988) The thermodynamics of short range order, in *Physical Properties and Thermodynamic Behaviour of Minerals*, (ed. E. K. H. Salje), NATO ASI series C 225, Reidel, Dordrecht, Boston, Lancaster, Tokyo, pp. 17–48.

McConnell, J. D. C. and Heine, V. (1985) Incommensurate structure and stability of mullite. *Physical Review B*, **31**, 6140–2.

Mirwald, P. W. (1976) A differential thermal analysis study of the high-temperature polymorphism of calcite at high pressure. *Contributions to Mineralogy and Petrology*, **59**, 33–40.

Müller, K. A. and Berlinger, W. (1971) Static critical exponents at structural phase transitions. *Physical Review Letters*, **26**, 13–16.

Müller, K. A. and von Waldkirch, T. (1973) Structural phase-transition studies in $SrTiO_3$ and $LaAlO_3$ by EPR. *IBM Research RZ620* (#20576) Solid State Physics.

Müller, K. A., Berlinger, W., and Waldner, F. (1968) Characteristic structural phase transition in perovskite-type compounds. *Physical Review Letters*, **21**, 814–17.

Navrotsky, A. and Loucks, D. (1977) Calculation of subsolidus phase relations in carbonates and pyroxenes. *Physics and Chemistry of Minerals*, **1**, 109–27.

Palmer, D. C., Bismayer, U., and Salje, E. K. H. (1990) Phase transitions in leucite: order parameter behaviour and the Landau potential deduced from Raman spectroscopy and birefringence studies. *Physics and Chemistry of Minerals*, **17**, 259–65.

Palmer, D. C., Salje, E. K. H. and Schmahl, W. W. (1989) Phase transitions in leucite: X-ray diffraction studies. *Physics and Chemistry of Minerals*, **16**, 714–19.

Putnis, A. (1980) The distortion index in anhydrous Mg–cordierite. *Contributions to Mineralogy and Petrology*, **74**, 135–41.

Putnis, A. and Holland, T. J. B. (1986) Sector trilling in cordierite and equilibrium overstepping in metamorphism. *Contributions to Mineralogy and Petrology*, **93**, 265–72.

Redfern, S. A. T., Graeme-Barber A., and Salje, E. (1988) Thermodynamics of plagioclase III:

spontaneous strain at the $I\bar{1} \rightleftharpoons P\bar{1}$ phase transition in Ca-rich plagioclase. *Physics and Chemistry of Minerals*, 16, 157–63.

Redfern, S. A. T. and Salje, E. (1987) Thermodynamics of plagioclase II: temperature evolution of the spontaneous strain at the $I\bar{1} \rightleftharpoons P\bar{1}$ phase transition in anorthite. *Physics and Chemistry of Minerals*, 14, 189–95.

Redfern, S. A. T. and Salje, E. (1988) Spontaneous strain and the ferroelastic phase transition in As_2O_5. *Journal of Physics C*, 21, 277–85.

Redfern, S. A. T., Salje, E., and Navrotsky, A. (1989) High-temperature enthalpy at the orientational order–disorder transition in calcite: implications for the calcite/aragonite phase equilibrium. *Contributions to Mineralogy and Petrology*, 101, 479–84.

Salje, E. (1985) Thermodynamics of sodium feldspar I: order parameter treatment and strain induced coupling effects. *Physics and Chemistry of Minerals*, 12, 93–8.

Salje, E. (1986) Raman spectroscopic investigation of the order parameter behaviour in hypersolvus alkali feldspar: displacive phase transition and evidence for Na–K site ordering. *Physics and Chemistry of Minerals*, 13, 340–6.

Salje, E. (1987a) Thermodynamics of plagioclases I: theory of the $I\bar{1} \rightleftharpoons P\bar{1}$ phase transition in anorthite and Ca-rich plagioclases. *Physics and Chemistry of Minerals*, 14, 181–8.

Salje, E. (1987b) Structural states of Mg-cordierite II: Landau theory. *Physics and Chemistry of Minerals*, 14, 455–60.

Salje, E. (1988a) Toward a thermodynamic understanding of feldspars: order parameters of Na-feldspar and the $I\bar{1} \rightleftharpoons P\bar{1}$ phase transition in anorthite, in *Advances in Physical Geochemistry*, vol 7: *Structural and Magnetic Phase Transitions in Minerals*, (eds S. Ghose, J. M. D. Coey, and E. Salje), Springer, New York, Berlin, Heidelberg, Tokyo, pp. 1–16.

Salje, E. (1988b) Structural phase transitions and specific heat anomalies, in *Physical Properties and Thermodynamic Behaviour of Minerals*, (ed. E. K. H. Salje), NATO ASI series C225, Reidel, Dordrecht, Boston, Lancaster, Tokyo, pp. 75–118.

Salje, E. (1988c) Kinetic rate laws as derived from order parameter theory I: theoretical concepts. *Physics and Chemistry of Minerals*, 15, 336–48.

Salje, E. (1989a) Characteristics of perovskite-related materials. *Philosophical Transactions of the Royal Society of London A*, 328, 409–16.

Salje, E. (1989b) Towards a better understanding of time-dependent geological processes: kinetics of structural phase transformations in minerals. *Terra Nova*, 1, 35–44.

Salje, E. (1990) *Phase Transitions in Ferroelastic and Co-elastic Crystals*, Cambridge University Press, Cambridge.

Salje, E., Bismayer, U., and Jansen, M. (1987) Temperature evolution of the ferroelastic order parameter of As_2O_5 as determined from optical birefringence. *Journal of Physics C*, 20, 3613–20.

Salje, E. and Devarajan, V. (1986) Phase transitions in systems with strain-induced coupling between two order parameters. *Phase Transitions*, 6, 235–48.

Salje, E. and Wruck, B. (1988) Kinetic rate laws as derived from order parameter theory II: interpretation of experimental data by Laplace-transformation, the relaxation spectrum, and kinetic gradient coupling between two order parameters. *Physics and Chemistry of Minerals*, 16, 140–7.

Salje, E., Kuscholke, B., Wruck, B., et al. (1985) Thermodynamics of sodium feldspar II: experimental results and numerical calculations. *Physics and Chemistry of Minerals*, 12, 99–107.

Schmahl, W. W. and Redfern, S. A. T. (1988) An X-ray study of coupling between acoustic and optic modes at the ferroelastic phase transition in As_2O_5. *Journal of Physics C*, 21, 3719–25.

Schmahl, W. W. and Salje, E. (1989) X-ray diffraction study of the orientational order/disorder transition in $NaNO_3$: evidence for order parameter coupling. *Physics and Chemistry of Minerals*, 16, 790–8.

Stokes, H. T. and Hatch, D. M. (1988) *Isotropy Subgroups of the 230 Crystallographic Space Groups*, World Scientific, Singapore.

Tolédano, P. and Tolédano, J.-C. (1976) Order-parameter symmetries for improper ferroelectric nonferroelastic transitions. *Physical Review B*, **14**, 3097–109.

Tolédano, P. and Tolédano, J.-C. (1977) Order-parameter symmetries for the phase transitions of nonmagnetic secondary and higher-order ferroics. *Physical Review B*, **16**, 386–407.

Tolédano, J.-C. and Tolédano, P. (1980) Order parameter symmetries and free-energy expansions of purely ferroelastic transitions. *Physical Review B*, **21**, 1139–72.

Tolédano, J.-C. and Tolédano, P. (1987) *The Landau Theory of Phase Transitions*, World Scientific, Singapore, New Jersey, Hong Kong.

Wruck, B. (1986) Einfluss des Na-Gehaltes und der Al,Si-Fehlordnung auf das thermodynamische Verhalten der Phasenumwandlung $P\bar{1} \rightleftharpoons I\bar{1}$ in Anorthit, doctoral dissertation, University of Hanover.

CHAPTER SIX
The stability of modulated structures

J. Desmond C. McConnell

6.1 Introduction

Considerable interest has been focused recently in physics and in mineralogy on the existence and origin of modulated crystal structures. These have the important and fundamental characteristic that they cannot properly be defined in terms of a conventional simple unit cell and space group.

Such structures invariably show additional intensity in diffraction experiments which, in terms of the conventional reciprocal lattice repeats, define spacings, or a wavelength, which are irrational. For this reason such modulated structures are frequently described as incommensurate structures, particularly when the additional intensity occurs as sharp additional maxima. The additional intensity has, in all cases that we wish to consider, the property that it is convoluted with the reciprocal lattice and hence appears in the same position, but not necessarily with the same intensity, in all cells throughout diffraction space.

It is usual to define the distribution of extra intensity, when it comprises sharp, additional diffraction maxima, in terms of a small reciprocal vector, Q, which serves to define the position of the additional intensity within a single unit cell in reciprocal space. In certain cases the intensity may occur as intensity streaks in the diffraction pattern.

Diffraction theory implies that, where the additional intensity can be described in this way in terms of a reduced vector Q, a function also exists in direct space with corresponding wavelength Q^{-1} associated with a modulation on the lattice. Where the small vector, Q, may be chosen with reference to the primary reciprocal lattice points, the modulation operates on the primary direct lattice and where the additional intensity lies close to the position of a potential superlattice maximum, it is useful to consider that the additional intensity relates to the modulation of the corresponding superlattice. In the formal definition of a modulated structure presented above, we have assumed that the value of the reduced vector, Q, is irrational. By definition this automatically excludes all simple superlattices with rational values of Q. This

The Stability of Minerals. Edited by G. D. Price and N. L. Ross.
Published in 1992 by Chapman & Hall, London. ISBN 0 412 44150 0

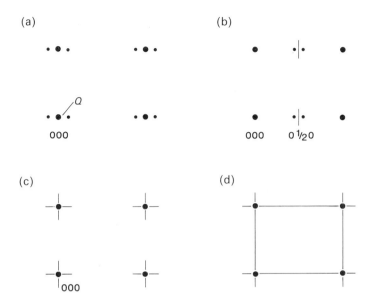

Figure 6.1 The distribution of extra diffracted intensity for modulated structures, (a) with Q located close to 000; (b) with Q located close to the superlattice point 0,1/2,0; (c) with diffuse streaks close to 000; and (d) with intensity streaks between Bragg diffraction peaks.

criterion, which may appear trivial at first sight, is extremely important since the theoretical principles involved in making this distinction are of fundamental importance in the analysis of the origin of incommensurate structures.

In many cases, as noted above, the additional intensity observed in modulated structures comprises sharp maxima with positions defined in terms of a precise value of Q. This is the case in a large class of modulated structures of direct interest to physicists that are currently described as incommensurate (IC) structures. More complicated intensity distributions also exist. In the present Chapter, for example, we will deal briefly with modulated structures which show additional streaks of intensity and hence a range of modulation wavelengths rather than a single preferred wavelength. The characteristics of the distribution of scattered intensity for the range of modulated structures discussed here are illustrated in Figure 6.1.

Modulated structures can arise in a number of different ways. In many cases it can be shown that they have a true field of thermodynamic stability, and hence play an important role in defining the relative stabilities of mineral phases within a multiphase assemblage. In other cases, the appearance of modulated structures can be related to the existence of purely kinetic constraints. An excellent example of the latter occurs in the spinodal modulations in alkali feldspar solid solutions that develop during relatively rapid cooling,

(McConnell, 1971). In this case the kinetic constraints are associated with the counter diffusion of K and Na ions in a process of incipient exsolution within the spinodal. The diffuse intensity present in this case is in the form of streaks and corresponds to a relatively wide range of possible wavelengths that reflect both free-energy criteria, and the statistics of the segregation process. A somewhat similar situation arises in relation to the development of the modulated structure in orthoclase and adularia (McConnell, 1965). In this case the modulation is associated with the transformation from monoclinic to triclinic symmetry, and the principal kinetic controls are associated with the difficulty of Al/Si diffusion during ordering in this order–disorder transformation.

As already noted, the implications of thermodynamically stable modulation structures in minerals are extremely important in relation to mineral stability. This is particularly so in the case of non-stoichiometric solid solutions where the onset of modulation may involve extensive ordering that results in a very substantial reduction in enthalpy and entropy. One of the most important examples of this phenomena in minerals is observed in the intermediate plagioclase feldspars. Here, accurate thermochemical data are available, (Carpenter *et al.*, 1985), on the enthalpy reduction and stabilization associated with the development of incommensurate structure. Such data are particularly relevant in discussing the stability relationships of the plagioclase feldspars in multiphase assemblages in regional metamorphism. Similar enthalpy criteria are involved in stabilization of the incommensurate structure of mullite (McConnell and Heine, 1985), and are also relevant in controlling the low-temperature structural state of natural nepheline. From this brief account it should be clear that it is unwise to ignore the implications of IC phase transformations in dealing with phase relationships among the common rock-forming minerals.

We now turn to considering the mode of origin of modulated and incommensurate structures in minerals. In general, the onset of modulation is associated with a phase transformation, generally of the second order, i.e. a transformation which is continuous in character. Further, it is usually possible to show that the origin of modulated structures is directly dependent on the prior existence of a normal-phase transformation in the same temperature range. Thus, in the case of quartz, we find that the modulated structure exists in a narrow temperature range in the immediate vicinity of the α to β transition at 573°C (Aslanyan and Levanyuk, 1979). Theory implies that this is not a coincidence, and that the origin of the incommensurate structure may indeed be directly related to the prior existence of the well-known structural phase transition in quartz involving a change from hexagonal to trigonal symmetry at this temperature.

The fact that incommensurate structures are almost invariably associated with conventional phase transformations implies that any profitable theoretical analysis of their origin must begin by considering how they relate to, and differ from, normal phase transformations. In this context the Landau theory of phase transformations provides the most useful general starting point (Landau and Lifshitz, 1958; Lifshitz, 1942a; 1942b). In establishing the theory of incommensurate structures in this chapter, we begin by reviewing those aspects of the Landau theory of phase transformations that are particularly relevant. Elementary Landau theory is then extended to deal with incommensurate structures and modulated structures generally. Having established the basis of a theory that links both normal and incommensurate phase transformations, the chapter concludes by illustrating the theory and general principles of stability in modulated structures with reference to carefully selected examples. Quartz is one of the examples chosen, and has the advantage that the structural changes in both the normal α to β transition, and the related incommensurate phase, can be readily demonstrated.

6.2 The theory of phase transformations

6.2.1 Special point transformations

Landau theory in its application to the physics of phase transformations depends essentially on symmetry principles (Landau and Lifshitz, 1958). First and foremost the theory depends on the existence of a symmetry change in a transformation. Later, we will discuss the principles that govern this symmetry change. Given that there is a discrete change in symmetry at the transformation temperature, T_c, the theory proceeds by classifying all structural aspects of the crystal into just two symmetry sets, one of which has the original, high-temperature symmetry and the other that possesses the reduced symmetry associated with the crystal below the transition temperature. This partition of all distribution functions (displacements, densities, etc.) for the single crystal into two discrete sets can be written provisionally as a simple sum:

$$\Xi_{total} = \Xi_{high} + \Xi_{low} \qquad (1)$$

Ξ_{high} has the full symmetry of the high-temperature structure and Ξ_{low} describes the set of functions of lower symmetry which appear only at the transition temperature.

We note now that the justification of this exercise depends on certain principles contained in group theory. Group theory has the characteristic that it deals in a purely abstract way with the properties of functions, and classifies them precisely and uniquely on abstract symmetry criteria. Thus, by using group theory, it is possible to make perfectly general statements about sets of functions which may be extremely complex but, which, at the same time need not be specified in any great detail. The reason why it is possible to make general statements of this kind on the basis of symmetry criteria alone is because group theory uses the global symmetry properties of a system to establish orthogonal relationships between functions. In this way any or all of the functions from the sets Ξ_{high} and Ξ_{low}, when multiplied together and integrated throughout space must yield exactly zero.

$$\int \Xi_{high} \cdot \Xi_{low} \cdot dV = 0 \tag{2}$$

This mathematical condition has the related important implication that we may also separate the free-energy contributions associated with the two sets of distribution functions, and hence write the total free energy of the system as a simple sum of contributions from the two orthogonal distributions. Thus we may write down:

$$G = G_{high} + G_{low} \tag{3}$$

In a polymorphic phase transformation, the contribution of G_{low} is necessarily zero above the transition temperature and has some finite value below the transition temperature where the crystal possesses lower symmetry. In the Landau theory of phase transitions G_{low} is expanded in a series which usually relies on the concept of a single parameter η, the order parameter, which is used to describe the amplitude of all functions in the set Ξ_{low}. This parameter η may be chosen in many different ways. It may describe a change in a lattice angle, or the extent of order on a specific structural site in an order–disorder transformation. It must always be defined so that it is identically zero above the transition temperature, implying that the contribution to the total free energy of the system is identically equal to zero there.

$$G_{low} = A\eta^2 + B\eta^3 + C\eta^4 + \tag{4}$$

Landau made the simple assumption that coefficients in this expansion were independent of temperature with the exception of the coefficient A which was given a temperature dependence of the form:

$$A = a(T - T_c) \tag{5}$$

This implies, when a is chosen to be positive, that the value of A is also positive above T_c, zero at the transition temperature T_c, and negative below the transition, leading to a potential reduction in the free energy of the system. The detailed thermodynamic characteristics of a system undergoing a phase transformation are determined in Landau theory, by the presence or absence of terms in the expansion (4) due to symmetry, and by the numerical values of the coefficients. We deal first with the question of the controls exercised by symmetry on the presence of the different terms in this free-energy expansion.

We have already used group-theoretical principles to separate the different components Ξ_{high} and Ξ_{low}, and their associated free-energy contributions in the transforming single crystal. Group theory can also be used to study the symmetry characteristics of the different terms in the free-energy expansion and hence the overall characteristics of the phase transformation. Thus the presence of a third-order term in the expansion, associated, for example, with the loss of a threefold axis, demands that the phase transformation must be first order in character.

In order to understand this use of group theoretical principles it is convenient to introduce the concept of group representation tables. The derivation and detailed use of these tables is discussed by Cotton (1971) and by Cracknell (1968). The usefulness of these tables is directly related to the fact that they provide, in a very simple form, the means of establishing and manipulating orthogonal functions associated with each of the possible subgroup symmetries (Ξ_{low}) of the crystal. Representation tables are used in this way in a vast range of problems where one wishes to separate orthogonal symmetry components. Examples include the definition of the several different modes of vibration of a molecule, or the possibility of identifying specific Raman and infra-red active models in a single crystal.

In applying group theory to the study of the characteristics of the free-energy expansion (4) we note that any subgroup symmetry may be associated with its corresponding representation, and that the representation itself may be used to characterize the symmetry properties of the powers of the order parameter η, within the separate terms in the free-energy expansion.

We note here that irreducible representations in general comprise a set of matrix elements, one for each symmetry element in the group, which, on matrix multiplication, behave exactly as the symmetry operations do themselves and, at the same time, provide, as an algebra, a unique description of each of the permissible subgroups of the parent symmetry group. All the irreducible representations in any symmetry group are orthogonal to each other as we have already noted, and their number and other properties are governed by simple group theoretical rules with which we need not be concerned.

Our principal interest here lies in the fact that the symmetry change at the transition temperature may be described in terms of a single and unique irreducible representation of the parent, or high-temperature symmetry group. Further, in order to characterize the nature of the phase transition in more detail, it is only necessary to determine whether or not the identity (full symmetry) representation is present within each term in the expansion in turn (Birman, 1966; McConnell, 1985). Thus, in proving that the third-order term must be included in the free-energy expansion for a phase transformation involving the loss of a threefold axis, it is only necessary to show that the identity representation of the high-temperature (full-symmetry) group is contained in the symmetrized cube of the relevant irreducible representation.

THE STABILITY OF MODULATED STRUCTURES

These rules do not depend on the complexity of the symmetry group itself, and are equally applicable to simple-point group representations or to the representations of extremely large groups such as the complete space group of the single crystal, where the number of group elements is very large indeed. Generally, in dealing with normal phase transformations only very simple group representations need be used. In the case of transformations involving modulated structures the representations are only slightly more complicated.

6.2.2 Landau theory for symmetry point structures

It will be convenient to illustrate the use of symmetry principles and representation theory for simple commensurate phase transformations by selecting a structure with four atoms per unit cell distributed over general positions in space group Pmm2. We will assume that the four atoms in the unit cell comprise two A atoms and two B atoms and we explore the possibilities for ordering them on a commensurate basis. This study forms an essential introduction to the theory of incommensurate phase transformations.

First, we consider the possibility of ordering the As and Bs without change in the lattice constants of the orthorhombic unit cell in space group Pmm2. This corresponds to the choice of Q at the origin of reciprocal space, at the symmetry point 000, and this ordering process will leave all the existing lattice points unchanged in the new regime. Consequently, we can ignore the existence of the lattice and its translations altogether (Bradley and Cracknell, 1972). (An adequate group-theoretical representation of the whole assembly of lattice points gives each lattice point the character 1, implying that all possible translation vectors are associated with the identity operation. In defining one lattice point, we fix all, and it is not necessary to deal further with the lattice translations.)

It remains to consider the role of the point group symmetry elements of which there are four: E, m_x, m_y, and C_{2z}. The group representation table for this small group of order four is set out in Table 6.1 and each of the four representations can be used to establish a different ordered structure for the A and B atoms. These are shown in Figure 6.2. The first representation is the trivial or identity representation which implies, from the presence of the identity element 1 throughout, that the A and B atoms cannot be distinguished in this representation. They may be represented in a density function by giving them mass (or scattering factor) $(A+B)/2$. It will now be apparent that the identity representation is incapable of representing anything except the completely disordered structure. Indeed, it describes the complete set of functions which we designated originally as Ξ_{high}.

The remaining three representations of point group mm2 relate to three possible ways of ordering the A and B atoms. Note that we must use antisymmetrical functions with average density zero in setting up suitable functions to associate with these three representations. It is convenient to use

Table 6.1 Derivative space groups for ordering vector 000 in space group Pmm2

	E	C_{2z}	m_x	m_y
Pmm2	1	1	1	1
Pm_x	1	−1	1	−1
Pm_y	1	−1	−1	1
P2	1	1	−1	−1

density functions of the form $\pm (A-B)/2$, in line with the fact that the characters in each of these three representations are also simply ± 1. It should now be apparent that the three additional ordering representations describe three independent functions of the type Ξ_{low}, as originally defined. Each of these three functions is anti-symmetrical and it is necessary to add them individually to the identity function $(A+B)/2$ in turn in order to obtain the three different ordering schemes for the A and B atoms in the unit cell.

It remains to note that all three functions Ξ_{low} are orthogonal. To prove this we need only multiply together the rows corresponding to any two of the

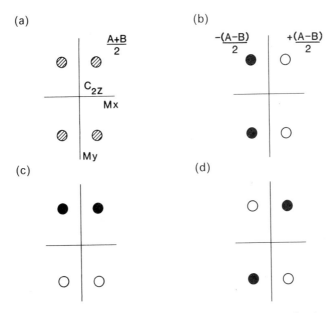

Figure 6.2 Ordering patterns associated with the derivative space groups for the vector 000 in space group Pmm2. Note that in (a), which is associated with the identity representation, all four items are equivalent, with weight $(A+B)/2$. In the remaining diagrams the ordering patterns are associated with weights $\pm(A-B)/2$, with sign corresponding to the characters of the representations in Table 6.1.

irreducible representations. Summation over such products is always identically equal to zero. Note that this operation is exactly equivalent, in its effect, to the enormous task of multiplying together two anti-symmetrical ordering functions over the whole of the single crystal, and then carrying out a complete integration, again over the whole of the single crystal.

We may now examine the characteristics of the three ordering representations in order to establish the likely nature of the ordering transformations. Since the representations are one-dimensional, and real, it follows that there can be no third-order invariant. It is also true that the anti-symmetrical square of all three representations does not exist implying that the corresponding phase transformation may be of second order (Landau and Lifshitz, 1958; Birman, 1966).

This concludes our investigation of the application of simple Landau theory to ordering associated with the symmetry point 000, i.e. at the Brillouin-zone centre. We now examine the possibilities for ordering of the A and B atoms in the same structure with space group Pmm2, while at the same time considering possible changes in the unit cell dimensions. We will assume a symmetry-point ordering process associated with the development of a simple superlattice in which the *b* cell edge is doubled, i.e., we assume that the transformation is associated with the vector 0,1/2,0. Since we have now introduced a change in the lattice, as distinct from the point group of the single crystal, we must examine the corresponding irreducible representations of the lattice associated with the doubling (Bradley and Cracknell, 1972).

The small representation associated with doubling the *b* axis belongs to the group of order two (Table 6.2), with symmetry elements E and t. It is anti-symmetrical, with character -1, in the initial translation operation *b* (t). The operation of doubling the cell edge yields the identity (2*b*) element with character $+1$. We note that all lattice translations belong to one of two sets and may be labelled with the characters -1 or $+1$ accordingly.

The extended group representations for ordering associated with change in the translations of the lattice are formed from the direct product of the original point group mm2, and the small group of order two in Table 6.2 associated with doubling the *b* cell dimension. In this new group, which is of order eight, only the four representations that are anti-symmetrical in the translation *b*, and have character -1, are relevant (Bradley and Cracknell, 1972). The complete ordering schemes associated with this doubling of the *b* cell edge are set out in Figure 6.3. Notice that the doubled cell contains eight atoms in all, and that there are four ordering schemes, and related space groups, that correspond to different ordering schemes compatible with the doubling operation.

This completes the theory associated with symmetry point transformations that occur both at the zone centre, i.e. without change in lattice constants, and in association with a change in cell dimensions (a zone-boundary transformation).

Table 6.2 The translation group of order two, and the derivative space groups for ordering vector 0, 1/2, 0 in space group Pmm2

					E/0	E/t		
					1	1		
					1	−1		
	E/0	C_{2z}/0	m_x/0	m_y/0	E/t	C_{2z}/t	m_x/t	m_y/t
Pmm2	1	1	1	1	−1	−1	−1	−1
Pmm2	1	−1	1	−1	−1	1	−1	1
Pbm2	1	−1	−1	1	−1	1	1	−1
Pbm2	1	1	−1	−1	−1	−1	1	1

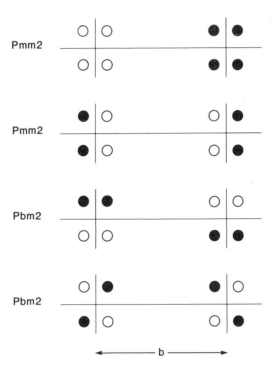

Figure 6.3 The ordering patterns associated with the vector 0,1/2,0 in space group Pmm2, matching the representations given in Table 6.2. Note that the space groups Pmm2 and Pbm2 each occur twice.

We have now reached the point where we must consider how these phase transitions differ from those that are incommensurate. The distinction was originally clearly stated by Landau (Landau and Lifshitz, 1958; Lifshitz, 1942a) in terms of the properties of special points (symmetry points) related to the choice of Q. First, we note that, by selecting Q in a totally arbitrary position in the reciprocal unit cell, it follows that it is necessary to include all the additional vectors that are related to Q by symmetry. In the general case this number (n) is equal to the order of the original point group of the crystal and defines an n-dimensional irreducible representation. It is therefore impossible to reduce the symmetry of the crystal by defining such a system of functions since the set must transform on translation as an invariant under the full symmetry, i.e. have the properties of Ξ_{high} as previously defined.

Landau indicated that a discrete change in symmetry in the crystal was possible where Q was chosen to lie on special points in the reciprocal unit cell. These special points correspond to centres of symmetry, or to the points of intersection of planes, or axes of symmetry. By choosing such points for Q, it is possible to define functions which are not invariant under the original translations and hence may be used to bring about a change in symmetry. We have already considered transformations of this kind in relation to the space group Pmm2 by selecting the special points 000 and 0,1/2,0.

The special point concept may be further illustrated with reference to the choice of Q, first close to, and then at, a centre of symmetry in the reciprocal unit cell. In this case two linearly independent functions associated with $\pm Q$, transforming as an invariant, make way for a single function associated with the choice of Q at the centre of symmetry (Lifshitz, 1942a; 1942b). The single function can no longer behave as an invariant under translation and its appearance therefore initiates a symmetry change. In developing the theory of IC structures the role of these special, symmetry-point structures is extremely important as we shall see in Section 6.3.

6.3 The theory of incommensurate (IC)-phase transformations

The theory of the origin of incommensurate structures hinges on the problem of deciding how we might describe a phase transformation based on a value of Q other than that for the Landau-type special points that we have been considering up to this point. Clues to the answer to this problem lie in something which we have already established, namely that for non-special-point values of Q we must consider several linearly independent functions, and a related n-dimensional irreducible representation. Now each representation in this set must have a unique symmetry specification (set of characters) since otherwise it would be possible to reduce the dimension of the representation. This leads one directly to the concept of the existence of multiple-order parameters for incommensurate structures since we have already indicated that each order parameter is associated with a specific irreducible representation, and a unique set of symmetry properties.

Thus, whereas it is always possible to describe a special, symmetry point phase transformation in terms of a single, one-dimensional irreducible representation, and an associated single-order parameter η, this is not true for the incommensurate and modulated structures of arbitrary Q that are of interest here (Heine and McConnell, 1981; 1984; Sannikov, 1981). In the simplest case we find that two irreducible representations, and two symmetrically inequivalent order parameters (η and ξ), are necessary. The theory which we will need to develop in this review is dependent on this simple theoretical argument.

At this point we should distinguish carefully between the existence of different physical events with the same symmetry, and those with different symmetry. In a symmetry-point phase transformation the Landau free-energy expansion may be rather more complicated than the simple expression given in equation (4). It may be necessary to consider interactions between different physical processes (functions) with the same symmetry (i.e. belonging to the same irreducible representation). This occurs in many mineral systems where it is necessary to distinguish between ordering, on the one hand, and a spontaneous shear on the other. In normal commensurate-phase transformations we may distinguish between these phenomena but since they may always be referred to the same irreducible representation, it is impossible to distinguish them on purely symmetry grounds. They give rise to product terms (invariants) of the form $\eta\eta'$ in the Landau free-energy expansion (Salje, 1985; 1988).

By contrast in IC crystal structures, at least two symmetrically distinct order parameters (η and ξ) are present corresponding to two different irreducible representations. In considering the nature of the possible interaction of these two order parameters it is necessary to note first that, since the product of two different irreducible representations of a group can never give the identity representation (an invariant), we may immediately eliminate all simple product terms of the form $\eta\xi$ in the Landau free-energy expansion.

In considering how two different symmetries might interact, i.e. yield an invariant, a further possibility arises. Gradient invariants may be present (Heine and McConnell, 1981; 1984; Sannikov and Golovko, 1989), meaning that it may be possible to find a situation in which the symmetry of the gradient of one function corresponds to the symmetry of a second function. In this case it will be possible to consider terms of the form $\eta \, d\xi/dx$, where the symmetry of the representation associated with η corresponds to the symmetry of the spatial derivative of the function associated with ξ.

Where two representations can be chosen in this way, valid terms involving gradients will appear in the Landau free-energy expansion, and we may assume that there may well be an interaction between two physical events with different symmetries. Of course, we must make the rather obvious point here that gradient terms in the free-energy expansion will not normally be present in the case of special point transformations since the related functions for Ξ_{low}

must be constant throughout the crystal. Gradient terms can only arise where we consider a modulation with some finite wavelength such as we may associate with some value of Q off a special point, or where the symmetry-point representation is itself degenerate due to the presence of an essential screw or glide operation (Sannikov and Golovko, 1989).

We have now reached the point where it is possible to say with confidence that we can distinguish between normal- and IC-phase transitions. For a modulated structure in the simplest case we must define pairs of irreducible representations that are related by a gradient relationship to one another. This principle is fundamental in attempting to establish both the origin of thermodynamic stability in modulated structures, and in unravelling their structures. The practical study of gradient invariants is very straightforward, and will be explained below by reference to normal crystal structures and symmetry operations.

Before leaving this part of the analysis we may note that we have by no means exhausted the possibilities for gradient interactions in writing down the quadratic terms $\eta.d\xi/dx$ and $\xi.d\eta/dx$. It is also possible to construct third-order invariants and the associated behaviour has been described as a triple product interaction (McConnell, 1988c). The search for gradient invariants of higher order is currently an important research topic in theoretical solid-stae physics (Sannikov and Golovko, 1987).

6.4 The symmetry of modulated structures

In discussing the properties of modulated structures it is necessary to deal with three types of symmetry operation. In the first place we deal with the lattice translation symmetry, which on its own involves a pair of complex conjugate representations associated with $\pm Q$, of the form:

$$\begin{array}{cccccc} 1 & \exp(iQ.r_1) & \exp(iQ.r_2) & \exp(iQ.r_3) & \exp(iQ.r_4) & \\ 1 & \exp(-iQ.r_1) & \exp(-iQ.r_2) & \exp(-iQ.r_3) & \exp(-iQ.r_4). & (6) \end{array}$$

It will convenient to rewrite these exponential functions as real functions:

$$\begin{array}{cccccc} 2 & 2\cos Q.r_1 & 2\cos Q.r_2 & 2\cos Q.r_3 & 2\cos Q.r_4 & \\ 0 & 2i\sin Q.r_1 & 2i\sin Q.r_2 & 2i\sin Q.r_3 & 2i\sin Q.r_4. & (7) \end{array}$$

Here we add and subtract the two original representations to yield real functions which are still linearly independent, and together form the basis for an irreducible, two-dimensional representation of the space group of the high-temperature structure.

In establishing suitable modulation functions, we must now consider the symmetry of the vector Q, i.e. the group of symmetry elements that leave Q

unchanged. This small group must be a subgroup of the group associated with the relevant symmetry point. We now observe that the function, or group of functions, that we use to describe the modulation, must be invariant under the complete set of symmetry operations in the group of Q, and we consider how such functions may be defined.

We may build up functions with all the necessary symmetry properties by selecting one of the local, symmetry-point ordered structures and modulating it with one of the translational functions written out in equation (7). Since the group of Q is a subgroup of the full symmetry group at the symmetry point it follows that our new and complex set of functions must automatically satisfy the symmetry of the group of Q.

We now examine the role of the remaining symmetry elements, namely those that take Q into $-Q$. Group theory requires that the character (trace) of the matrices for each of these elements in a full two-dimensional irreducible representation must be zero since they originally relate two linearly independent functions associated respectively with Q and $-Q$. It follows from this that, in defining suitable functions associated with the modulation, suitable representations must be associated in pairs, i.e. there must be not one but two symmetry-point ordering schemes with the property that they have characters of opposite sign, i.e. ± 1, for all symmetry elements that take Q into $-Q$. Thus we must first choose and modulate *two* symmetry point structures in order to create a full two-dimensional irreducible representation associated with $\pm Q$.

We now illustrate these principles with a simple example taken from our previous study of ordering in the space group Pmm2. Here we define a symmetry point and write down the set of representations associated with it. It will be convenient to use the symmetry point 0,1/2,0 illustrated in Table 6.2, and establish Q parallel to b^*. The point group symmetry of this vector Q, at 0,1/2,0 $\pm \Delta b^*$, is the group of order 2 containing the symmetry elements E and m_x, and contains just two representations in which m_x is either even ($+1$) or odd (-1).

On returning to the full representation table for the symmetry point 0,1/2,0 provided in Table 6.2, we find that the four representations associated with the symmetry point 0,1/2,0, break into two sets of two on the basis of the representations of the little group of Q comprising E and m_x. Each pair may be used to establish a full, two-dimensional irreducible representation associated with the vectors $\pm Q$.

6.5 Gradient invariants

In order to appreciate the role of gradient invariants it is convenient to re-examine the representations of the space group Pmm2 that we associated with the vectors $\pm Q$ at 0,1/2,0 $\pm \Delta b^*$. In Figure 6.4 we compare the symmetry point structures incorporated in each of the two possible two-dimensional irreducible representations of $\pm Q$. These pairs have been labelled (a,b) and

THE STABILITY OF MODULATED STRUCTURES

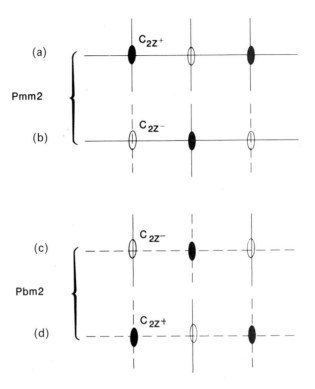

Figure 6.4 Symmetry properties of the representations listed in Table 6.2 for Q parallel to b. Note that the ordered structures of Figure 6.3 are now related to one another in pairs in the context of: (a) the symmetry elements in the group of Q, where they have the same symmetry; and (b) the symmetry elements which turn Q into $-Q$, where their symmetries are related gradientwise. Note that symmetry planes and glide planes with character $+1$ are shown with full lines and with character -1 are shown with dashed lines. Diad axes with character $+1$ are shown as full symbols and with character -1 as open symbols.

(c,d) in Figure 6.4. Notice that both members of a pair correspond with one of the representations of the little group of Q, i.e. they have the same characters for E and m_x. Next, we examine the symmetry elements from the group that takes Q into $-Q$ in each case. Here the symmetry operations for the paired symmetry-point structures have characters of opposite sign. In the case of the twofold axis C_{2z}, for example, where the structure labelled (a) has character $+1$ the associated structure labelled (b) has character -1, and a similar relationship exists for the pair of structures labelled (c) and (d). A similar situation exists for the symmetry element m_y. The character labels imply, in each case, that the paired structures are related to one another by a gradient operation for a modulation wave vector parallel to b^*. Thus, if the members of a pair of structures are assigned order parameters η and ξ, the products $\eta\, d\xi/dy$ and $\xi\, d\eta/dy$ are clearly invariants.

230

So far we have concentrated entirely on the symmetry aspects of modulated structures. We now consider the physics of the origin of modulated structures, and the conditions under which it becomes necessary to include related gradient terms in the Landau free-energy expansion. In normal circumstances the free-energy function in Q space must be governed by symmetry. Hence the free energy will normally be a minimum (extremum) at the symmetry point (where Q is equal to zero) for a symmetry-point transformation. If we do propose a modulation on this symmetry-point structure, this will normally involve an increase in energy with increasing Q as indicated in Figure 6.5. This energy increase we associate with the fact that regions of maximum gradient, and incorrect structure, will contribute extra energy in the way that boundaries normally do.

This argument fails where the second representation of the pair associated with $\pm Q$ is capable of lowering the energy of the system on incorporation at the points of maximum gradient of the initial modulation. We must now decide how this will affect the free energy of the modulation. In the first instance we note that the gradient interaction of the form $\eta\, d\xi/dx$ must be linear in Q and, to be effective, must provide a negative term in the free-energy expansion. If we assume simple quadratic behaviour of the free energy as a function of increasing Q for the modulation of the dominant symmetry-point structure, and a linear, gradient interaction term of the form $-bQ$ for interaction with a second structure, we may write for the free energy:

$$G = G_0 - bQ + aQ^2 \tag{8}$$

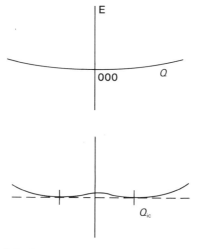

Figure 6.5 Comparison of the free energy as a functon of Q for: (a) a normal special point transformation; and (b) for an IC phase based on the same symmetry-point transformation where a gradient term has been included.

yielding a minimum of the free-energy function at $Q=b/2a$. The effect of including this additional, linear interaction term in Q is illustrated in Figure 6.5.

General study of incommensurate structures indicates that the value of Q for incommensurate structures is generally small, corresponding, therefore, to a wavelength of many unit cell repeats. Thus, in general, the gradient term $-bQ$ must be relatively small in comparison to the free energy increase likely to be associated with the modulation of the dominant stucture at wavelengths corresponding to several unit cell repeats.

The picture which we present of an IC structure implies that it is always likely to be associated with a dominant symmetry-point transformation. Close to the transition temperature the existence of a second, or servient, structure acts to reduce the free energy for some small value of Q so that the transformation takes place to a modulated structure at some temperature (T_{ic}) above that defined for the dominant phase alone at T_c.

This lowering of the free energy must be incorporated in the main Landau expansion where the free energy is lowered and the temperature of the transformation, T_c, is raised due to the presence of the small negative term associated with a gradient interaction (Sannikov and Golovko, 1989):

$$G = G_0 + a(T-T_c)\eta^2 + b\eta^4 \ldots - g(\eta\, d\xi/dx - \xi\, d\eta/dx) \qquad (9)$$

Since in many cases the increase in T_c is quite small, and of the order of a few degrees centigrade it would seem that the magnitude of the gradient term is generally quite small.

In the original discussion of gradient invariants of the kind we are using here, Lifshitz concluded that they could not lead to a phase transformation involving a change in symmetry (Lifshitz, 1942a), and a similar conclusion was tacitly inferred early in this article. This conclusion is false for the following reason. If a gradient interaction occurs between a pair of appropriate symmetry-point structures, there is a choice in how their phase relationships can be chosen. Thus in dealing with the transition components of the symmetry written out in equation (7) it is permissible to take either of the two solutions:

$$a \cos Q.r \pm ib \sin Q.r. \qquad (10)$$

Physically, these alternatives are different and involve a free-energy splitting term such that the structure may adopt with advantage one choice rather than the other. But in making this choice we must accept one of a pair of irreducible representations rather than the other, with concomitant loss of symmetry. In general, this choice will lead to a second-order incommensurate phase transformation as Landau was at pains to demonstrate in the case of symmetry-point transformations generally.

The existence of this choice of phase leads directly to a useful ploy in the study of the interaction of the component structures in modulated phases (Heine and McConnell, 1981; 1984). This has been described as the gradient ploy and involves examining the nature of a sharp anti-phase boundary in either the dominant or the servient structures. From such an analysis it is usually possible to decide on the choice of phase relationship for the two structural components of the modulation that will be most acceptable structurally and energetically.

One final feature of the relationship between dominant and servient symmetry-point structures should be noted here. In the product of their representations the final characters for all point group elements that turn Q into $-Q$ must be -1 since they are initially unequal (± 1). This implies that the product representation as a whole must transform as the vector Q (Sannikov, 1981). The same rule may be applied to pairs of irreducible representations that are degenerate (directly related by symmetry) at symmetry points on the Brillouin-zone boundary as determined by the presence of an essential screw axis or glide plane.

6.6 Examples of modulated structures

6.6.1 The incommensurate (IC) structure of quartz

In this, the final section of the chapter, we demonstrate, by means of examples, the general principles of the theory of IC or modulated structures. The first example chosen for detailed study is that of quartz which has a symmetry-point transformation from the β to the α form on cooling at 573°C. In a narrow temperature interval around this temperature, quartz develops an IC structure which has been the subject of considerable theoretical study (Axe and Shirane, 1970; Boysen et al., 1980; McConnell, 1988b). The IC phase has also been imaged, at temperature, by electron microscopy (Van Landuyt et al., 1985; Van Tendeloo et al., 1976).

In line with the general theory as developed here we first of all examine the high–low transition in quartz as a simple symmetry-point transformation and then consider the origin of the closely related IC phase. Since the transformation in quartz is a zone-centre transformaton (000), and takes place without change in the unit-cell dimensions, it is only necessary to examine the representations of the point-group 62 as set out in Table 6.3, where all the one-dimensional representations have been listed with their corresponding space groups. The first representation, R_1, which is the identity representation, corresponds to the high temperature or β structure with space group $P6_222$. The structure of this phase is illustrated in Figure 6.6. The remaining three one-dimensional representations R_2, R_3 and R_4 accord with potential phase transformations to each of the space groups $P6_2$, $P3_212$, and $P3_221$, respectively. Of these the low-temperature, α, structure for quartz has space group $P3_221$ with structure as illustrated in Figure 6.7. Note that the transformation

THE STABILITY OF MODULATED STRUCTURES

Table 6.3 Derivative space groups for vector 000 of space group $P6_222$, β quartz

	E	$2C_6$	$2C_3$	C_{2z}	$3C_{2x}$	$3C_{2y}$	
R_1	1	1	1	1	1	1	$P6_222$
R_2	1	1	1	1	−1	−1	$P6_2$
R_3	1	−1	1	−1	1	−1	$P3_212$
R_4	1	−1	1	−1	−1	1	$P3_221$
	2	1	−1	−2	0	0	
	2	−1	−1	2	0	0	

involves the loss of the diad axes parallel to a^* and c in the low-temperature structure.

The application of Landau theory to this β- to α-symmetry-point transformation is straightforward. There is no third-order invariant term in the free-energy expansion, and the anti-symmetrical square does not exist (Birman, 1966). This implies that the transformation fulfils the criteria for possible second-order behaviour. In practice we know that an incommensurate phase exists in the temperature range of the β to α transition so that the question of second-order behaviour, as implied by Landau theory, is irrelevant.

In order to establish the nature of the incommensurate transformation we now set out the details of the modulation. Diffraction data imply that the modulation wave vectors lie in the (0001) plane. The satellite intensity maxima are six in number, with the vector Q located at the zone centre point 000 and orientated parallel to a^* (indicated as the direction x in Figure 6.8).

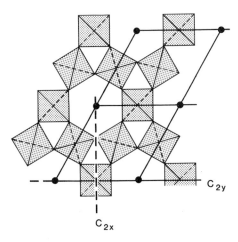

Figure 6.6 The structure of high quartz. The tetrahedra have been shaded and the positions of the diads parallel to x and y have been indicated.

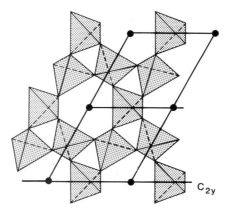

Figure 6.7 The structure of low quartz which may be compared with Figure 6.6. Note the loss of the diads parallel to x and z associated with rotation of the SiO$_4$ tetrahedra about the y axis.

Since the vector, Q, is located near to 000, it follows that we may discuss the nature of the modulation in terms of the point-group representations set out in Table 6.3. From this Table we must choose a second representation, (space group), associated with a second structure which has the correct symmetry relationship to P3$_3$21 both in respect of the symmetry of Q, and the symmetry of elements which turn Q into $-Q$.

The symmetry group of Q is the group of order two shown in Table 6.4. This group includes the symmetry elements E and C$_{2x}$ where we have chosen orthogonal axes with x parallel to a^* and z parallel to c. Assuming that low quartz (the α form) must be the dominant structure in the modulation, we note that the relevant representation in the group of Q is anti-symmetrical in C$_{2x}$ i.e. has character -1. The second structure associated with the modulation must therefore also have C$_{2x} = -1$. Returning to Table 6.3 we find that there is only one other possible space group which also has character -1 for the symmetry operation C$_{2x}$, namely P6$_2$. We deduce that the second, or servient, structure in the modulation must therefore be P6$_2$. As a further check on this conclusion we now examine the operation of symmetry elements that take Q

Table 6.4 The symmetry group of Q for the incommensurate structure of quartz

E	C$_{2x}$
1	1
1	-1

THE STABILITY OF MODULATED STRUCTURES

Table 6.5 The symmetry group for the incommensurate structure of quartz. Note that characters in the group of Q, E, C_{2x}, shown bold, are identical, and that characters in the group which turn Q into $-Q$, C_{2y}, and C_{2z}, have been underlined

	E	C_{2x}	C_{2y}	C_{2z}
$P6_222$	**1**	**1**	1	1
$P3_221$	**1**	**−1**	1	−1
$P6_2$	**1**	**−1**	−1	1
$P3_212$	**1**	**1**	−1	−1

into $-Q$ (see Table 6.5). In the case of C_{2z}, $P3_221$, and $P6_2$ have characters −1 and 1 respectively as required by the theory. The same conclusion holds for C_{2y} which also turns Q into $-Q$. Here the characters are 1 and −1 associated with $P3_221$ and $P6_2$, respectively. This set of relationships is illustrated in Figure 6.8, which has been labelled with the character data for all three structures involved.

From this brief analysis we conclude that the incommensurate structure in quartz must be based on a dominant structure (low quartz) with space group

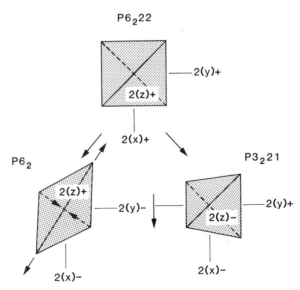

Figure 6.8 Diagrammatic representation of the data contained in Table 6.5 indicating the loss of the diad parallel to x, the modulation wave vector, in both the structures $P3_221$ and $P6_2$ in the little group of Q, and the fact that these two structures are also gradient related by the operation of the remaining diad axes C_{2y} and C_{2z}.

$P3_221$ and a second or servient structure with space group $P6_2$. It remains to demonstrate that these conclusions are physically acceptable and in the course of examining the actual structure of the phases it should also be possible to establish the correct phase relationships for the coexistence of the two low-temperature structures in the modulation. This we do in order to provide the best fit and minimum enthalpy by applying the gradient ploy discussed earlier.

The characteristics of the structure $P6_2$ have been illustrated in Figure 6.9. It is interesting to note that this structure involves some distortion of the individual tetrahedral bond lengths in addition to rigid body rotation of the tetrahedra. The correct phase relationship of this structure to the dominant structure $P3_221$ has been illustrated in Figure 6.10. It will be obvious from this diagram that there is a very clear reason for accepting the $+P6_2$ orientation rather than $-P6_2$, in other words the gradient ploy indicates clearly which of the two alternatives is correct, and is likely to provide a favourable gradient interaction term in the free energy of the modulated phase.

The theory of the incommensurate structure of quartz that has been presented here differs somewhat in its approach from current treatments in theoretical solid-state physics where emphasis is placed on establishing the phonon models at Q (Axe and Shirane, 1970; Boysen et al., 1980) that are thought to participate in the modulation. In using the theory of symmetry-point structures, one arrives directly at the specification of a complete and internally consistent suite of eigenvectors for displacements drawn from all the relevant modes. This follows from the fact that we deal directly with a second structure rather than an arbitrary set of soft modes. Clearly, we must deal with complete structures since these make a volume contribution to the effect and it is quite unsatisfactory to deal with boundaries alone. Electron optical data on contrast also indicate that there are two component structures

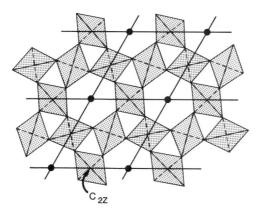

Figure 6.9 The structure of the $P6_2$ form of quartz. In this structure only the diad axis parallel to c^* (C_{2z}) is preserved.

THE STABILITY OF MODULATED STRUCTURES

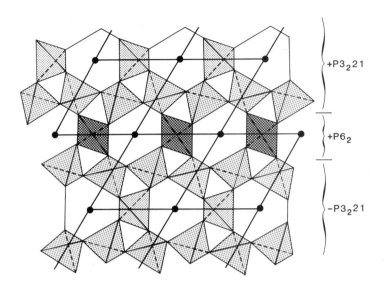

Figure 6.10 Nature of the incommensurate structure of quartz showing the presence of slabs of the structures P3$_2$21 and P6$_2$ with the correct phase relationships. In the modulated structure the change from one structure to the other is necessarily gradual.

(Van Tendeloo et al., 1976). Finally, we note that the method used here lays some considerable emphasis on the existence of tetrahedral distortion in addition to rigid body rotations of the tetrahedra and this is something which has not been included in alternative treatments of the problem.

6.6.2 The modulated structure in K (potassium) feldspar

So far we have considered an example of a modulated structure based on the existence of a quadratic invariant of the form $\eta \, d\xi/dx$ which is responsible for the appearance of simple incommensurate structures in quartz and other minerals. We turn now to an analysis of the modulated structure observed in K feldspar. This modulated structure is associated with the monoclinic to triclinic (C2/m to C$\bar{1}$) transformation in this material.

The diffraction data observed in this case comprise two sets of streaks through the Bragg maxima, the first streak in the mirror plane 010, and the second streak parallel to [010] (McConnell, 1965; 1971). The actual distribution of intensity in the streaks, when examined throughout reciprocal space, indicates that the intensity is associated with a system of transverse wave displacements with displacement vectors, ε_y and ε_x lying in the plane defined by the two systems of wave vectors Q_x and Q_y respectively. These criteria are illustrated in Figure 6.11.

This system of substantial displacements can only reasonably be interpreted

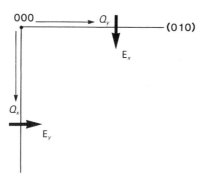

Figure 6.11 The geometry of the displacement waves, and associated amplitude vectors, for the orthogonal transverse wave system observed in potassium feldspar. The full symmetry of the displacements is given in Table 6.6, where the product representation transforms as a shear xy, or the rotation R_x. The shear xy also has the symmetry of the ordering pattern of the low-temperature structure.

if it is assumed that the resulting microstructure is dependent on the existence of local Al/Si ordering associated with the low-temperature triclinic structure for potassium feldspar. This is confirmed by studying the symmetry properties of the two orthogonal transverse waves in relation to the symmetry of the triclinic ordered phase.

The fact that the streaks of intensity are centred on the primary Bragg maxima implies that the transformation is associated with the zone centre 000 and endorses the hypothesis presented above. It is possible to write down the symmetry characteristics of the displacements ε_y and ε_x associated with the two orthogonal transverse waves, by utilizing both the symmetry elements in the groups of Q_x and Q_y, and the symmetry elements that turn Q into $-Q$ in each case. The corresponding representations are shown in Table 6.6. Since the transverse waves overlap in space, we may now consider the nature of their interaction with each other. This is achieved by taking the product of the two representations associated with ε_y and ε_x in Table 6.6. The product representation corresponds to the spontaneous shear (xy) of the ordered triclinic structure, implying that the product of all three representations, comprising both the displacements and the ordering, will yield an invariant. We may

Table 6.6 The symmetry of transverse displacements, $\varepsilon_x, \varepsilon_y$ and ordering, in 2/m

	E	C_{2y}	i	m_y		
	1	1	1	1		
	1	−1	1	−1	R_x, R_z	xy, zy
ε_x	1	−1	−1	1		x, z
ε_y	1	1	−1	−1		y

conclude from this analysis that there should be an extremely favourable interaction between the pair of displacement waves and the ordered structure of the low-temperature phase. This triple interaction yields a third-order Landau term and has been described as an order–phonon interaction (McConnell, 1988a).

A more detailed analysis of the problem indicates that there are actually two symmetrically distinct components in the resulting microstructure in K feldspar, and that both belong to the irreducible representation of the low-temperature ordered phase. The first of these corresponds to the spontaneous shear (xy) of the ordered triclinic phase, and the second corresponds to a pure rotation, (R_x), implying that a second, disordered component structure is also present. Our analysis therefore suggests that the coexistence of disordered and ordered structures leads to a favourable interaction, with concomitant lowering of the free energy of the modulated phase as found in simple IC structures such as $NaNO_2$ (Heine and McConnell, 1981; Sannikov, 1981). This dual aspect of the structure of order–phonon microstructures is illustrated in Figure 6.12.

Modulated structures of this type are observed in practice both as a purely kinetic phenomena as in K feldspars and in the mineral cordierite (Putnis, 1980a; 1980b), and also as a stable intermediate phase in certain alloy systems (Tanner *et al.*, 1982). The theory of these modulated structures has not been studied exhaustively as yet and is likely to be the subject of much more detailed study in the future.

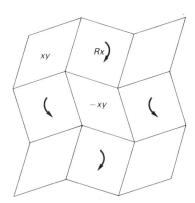

Figure 6.12 Diagram showing the presence of the two representations involving both rotation, R_x, and shear, xy, in the tweed texture observed in potassium feldspar.

References

Aslanyan, T. A. and Levanyuk, A. P. (1979) On the possibility of the incommensurate phase near α–β transition point in quartz. *Solid State Communications*, **31**, 547–50.

Axe, J. D. and Shirane, G. (1970) Study of the α–β quartz phase transformation by inelastic neutron scattering. *Physical Review*, **B1**, 342–8.

Birman, J. L., (1966) Simplified theory of symmetry change in second order phase transitions: application to V_3Si. *Physical Review Letters*, **17**, 1216–19.

Boysen, H., Dorner, B., Frey, F., *et al.* (1980) Dynamic structure determination of two interacting modes at the M point in α and β quartz by inelastic neutron scattering. *Journal of Physics C: Solid State Physics*, **13**, 6127–46.

Bradley, C. J. and Cracknell, A. P. (1972) *The Mathematical Theory of Symmetry in Crystals*, Clarendon Press, Oxford.

Carpenter, M. A., McConnell, J. D. C., and Navrotsky, A. (1985) Enthalpies of ordering in the plagioclase feldspar solid solution. *Geochimica et Cosmochimica Acta*, **49**, 946–66.

Cotton, F. A. (1971) *Chemical Applications of Group Theory*, John Wiley and Sons, New York.

Cracknell, A. P. (1968) *Applied Group Theory*, Pergamon Press, Oxford.

Heine, V. and McConnell, J. D. C. (1981) Origin of modulated incommensurate phases in insulators. *Physical Review Letters*, **46**, 1092–5.

Heine, V. and McConnell, J. D. C. (1984) The origin of incommensurate structures in insulators. *Journal of Physics C: Solid State Physics*, **17**, 1199–1220.

Landau, L. D. and Lifshitz, E. M. (1958) *Statistical Physics*, Addison Wesley, Reading, Massachusetts.

Lifshitz, E. M. (1942a) On the theory of phase transitions of second order. I. Changes in the elementary cell in phase transitions of second order. *Journal of Physics, Moscow*, **6**, 61–74.

Lifshitz, E. M. (1942b) On the theory of phase transitions of second order. II. Phase transitions of the second order in alloys. *Journal of Physics, Moscow*, **6**, 251–63.

McConnell, J. D. C. (1965) Electron optical study of effects associated with partial inversion of a silicate phase. *Philosophical Magazine*, **11**, 1289–1301.

McConnell, J. D. C. (1971) Electron optical study of phase transitions. *Mineralogical Magazine*, **38**, 1–20.

McConnell, J. D. C. (1985) Symmetry aspects of order–disorder and the application of Landau theory, in *Reviews in Mineralogy* **14**: *Microscopic to Macroscopic*, (eds S. W. Keiffer and A. Navrotsky), American Mineralogical Society, pp. 165–86.

McConnell J. D. C. (1988a) The thermodynamics of short range order, in *Physical Properties and Thermodynamic Behaviour of Minerals*, (ed. E. K. H. Salje), Reidel, Dordrecht, Boston, Lancaster, Tokyo, pp. 17–48.

McConnell, J. D. C. (1988b) The structure of incommensurate materials, in *Electron Beam Imaging*, (ed. K. Knowles), Institute of Physics, Bristol, pp. 1–18.

McConnell, J. D. C. (1988c) Symmetry aspects of pretransformation behaviour in metallic alloys. *Metallurgical Transactions. AIME*, **A19** (2), 159–67.

McConnell, J. D. C. and Heine, V. (1985) Incommensurate structure and stability of mullite. *Physics Review*, **B31**, 6140–2.

Putnis, A. (1980a) Order modulated structures and the thermodynamics of cordierite reactions. *Nature*, **287**, 128–31.

Putnis, A. (1980b) The distortion index in anhydrous Mg-cordierite. *Contributions to Mineralogy and Petrology*, **74**, 135–41.

Salje, E. K. H. (1985) Thermodynamics of sodium feldspar I: order parameter treatment and strain induced coupling effects. *Physics and Chemistry of Minerals*, **12**, 93–8.

Salje, E. K. H. (1988) Structural phase transitions and specific heat anomalies, in *Physical Properties and Thermodynamic Behaviour of Minerals*, (ed. E. K. H. Salje), Reidel, Dordrecht, Boston, Lancaster, Tokyo, pp. 75–118.

Sannikov, D. G. (1981) Thermodynamic theory of incommensurate phase transitions in the vicinity of the Lifshitz point illustrated for the case of ferroelectric $NaNO_2$. *Soviet Physics Solid State*, **23** (10), 1827–30.

Sannikov, D. G. and Golovko, V. A. (1987) New type of incommensurate phase transitions. *Ferroelectrics Letters*, **8**, 15–18.

Sannikov, D. G. and Golovko, V. A. (1989) Role of gradient invariants in the theory of incommensurate phase. *Soviet Physics Solid State*, **31** (1), 137–40.

Tanner, L. E., Pelton, A. R., and Gronsky, R. (1982) The characterization of pretransformation morphologies: periodic strain modulations. *Journal de Physique*, **C4 43**, 169–72.

Van Landuyt, J., Van Tendeloo, G., Amelinckx, S., *et al.* (1985) Interpretation of Dauphine-twin-domain configurations resulting from the α–β phase transition in quartz and aluminium phosphate. *Physics Review*, **31B**, 2986–92.

Van Tendeloo, G., Van Landuyt, J., and Amelinckx, S. (1976) The α–β phase transition in quartz and $AlPO_4$ as studied by electron microscopy and diffraction. *Physics Status Solidi* **(a) 33**, 723–35.

CHAPTER SEVEN
Thermochemistry of tetrahedrite–tennantite fahlores

Richard O. Sack

7.1 Introduction

In this chapter some of the methods used in constructing models for the thermodynamic mixing properties of multicomponent solid solutions will be illustrated using the mineral tetrahedrite–tennantite as an example. In the case of this complex sulphosalt, or the more common oxide and silicate solid solutions (e.g. spinels, rhombohedral oxides, and pyroxenes), accurate models for activity–composition relations are required to construct internally consistent thermodynamic databases for the common rock-forming minerals (cf Berman, 1988; Sack and Ghiorso, 1989; 1991a; 1991b; Ghiorso, 1990a). It is usually not possible to construct such a database from considerations of end-member properties alone, because there are insufficient, and often conflicting constraints on the standard state properties of all of the end-member components that govern exchange reactions involving solid solutions. Moreover, practical considerations require that constraints on mixing and standard-state properties be evaluated in concert, if consistent interpretations of natural phenomena are to be achieved.

The mineral tetrahedrite–tennantite provides a case in point, because there are virtually no experimental constraints on standard-state properties of its end-member components and only partial constraints on the thermodynamic properties of the simpler sulphosalts, sulphides, and alloys with which it coexists. There is the added complication that its binary substitutions (Ag for Cu; Zn for Fe; and As for Sb) may be energetically coupled, a phenomenon that is of interest in evaluating the distribution of elements in zoned, polymetallic base-metal sulphide deposits of hydrothermal-vein type. Although it has been a common practice in economic petrology to ascribe element zoning trends that are often prominent in such deposits to a host of operators involving the vanished fluid phase (e.g. boiling, dilution, cooling, pH changes associated with wall-rock alteration, etc.), such zoning may instead largely reflect non-ideal interactions between elements within mineral solutions. In this context models of tetrahedrite–tennantite thermochemistry may be useful as guides to calculation of ore reserves and to the properties of semimetal ions in hydrothermal fluids.

The Stability of Minerals. Edited by G. D. Price and N. L. Ross.
Published in 1992 by Chapman & Hall, London. ISBN 0 412 44150 0

7.2 Systematics

Tetrahedrite–tennantites are the most common representatives of minerals of the fahlore group (e.g. Spiridonov, 1984). They exhibit the most extensive solid solution and widespread occurrence of all sulphosalts (complex sulphides containing semimetal + sulphur pyramids). Despite the considerable flexibility in stoichiometry that is evident in some synthetic varieties in chemical systems of few components (e.g. Tatsuka and Morimoto, 1977; Makovicky and Skinner, 1978; 1979), natural tetrahedrite–tennantites typically closely approximate the simple formula $(Cu, Ag)_{10}(Fe, Zn)_2(Sb, As)_4S_{13}$, with trace to minor amounts of Mn, Cd, and Hg substituting for Fe and Zn, and Bi substituting for As and Sb (e.g. Springer, 1969; Charlat and Levy, 1974; Pattrick, 1978; Johnson and Jeanloz, 1983; Johnson et al., 1986). The names tetrahedrite and tennantite are usually reserved for varieties with $Sb/(Sb+As)$ ratios near unity and zero, respectively.

In tetrahedrite–tennantites the metals are equally distributed between two sites that are three- and fourfold coordinated by sulphurs in a sphalerite derivative structure of space group $I\bar{4}3m$, with two formula units per unit cell (e.g. Pauling and Neuman, 1934; Wuensch, 1964; Wuensch et al., 1966; Kalbskopf, 1972; Johnson and Burnham, 1985; Peterson and Miller, 1986). In this derivative structure, semimetals replace Zn atoms of the sphalerite structure at the midpoints of the half-diagonals of the unit cell ($\frac{1}{4}, \frac{1}{4}, \frac{1}{4}$, etc.). Sulphur atoms are added at its corners and centre, and eight sulphur atoms are removed at $\frac{1}{8}, \frac{1}{8}, \frac{1}{8}$, etc. The results of most structural studies are consistent with the inference that the divalent metals occupy one-third of the tetrahedral metal sites, although Peterson and Miller (1986) conclude that Fe^{2+} occupies trigonal–planar metal sites in a natural tetrahedrite with an $Ag/(Ag+Cu)$ ratio of about 0.4. Cu and Ag occupy both the remaining two-thirds of the tetrahedral metal sites and the trigonal-planar metal sites. The sulphur atoms at the corners and centre of the unit cell are octahedrally coordinated by the metals in these latter sites, and, the lone-pairs of the unbounded valence electrons of the semi-metals extend towards the sites of the sulphur vacancies (cf. Wuensch, 1964).

Spectroscopic studies support the inference that Ag enters trigonal sites in preference to tetrahedral metal sites, at least in iron-rich tetrahedrite–tennantites with $Ag/(Ag+Cu)$ ratios less than 0.4 (e.g. Kalbskopf, 1972; Johnson and Burnham, 1985; Peterson and Miller, 1986; Charnock et al., 1988; 1989). However, the structural role of Ag and its thermal dependence is less well understood both in synthetic tetrahedrite–tennantites and natural tetrahedrites with more Ag and/or Zn. Tetrahedrites that are fully Ag substituted have the same basic I-type cell structure (Criddle, pers. comm.), and thus must have substantial tetrahedral Ag. Because synthetic argentian tetrahedrites have unit-cell volumes that are roughly consistent with the linear trend defined by fully Cu- and Ag-substituted tetrahedrites, it is possible that Ag and Cu have fairly disordered distributions between trigonal and tetrahedral sites at high temperatures (e.g. Pattrick and Hall, 1983). In contrast, natural iron-rich tetrahedrites, presumably with low-temperature, strongly ordered cation distributions, display a pronounced local maximum in unit-cell volume at an Ag/(Ag+Cu) ratio of about 0.35 (Indolev et al., 1971; Riley, 1974). The sigmoid volume–composition systematics exhibited by these tetrahedrites may indicate that Ag undergoes a change in site preference with increasing substitution of Ag for Cu (e.g. O'Leary and Sack, 1987; Sack et al., 1987) analogous to that exhibited by Fe^{3+} in Fe_3O_4–$FeCr_2O_4$ solid solutions (e.g. Robbins et al., 1971). Alternatively, the collapsed unit cell exhibited by natural (low-temperature) tetrahedrites with intermediate Ag/(Ag+Cu) ratios could be the result of polyhedral rotations induced by the stretching of metal-sulphur bonds in trigonal sites as a consequence of increasing substitution of Ag for Cu. Stretching these bonds or substituting As for Sb causes tetrahedral rotations that bring the semimetals closer to the trigonal sites, the extent of this approach being limited by Van der Waals forces between semimetals and octahedrally coordinated sulphurs (cf Wuensch et al., 1966; Johnson and Burnham, 1985; Peterson and Miller, 1986). Assuming that Ag occupies only trigonal sites, substitution of Ag for Cu beyond Ag/(Ag+Cu) ratios of about 0.35 could conceivably lead to contraction in the unit cell, if these Van der Waals forces are overcome (cf Charnock et al., 1989).

One of the principal occurrences of tetrahedrite–tennantite is in epithermal–mesothermal, polymetallic base–metal sulphide deposits of fissure-vein type. In such veins tetrahedrite–tennantite varies from a minor constituent to the dominant phase, as in the Coeur d'Alene district in Idaho (e.g. Fryklund, 1964). Tetrahedrite–tennantites in these deposits are typically zoned in Ag–Cu, Zn–Fe, and As–Sb ratios in both time and space (e.g. Goodell and Petersen, 1974; Wu and Petersen, 1977). The marked compositional discontinuities recorded in many crystals in such deposits are consistent with the interpretation that initial compositions are commonly retained, at least at the lower temperatures of epithermal mineralization, 200–300°C (e.g. Yui, 1971; Wu and Petersen, 1977; Raabe and Sack, 1984). Tetrahedrite–tennantites deposited first are typically richer in Ag and Sb than those deposited in the centre of such veins at the end of the hydrothermal cycle. Ag and Sb are also enriched along fluid flow paths distal to the hydrothermal source of the mineralizing fluids. Along such paths they typically exhibit positive correlations between Sb/(Sb+As) and Ag/(Ag+Cu) ratios that are strongly concave with respect to the Ag/(Ag+Cu) axis (e.g. Wu and Petersen, 1977; Hackbarth and Petersen, 1984). Mass-balance considerations require that, along such paths, there is a strong composition dependence to the distribution coefficients for the exchange of As and Sb, and Ag and Cu between tetrahedrite–tennantites and hydrothermal fluids (e.g. Hackbarth and Petersen, 1984). The trajectory of these paths requires that the As/(As+Sb) and Ag/(Ag+Cu) ratios of tetrahedrite–tennantites are respectively larger and smaller than those of hydrothermal fluids nearest to the hydrothermal source, but these differences diminish along the direction of flow of the hydrothermal fluid.

It is difficult to understand the strong composition dependence inferred for these distribution coefficients along flow paths distal to hydrothermal sources unless As and Ag are strongly incompatible in the tetrahedrite structure. Such an incompatibility would require that the reciprocal reaction between end-member tetrahedrite–tennantite components

$$Cu_{10}Fe_2As_4S_{13} + Ag_{10}Fe_2Sb_4S_{13} = Cu_{10}Fe_2Sb_4S_{13} + Ag_{10}Fe_2As_4S_{13} \quad (1)$$

has a large positive Gibbs energy. I will summarize some of the observations that require that the Gibbs energies of this reaction and the other two reciprocal reactions characteristic of $(Cu, Ag)_{10}(Fe, Zn)_2(Sb, As)_4S_{13}$ tetrahedrite–tennantites

$$Cu_{10}Zn_2Sb_4S_{13} + Cu_{10}Fe_2As_4S_{13} = Cu_{10}Fe_2Sb_4S_{13} + Cu_{10}Zn_2As_4S_{13} \quad (2)$$

and

$$Cu_{10}Zn_2Sb_4S_{13} + Ag_{10}Fe_2Sb_4S_{13} = Cu_{10}Fe_2Sb_4S_{13} + Ag_{10}Zn_2Sb_4S_{13} \quad (3)$$

are each positive. Values for the energies of these reactions may be established from constraints on the Ag–Cu, Zn–Fe, and As–Sb exchange reactions between tetrahedrite–tennantites and assemblages consisting of simpler sulphosalts, sulphides, and alloys (cf Raabe and Sack, 1984; Sack and Loucks, 1985; O'Leary and Sack, 1987; Sack et al., 1987; Ebel and Sack, 1989). Results of these studies may be used to constrain the shape of the Gibbs energy surface of natural tetrahedrite–tennantites in the context of a simple formulation for thermodynamic properties (cf Sack et al., 1987).

7.3 Thermodynamics

7.3.1 Ag-poor tetrahedrite–tennantites

The model for the thermodynamic properties of tetrahedrite–tennantites may be developed starting with tetrahedrite–tennantites in the Ag-free subsystem which approximate the formula $Cu_{10}(Fe, Zn)_2(Sb, As)_4S_{13}$. Only two composition variables

$$X_2 = Zn/(Zn + Fe)$$

and

$$X_3 = As/(As + Sb) \quad (4)$$

are sufficient to specify the thermodynamic state of such tetrahedrite–tennantites at a given temperature and pressure. The molar Gibbs energy may be defined by the equation

$$\bar{G} = \bar{G}^* - T\bar{S}^{\text{IC}} \tag{5}$$

where $\bar{G}^*$ is the molar *vibrational* Gibbs energy and $\bar{S}^{\text{IC}}$ is the molar configurational entropy. Assuming that the metals and semimetals are randomly distributed on sites, an expression for the molar configurational entropy is readily deduced by substituting the definitions of site metal and semimetal fractions in terms of composition variables into the relation

$$\bar{S}^{\text{IC}} = -R \sum_r \sum_a \bar{r} X_{a,r} \ln X_{a,r} \tag{6}$$

where $\bar{r}$ is the number of sites of type r per formula unit, $X_{a,r}$ is the atom fraction of a on site r, and R is the universal gas constant. For Ag-free tetrahedrite–tennantites $\{\approx(\text{Cu})_6^{\text{III}}(\text{Cu}_{2/3},[\text{Fe},\text{Zn}]_{1/3})_6^{\text{IV}}(\text{Sb},\text{As})_4^{\text{SM}} S_{13}\}$ the relevant relations between fractions of atoms on sites and composition variables are

$$X_{\text{Cu}}^{\text{TET}} = \tfrac{2}{3},$$

$$X_{\text{Fe}}^{\text{TET}} = \tfrac{1}{3}(1 - X_2),$$

$$X_{\text{Zn}}^{\text{TET}} = \tfrac{1}{3}X_2,$$

$$X_{\text{As}}^{\text{SM}} = X_3,$$

and
$$X_{\text{Sb}}^{\text{SM}} = (1 - X_3). \tag{7}$$

When these relations are substituted into equation (6), the following expression is obtained for the molar configurational entropy:

$$\bar{S}^{\text{IC}} = -R(2X_2 \ln([X_2]/3) + 2(1 - X_2)\ln([1 - X_2]/3))$$
$$- R(4\ln(2/3) + 4X_3 \ln X_3 + 4(1 - X_3)\ln(1 - X_3)). \tag{8}$$

In a first approximation, we may consider that the *vibrational* Gibbs energy of such tetrahedrite–tennantites may be satisfactorily described with a Taylor expansion of only second degree in these variables

$$\bar{G}^* = g_0 + g_{x_2}(X_2) + g_{x_3}(X_3) + g_{x_2 x_2}(X_2)^2 + g_{x_3 x_3}(X_3)^2 + g_{x_2 x_3}(X_2)(X_3). \tag{9}$$

A second-degree dependence of the vibrational Gibbs energy on composition variables is a minimum requirement for a thermodynamic model that makes explicit provision for non-ideality in the substitution of the metals and semimetals on the different crystallographic sites, and for a non-zero energy of the reciprocal reaction expressing the incompatibility of Zn and As in tetrahedrite–tennantite (equation (2)). Coefficients of the Taylor expansion correspond to three types of thermodynamic parameters which may be identified by setting the composition variables to the values they assume in end-member and in binary tetrahedrite–tennantites (cf Thompson, 1969; Sack, 1982). These are:

1. vibrational Gibbs energies of the end-member components $Cu_{10}Fe_2Sb_4S_{13}(\bar{G}_1^*)$, $Cu_{10}Zn_2Sb_4S_{13}(\bar{G}_2^*)$, and $Cu_{10}Fe_2As_4S_{13}(\bar{G}_3^*)$;
2. regular-solution-type parameters which describe the non-ideality associated with the substitution of Zn for Fe in $Cu_{10}(Fe, Zn)_2Sb_4S_{13}$ and $Cu_{10}(Fe, Zn)_2As_4S_{13}$ (W_{FeZn}^{TET}) and for the substitution of As for Sb in $Cu_{10}Fe_2(Sb, As)_4S_{13}$ and $Cu_{10}Zn_2(Sb, As)_4S_{13}(W_{AsSb}^{SM})$; and
3. the Gibbs energy of the reciprocal reaction for Ag-free tetrahedrite–tennantites ($\Delta\bar{G}_{23}^0$, equation (2)).

The procedure for identifying the coefficients of the Taylor expansion with the thermodynamic parameters is readily illustrated by considering the binary join $Cu_{10}Fe_2Sb_4S_{13}$–$Cu_{10}Fe_2As_4S_{13}$. The vibrational Gibbs energy of tetrahedrite–tennantites on this join is given by the expression

$$\bar{G}^* = g_0 + g_{x_3}(X_3) + g_{x_3x_3}(X_3)^2. \tag{10}$$

Taking into account the definitions for the end-members on this join

$$\bar{G}^*_{Cu_{10}Fe_2Sb_4S_{13}} = \bar{G}_1^* = g_0$$

and

$$\bar{G}^*_{Cu_{10}Fe_2As_4S_{13}} = \bar{G}_3^* = g_0 + g_{x_3} + g_{x_3x_3},$$

equation (10) may be rewritten as

$$\bar{G}^* = \bar{G}^*_{Cu_{10}Fe_2Sb_4S_{13}}(1-X_3) + \bar{G}^*_{Cu_{10}Fe_2As_4S_{13}}(X_3) - g_{x_3x_3}(1-X_3)(X_3). \tag{11}$$

It may be readily verified that the term $-g_{x_3x_3}$ is equivalent to a regular-solution-type parameter for this binary by comparing the form of equation (11) with that for a symmetric, binary regular solution (e.g. Thompson, 1967). Taking into account these and the remaining relationships between Taylor expansion coefficients and thermodynamic parameters (Table 7.1), we may write the following expression or the vibrational Gibbs energy of Ag-free tetrahedrite–tennantites:

$$\bar{G}^* = \bar{G}^*_{Cu_{10}Fe_2Sb_4S_{13}}(1-X_2-X_3) + \bar{G}^*_{Cu_{10}Zn_2Sb_4S_{13}}(X_2)$$
$$+ \bar{G}^*_{Cu_{10}Fe_2As_4S_{13}}(X_3) + \Delta\bar{G}_{23}^0(X_2)(X_3)$$
$$+ W_{FeZn}^{TET}(1-X_2)(X_2) + W_{AsSb}^{SM}(1-X_3)(X_3). \tag{12}$$

Table 7.1 Thermodynamic parameters for Ag-free tetrahedrite–tennantites, equivalent Taylor coefficients, and estimated values (kcal/gfw)

Parameter	Equivalent Taylor coeff.	Value (kcal/gfw)
$\bar{G}_1^* = \bar{G}_{Cu_{10}Fe_2Sb_4S_{13}}^*$	g_0	*
$\bar{G}_2^* = \bar{G}_{Cu_{10}Zn_2Sb_4S_{13}}^*$	$g_0 + g_{x_2} + g_{x_2x_2}$	*
$\bar{G}_3^* = \bar{G}_{Cu_{10}Fe_2As_4S_{13}}^*$	$g_0 + g_{x_3} + g_{x_3x_3}$	*
W_{FeZn}^{TET}	$-g_{x_2x_2}$	0.0 ± 0.2
W_{AsSb}^{SM}	$-g_{x_3x_3}$	4.0 ± 0.6
$\Delta \bar{G}_{23}^0$	$g_{x_2x_3}$	2.59 ± 0.14

*No estimates given for end-member components.

Chemical potentials of end-member components are evaluated by application of an extended form of the Darken equation (tangent intercept rule)

$$\mu_j = \bar{G} + n_1(1-X_1)(\partial \bar{G}/\partial X_1)_{X_2/X_3} + n_2(1-X_2)(\partial \bar{G}/\partial X_2)_{X_1/X_3}$$
$$+ n_3(1-X_3)(\partial \bar{G}/\partial X_3)_{X_1/X_2} \tag{13}$$

where the $n_{ij}(n_1, n_2,$ and $n_3)$ are the numbers of moles of the linearly independent components $Cu_{10}Fe_2Sb_4S_{13}$, $Cu_{10}Zn_2Sb_4S_{13}$, and $Cu_{10}Fe_2As_4S_{13}$ in one mole of component j (e.g. $n_1 = -1$, $n_2 = +1$, and $n_3 = +1$ in the j component $Cu_{10}Zn_2As_4S_{13}$) and X_1 is the dependent mole fraction given by the relation

$$X_1 = 1 - X_2 - X_3.$$

This relation may be written in the more general form as

$$\mu_j = \bar{G} + \sum_i n_{ij}(1-X_i)(\partial \bar{G}/\partial X_i)_{X_k/X_l}(k \neq l \neq i) \tag{14}$$

(cf. Sack, 1982; Ghiorso, 1990b).

Several of the parameters of this formulation may be evaluated from the results of Fe–Zn exchange experiments between sphalerites and Ag-free tetrahedrites (Sack and Loucks, 1985) and studies of ore deposits bearing the sphalerite + Ag-poor tetrahedrite–tennantite subassemblage (Raabe and Sack, 1984; O'Leary and Sack, 1987). The condition of equilibrium for the Fe–Zn exchange reaction between sphalerites and Ag-free tetrahedrite–tennantites

$$\begin{array}{ccccc} ZnS & + \tfrac{1}{2}Cu_{10}Fe_2Sb_4S_{13} & = FeS & + \tfrac{1}{2}Cu_{10}Zn_2Sb_4S_{13}, \\ SPH & TD\text{-}TN & SPH & TD\text{-}TN \end{array} \tag{15}$$

$$0 = (\partial \bar{G}^{SPH}/\partial X_{FeS}^{SPH}) + \tfrac{1}{2}(\partial \bar{G}^{TD\text{-}TN}/\partial X_2), \tag{16}$$

may be expressed as

$$\bar{Q}_1^* \equiv RT \ln\left[\left(\frac{X_{Fe}^{TET}}{X_{Zn}^{TET}}\right)^{TD-TN}\left(\frac{X_{ZnS}^{SPH}}{X_{FeS}^{SPH}}\right)\right] + RT \ln\left(\frac{\gamma_{ZnS}^{SPH}}{\gamma_{FeS}^{SPH}}\right)$$
$$= \Delta\bar{G}_{Zn(Fe)_{-1}}^0 + \tfrac{1}{2}\Delta\bar{G}_{23}^0(X_3) + \tfrac{1}{2}W_{FeZn}^{TET}(1-2X_2) \quad (17)$$

where $\Delta\bar{G}_{Zn(Fe)_{-1}}^0$ is the standard state Gibbs energy of the Fe–Zn exchange reaction and the activity coefficient ratio ($\gamma_{ZnS}^{SPH}/\gamma_{FeS}^{SPH}$) may be defined from the data of Barton and Toulmin (1966). Sack and Loucks (1985) have demonstrated that Fe and Zn mix ideally in the tetrahedrite–tennantite structure (i.e. $W_{FeZn}^{TET} = 0$), because the correction for the non-ideality associated with the substitution of Fe for Zn in sphalerite, $RT \ln(\gamma_{ZnS}^{SPH}/\gamma_{FeS}^{SPH})$, removes any dependence of $\bar{Q}_1^*$ on FeS content in sphalerites in the 500°C isotherms for Fe–Zn exchange between sphalerites and tetrahedrite–tennantites with As/(As+Sb) ratios of 0.00, 0.65, and 1.00. Because sphalerite + tetrahedrite–tennantite assemblages from the experimental and petrological studies define similar values of $\bar{Q}_1^*$, and these values of $\bar{Q}_1^*$ are a linear function of X_3 (Fig. 7.1), it is concluded that:

1. the formulation is both necessary and sufficient; and
2. the entropies of the exchange and reciprocal reactions are negligible.

The remaining parameter pertinent to the description of the thermodynamic mixing properties of $Cu_{10}(Fe, Zn)_2(Sb, As)_4 S_{13}$ tetrahedrite–tennantites, W_{AsSb}^{SM}, is only loosely constrained by available data. W_{AsSb}^{SM} must be less than about 7.5 kcal/gfw to be consistent with the observation that Ag-poor tetrahedrite–tennantites span the entire range of As/(As+Sb) ratios in epithermal ore deposits ($T_C = W_{AsSb}^{SM}/8RT \leqslant 200°C$, e.g. Springer, 1969; Raabe and Sack, 1984; Johnson et al., 1986). Ideal mixing of As and Sb ($W_{AsSb}^{SM} = 0$) may be ruled out based on size considerations and petrological constraints on the partitioning of As and Sb between Ag-poor tetrahedrite–tennantite and bournonite [CuPb(As, Sb)S_3] solid solutions in the Rajpura–Dariba polymetallic deposits, Poona, India (Mishra and Mookherjee, 1986). The data of Mishra and Mookherjee (1986) require that the substitution of As for Sb is substantially non-ideal. Decreases in X_3 that are associated with increases in X_2 for similar values of As/(As+Sb) in coexisting bournonites are greater than those than can be attributed to the incompatibility between As and Zn in tetrahedrite–tennantite (equation (2)). Values for W_{AsSb}^{SM} near this upper bound are suggested from analysis of the condition of As–Sb exchange equilibrium for these phases,

$$0 = (\partial\bar{G}^{BRN}/\partial X_{CuPbSbS_3}^{BRN}) + \tfrac{1}{4}(\partial\bar{G}^{TD-TN}/\partial X_3), \quad (18)$$

for various assumed temperatures of equilibration (cf Ebel and Sack, 1989), but preliminary results of As–Sb exchange experiments between them (400°C) permit values of W_{AsSb}^{SM} only slightly greater than one half of it (Table 7.1).

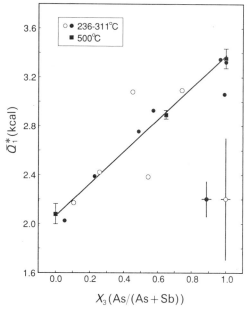

Figure 7.1 Calibration of equation (17) compared with experimental and natural constraints for Ag-poor tetrahedrite–tennantites. Diagonal line represents the calibration of equation (16) for $\Delta \bar{G}^0_{Zn(Fe)_{-1}} = 2.0718$, $\Delta \bar{G}^0_{23} = 2.59$, and $W^{TET}_{FeZn} = 0.0$ kcal/gfw. Solid squares represent 500°C reversal brackets of Sack and Loucks (1985) for sphalerites with $X^{SPH}_{FeS} > 0.04$ calculated utilizing the following expression for $\ln(\gamma^{SPH}_{ZnS}/\gamma^{SPH}_{FeS})$:

$$\ln\left\{\frac{\gamma^{SPH}_{ZnS}}{\gamma^{SPH}_{FeS}}\right\} = \{0.7285 - 0.9186 X_{FeS} - 0.5295 X^2_{FeS} - 0.1772 X^3_{FeS}\}(1 - 2X^{SPH}_{FeS})$$

(cf Sack and Loucks, 1985.) Attached error bars indicate the 1σ errors associated with averaging these brackets. Open and solid circles are constraints determined by O'Leary and Sack (1987) for natural sphalerite + Ag-poor tetrahedrite–tennantite assemblages with sphalerites with $X^{SPH}_{FeS} > 0.006$ (solid) and $X^{SPH}_{FeS} < 0.006$ (open). A correction of $-2.5(X_4)$ kcal/gfw $\{\approx \{-\frac{1}{2}\Delta \bar{G}^0_{24}(X_4) - \frac{1}{2}\Delta \bar{G}^*_{2s}(s)\}$ for Ag-poor compositions} is applied to $\bar{Q}^*_1$ to correct for minor amounts of Ag ($X_4 < 0.15$) (see Figure 7.2). Representative 1σ error bars for these estimates are given in the lower right-hand corner.

7.3.2 Argentian tetrahedrite–tennantites

The analysis of thermodynamic mixing properties may be readily extended to argentian tetrahedrite–tennantites. To do so it is necessary to consider one new composition variable,

$$X_4 \equiv Ag/(Ag + Cu), \tag{19}$$

and at least one order variable in the expressions for the Gibbs energy. In the absence of evidence for any change in space group in argentian tetrahedrite–

tennantites at the temperatures of interest (i.e. all trigonal and tetrahedral metal sites are crystallographically equivalent), it is assumed that only one such ordering variable is sufficient to describe the average distribution of Ag and Cu between trigonal and tetrahedral metal sites,

$$s \equiv X_{Ag}^{TRG} - \tfrac{3}{2} X_{Ag}^{TET}. \tag{20}$$

For the choice of composition and ordering variables given by equations (4), (19), and (20), we may write the following relations between fractions of atoms on sites and these variables:

$$X_{Fe}^{TET} = \tfrac{1}{3}(1 - X_2),$$
$$X_{Zn}^{TET} = \tfrac{1}{3} X_2,$$
$$X_{As}^{SM} = X_3,$$
$$X_{Sb}^{SM} = (1 - X_3),$$
$$X_{Cu}^{TRG} = (1 - X_4) - \tfrac{2}{5}s,$$
$$X_{Ag}^{TRG} = X_4 + \tfrac{2}{5}s,$$
$$X_{Cu}^{TET} = \tfrac{2}{3}(1 - X_4) + \tfrac{2}{5}s,$$

and
$$X_{Ag}^{TET} = \tfrac{2}{3} X_4 - \tfrac{2}{5}s. \tag{21}$$

Following the same procedures already outlined we may equate thermodynamic parameters with coefficients of a Taylor expansion of second degree in the variables X_2, X_3, X_4, and s (Table 7.2) to develop the following expression for the vibrational Gibbs energy:

$$\begin{aligned}
\bar{G}^* &= \bar{G}^*_{Cu_{10}Fe_2Sb_4S_{13}}(X_1) + \bar{G}^*_{Cu_{10}Zn_2Sb_4S_{13}}(X_2) + \bar{G}^*_{Cu_{10}Fe_2As_4S_{13}}(X_3) \\
&+ \bar{G}^*_{Ag_{10}Fe_2Sb_4S_{13}}(X_4) + \Delta \bar{G}^*_s(s) + \Delta \bar{G}^0_{23}(X_2)(X_3) + \Delta \bar{G}^0_{24}(X_2)(X_4) \\
&+ \Delta \bar{G}^0_{34}(X_3)(X_4) \\
&+ \Delta \bar{G}^*_{2s}(X_2)(s) + \Delta \bar{G}^*_{3s}(X_3)(s) \\
&+ \tfrac{1}{10}(\Delta \bar{G}^*_{4s} + 6W_{AgCu}^{TET} - 4W_{AgCu}^{TRG})(2X_4 - 1)(s) \\
&+ W_{FeZn}^{TET}(1 - X_2)(X_2) + W_{AsSb}^{SM}(1 - X_3)(X_3) \\
&+ (\Delta \bar{G}^*_{4s} + W_{AgCu}^{TRG} + W_{AgCu}^{TET})(X_4)(1 - X_4) \\
&+ \tfrac{1}{25}(6\Delta \bar{G}^*_{4s} - 9W_{AgCu}^{TET} - 4W_{AgCu}^{TRG})(s)^2
\end{aligned} \tag{22}$$

The new parameters appearing in this relation correspond to:

1. the vibrational Gibbs energy of $Ag_{10}Fe_2Sb_4S_{13}$;

Table 7.2 Additional thermodynamic parameters for argentian tetrahedrite–tennantite, equivalent Taylor coefficients, and estimated values

Parameter	Equivalent Taylor coeff.	Value (kcal/gfw)
$\bar{G}_4^* = \bar{G}^*_{Ag_{10}Fe_2Sb_4S_{13}}$	$g_0 + g_{x_4} + g_{x_4x_4}$	*
$\Delta \bar{G}_s^{*\P}$	$+g_s + \frac{1}{2}g_{x_4s}$	-0.40
W_{AgCu}^{TRG}	$-\frac{9}{25}g_{x_4x_4} - g_{ss} - \frac{3}{5}g_{x_4s}$	0.00
W_{AgCu}^{TET}	$-\frac{4}{25}g_{x_4x_4} - g_{ss} - \frac{2}{5}g_{x_4s}$	6.93
$\Delta \bar{G}_{24}^0$	$g_{x_2x_4}$	2.30
$\Delta \bar{G}_{34}^0 \dagger$	$g_{x_3x_4}$	12.25 ± 1.92
$\Delta \bar{G}_{2s}^* \P$	g_{x_2s}	2.60
$\Delta \bar{G}_{3s}^* \P$	g_{x_3s}	0.00
$\Delta \bar{G}_{4s}^* \P$	$-\frac{12}{25}g_{x_4x_4} + 2g_{ss} + \frac{1}{5}g_{x_4s}$	-2.60

*No estimates given for end-member components.
†Uncertainties stated only for estimates derived from experimental data.
¶Gibbs energies of the reciprocal ordering reactions defined in Table 7.3.

2. regular-solution-type parameters which describe the non-ideality associated with the substitution of Ag for Cu in trigonal metal sites in $(Cu, Ag)_6^{III}(Cu_{2/3}, [Fe, Zn]_{1/3})_6^{IV}(Sb, As)_4S_{13}$ and $(Cu, Ag)_6^{III}(Ag_{2/3}, [Fe, Zn]_{1/3})_6^{IV}(Sb, As)_4S_{13}$ tetrahedrite–tennantites (W_{AgCu}^{TRG}), and in tetrahedral metal sites in $Cu_6^{III}([Cu, Ag]_{2/3}, [Fe, Zn]_{1/3})_6^{IV}(Sb, As)_4S_{13}$ and $Ag_6^{III}([Cu]_{2/3}, [Fe, Zn]_{1/3})_6^{IV}(Sb, As)_4S_{13}$ tetrahedrite–tennantites (W_{AgCu}^{TET});
3. the standard state Gibbs energy of the reciprocal reactions given by reactions (1) and (3) ($\Delta \bar{G}_{34}^0$ and $\Delta \bar{G}_{24}^0$, respectively); and
4. the vibrational Gibbs energies of the ordering and reciprocal ordering reactions given in Table 7.3.

Taking into account the additional relations between fractions of atoms on

Table 7.3 Definitions of reciprocal ordering reactions

$\frac{1}{2}(Cu)_6^{III}(Ag_{2/3}, Fe_{1/3})_6^{IV}Sb_4S_{13} + \frac{1}{10}Ag_{10}Fe_2Sb_4S_{13} \Leftrightarrow \frac{1}{2}(Ag)_6^{III}(Cu_{2/3}, Fe_{1/3})_6^{IV}Sb_4S_{13}$ $+ \frac{1}{10}Cu_{10}Fe_2Sb_4S_{13}$	$[\Delta \bar{G}_s^*]$
$\frac{1}{2}(Cu)_6^{III}(Ag_{2/3}, Zn_{1/3})_6^{IV}Sb_4S_{13} + \frac{1}{10}Cu_{10}Fe_2Sb_4S_{13} + \frac{1}{2}(Ag)_6^{III}(Cu_{2/3}, Fe_{1/3})_6^{IV}Sb_4S_{13}$ $+ \frac{1}{10}Ag_{10}Zn_2Sb_4S_{13} \Leftrightarrow \frac{1}{2}(Ag)_6^{III}(Cu_{2/3}, Zn_{1/3})_6^{IV}Sb_4S_{13} + \frac{1}{10}Ag_{10}Fe_2Sb_4S_{13}$ $+ \frac{1}{2}(Cu)_6^{III}(Ag_{2/3}, Fe_{1/3})_6^{IV}Sb_4S_{13} + \frac{1}{10}Cu_{10}Zn_2Sb_4S_{13}$	$[\Delta \bar{G}_{2s}^*]$
$\frac{1}{2}(Cu)_6^{III}(Ag_{2/3}, Fe_{1/3})_6^{IV}As_4S_{13} + \frac{1}{10}Cu_{10}Fe_2Sb_4S_{13} + \frac{1}{2}(Ag)_6^{III}(Cu_{2/3}, Fe_{1/3})_6^{IV}Sb_4S_{13}$ $+ \frac{1}{10}Ag_{10}Fe_2As_4S_{13} \Leftrightarrow \frac{1}{2}(Ag)_6^{III}(Cu_{2/3}, Fe_{1/3})_6^{IV}As_4S_{13} + \frac{1}{10}Ag_{10}Fe_2Sb_4S_{13}$ $+ \frac{1}{2}(Cu)_6^{III}(Ag_{2/3}, Fe_{1/3})_6^{IV}Sb_4S_{13} + \frac{1}{10}Cu_{10}Fe_2As_4S_{13}$	$[\Delta \bar{G}_{3s}^*]$
$Cu_{10}Fe_2Sb_4S_{13} + Ag_{10}Fe_2Sb_4S_{13} \Leftrightarrow (Ag)_6^{III}(Cu_{2/3}, Fe_{1/3})_6^{IV}Sb_4S_{13}$ $+ (Cu)_6^{III}(Ag_{2/3}, Fe_{1/3})_6^{IV}Sb_4S_{13}$	$[\Delta \bar{G}_{4s}^*]$

sites and thermodynamic variables (equation (2.1)), we may readily derive the following relation for the molar configurational Gibbs energy:

$$
\begin{aligned}
-T\bar{S}^{\mathrm{IC}} =\ & +RT(2X_2\ln(X_2/3)+2(1-X_2)\ln([1-X_2]/3)+4X_3\ln X_3) \\
& +RT(4(1-X_3)\ln(1-X_3)+6(1-X_4-\tfrac{2}{5}s)\ln(1-X_4-\tfrac{2}{5}s)) \\
& +RT(6(X_4+\tfrac{2}{5}s)\ln(X_4+\tfrac{2}{5}s)+6(\tfrac{2}{3}X_4-\tfrac{2}{5}s)\ln(\tfrac{2}{3}X_4-\tfrac{2}{5}s) \\
& +RT(6(\tfrac{2}{3}[1-X_4]+\tfrac{2}{5}s)\ln(\tfrac{2}{3}[1-X_4]+\tfrac{2}{5}s)).
\end{aligned} \quad (23)
$$

Utilizing equation (5) to combine equations (22) and (23), the order variable s is evaluated by minimizing the molar Gibbs energy with respect to it:

$$
\begin{aligned}
(\partial\bar{G}/\partial s)_{T,X_2,X_3,X_4} &= 0 \\
&= RT\ln\left(\frac{(X_4+\tfrac{2}{5}s)(\tfrac{2}{3}[1-X_4]+\tfrac{2}{5}s)}{(1-X_4-\tfrac{2}{5}s)(\tfrac{2}{3}X_4-\tfrac{2}{5}s)}\right)+\tfrac{5}{12}(\Delta\bar{G}_s^*+\Delta\bar{G}_{2s}^*(X_2)+\Delta\bar{G}_{3s}^*(X_3)) \\
&\quad +\tfrac{1}{24}(\Delta\bar{G}_{4s}^*+6W_{\mathrm{AgCu}}^{\mathrm{TET}}-4W_{\mathrm{AgCu}}^{\mathrm{TRG}})(2X_4-1) \\
&\quad +\tfrac{1}{30}(6\Delta\bar{G}_{4s}^*-9W_{\mathrm{AgCu}}^{\mathrm{TET}}-4W_{\mathrm{AgCu}}^{\mathrm{TRG}})(s).
\end{aligned} \quad (24)
$$

For argentian tetrahedrite–tennantites the appropriate extended form of the Darken equation required to evaluate component chemical potentials is

$$\mu_j = \bar{G}+\sum_i n_{ij}(1-X_i)(\partial\bar{G}/\partial X_i)_{X_k/X_l}+(q_i-s)(\partial\bar{G}/\partial s)_{T,X_2,X_3,X_4} \quad (25)$$

where q_i is the value of s in the component of interest.

Following the same method of analysis employed for Ag-poor tetrahedrite–tennantites, we may write the following condition of equilibrium for the Fe–Zn exchange reaction between sphalerites and argentian tetrahedrites (i.e. $X_3 = \mathrm{As}/(\mathrm{As}+\mathrm{Sb})\approx 0$)

$$
\begin{aligned}
\bar{Q}_2^* &\equiv RT\ln\left[\left(\frac{X_{\mathrm{Fe}}^{\mathrm{TET}}}{X_{\mathrm{Zn}}^{\mathrm{TET}}}\right)^{\mathrm{TD-TN}}\left(\frac{X_{\mathrm{ZnS}}^{\mathrm{SPH}}}{X_{\mathrm{FeS}}^{\mathrm{SPH}}}\right)\right]+RT\ln\left(\frac{\gamma_{\mathrm{ZnS}}^{\mathrm{SPH}}}{\gamma_{\mathrm{FeS}}^{\mathrm{SPH}}}\right) \\
&= \Delta\bar{G}_{\mathrm{Zn(Fe)}-1}^0+\tfrac{1}{2}\Delta\bar{G}_{24}^0(X_4)+\tfrac{1}{2}\Delta\bar{G}_{2s}^*(s).
\end{aligned} \quad (26)
$$

In this equation $\Delta\bar{G}_{\mathrm{Zn(Fe)}-1}^0$ is the standard state Gibbs energy of reaction (15) and the term $\tfrac{1}{2}W_{\mathrm{FeZn}}^{\mathrm{TET}}(1-2X_2)$ is omitted because $W_{\mathrm{FeZn}}^{\mathrm{TET}}$ is negligible. Taking this modification into account, equation (26) differs from that for Ag-free tetrahedrite–tennantites (equation (17)) in that the $\bar{Q}^*$ function is linear in both the composition variable X_4 and the ordering variable s. It is necessary to make explicit provision for both dependencies, because constraints derived from nature (O'Leary and Sack, 1987) demonstrate that $\bar{Q}_2^*$ is a sigmoid function of Ag/(Ag+Cu) ratio at a given temperature and composition of coexisting sphalerite (Fig. 7.2). With increasing X_4, $\bar{Q}_2^*$ first rises, and then falls below the line formed by $\bar{Q}_2^*$ values for Ag-rich and Ag-poor tetrahedrites, a feature readily illustrated for argentian tetrahedrites in the composition range $0.08 \leqslant X_{FeS}^{SPH} \leqslant 0.12$ (Fig. 7.2). For sphalerites with lower X_{FeS}^{SPH}'s, the sigmoid $\bar{Q}_2^* - X_4$ function is displaced to lower values of $\bar{Q}_2^*$ at a given X_4.

Despite this added complexity, the Gibbs energy of the reciprocal reaction expressing the incompatibility between Zn and Ag in argentian tetrahedrites ($\Delta \bar{G}_{24}^0$, equation (3)) may still be determined directly from differences in Q_2^* between Ag and Cu tetrahedrites, because the order variable s is inoperative in them. A minimum bound on ΔG_{24}^0 of 1.8 ± 0.4 kcal/gfw may be obtained by comparing $\bar{Q}_2^*$ data for Ag-rich tetrahedrites with the previously derived constraints for Cu-tetrahedrites ($\bar{Q}_2^* = 2.07 \pm 0.07$ kcal/gfw). The parameter $\Delta \bar{G}_{2s}^*$ must also be positive, because the order variable is positive in tetrahedrites with low-intermediate values of X_4 (e.g. Kalbskopf, 1972; Johnson and Burnham, 1985; Peterson and Miller, 1986), and these tetrahedrites have values of $\bar{Q}_2^*$ above the line defined by Ag-rich and Ag-poor tetrahedrites. In the

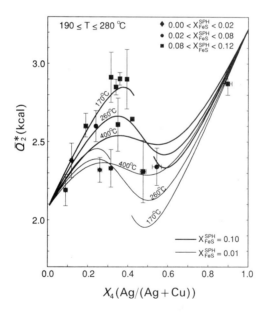

Figure 7.2 Calibration of equation (26) compared with constraints for argentian tetrahedrites given by O'Leary and Sack (1987). Curves are calculated for 170, 260, and 400°C for $X_{FeS}^{SPH} = 0.01$ and $X_{FeS}^{SPH} = 0.10$ using the parameter values given in Table 7.3, $\Delta \bar{G}_{Zn(Fe)_{-1}}^0 = 2.0718$ kcal/gfw, and the expression for $\ln(\gamma_{ZnS}^{SPH}/\gamma_{FeS}^{SPH})$ given in the caption to Figure 7.1. A correction of $-1.295(X_3)$ kcal/gfw $\{-\frac{1}{2}\Delta \bar{G}_{23}^0(X_3)\}$ is applied to values $\bar{Q}_2^*$ calculated from the composition–temperature data summarized by O'Leary and Sack (1987, Tables 1, 3) to correct for minor amounts of As ($X_3 < 0.05$). Error bars indicate 1σ uncertainties of these estimates.

context of this formulation, acceptable solutions for these and the other relevant parameters in equation (24) ($\Delta \bar{G}_s^*$, $\Delta \bar{G}_{4s}^*$, W_{AgCu}^{TRG}, and W_{AgCu}^{TET}) must yield values of s that have the same form of sigmoid dependence on X_4 as that exhibited by $\bar{Q}_2^*$, because $\bar{Q}_2^*$ is linear in the order variable s. Assuming that all regular-solution-type parameters are either negligible or positive (cf Lawson, 1947), solutions for the parameters in equation (24) which have this feature require that argentian tetrahedrites with intermediate Ag/(Ag+Cu) ratios have miscibility gaps that extend to temperatures of at least 190°C. It is unlikely that any such gaps extend to temperatures much above 200°C, because natural argentian tetrahedrites crystallized in epithermal veins display continuous composition trends that cross any such hypothesized gaps (Fig. 7.3). On the other hand, it is probable that such gaps are fairly extensive at temperatures

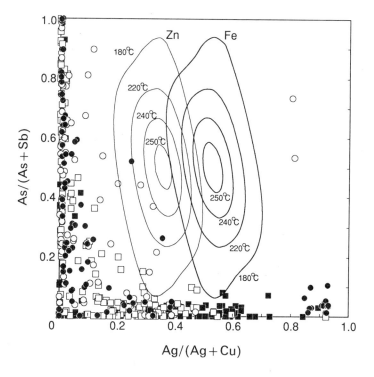

Figure 7.3 Calculated 180, 220, 240, and 250°C miscibility gaps for $(Ag, Cu)_{10}Zn_2(As, Sb)_4S_{13}$ (Zn) and $(Ag, Cu)_{10}Fe_2(As, Sb)_4S_{13}$ (Fe) tetrahedrite–tennantites compared with the composition data from nature. Ranges of Zn/(Zn+Fe) ratios of tetrahedrite-tennantites are indicated as follows: solid circles, $1 \leqslant Zn/(Zn+Fe) \leqslant 0.8$; open circles $0.8 < Zn/(Zn+Fe) \leqslant 0.5$; open squares, $0.5 < Zn/(Zn+Fe) \leqslant 0.2$; solid squares, $0.2 < Zn/(Zn+Fe) \leqslant 0.0$. Sources of the composition data are cited in Sack et al. (1987, Fig. 5) except for that reported by Bishop et al. (1977), Imai and Lee (1980), Kvacek et al. (1974), and Paar et al. (1978).

below 200°C, because coexisting Ag-rich and Ag-poor argentian tetrahedrites in apparent chemical equilibrium (e.g. Indolev et al., 1971) have differences in Fe/Zn ratios that are consistent with temperatures of at least 160°C. Moreover, gaps calculated from calibrations of the parameters in equations (24) and (26) that have critical temperatures below 200°C are in general accord with this composition data (cf O'Leary and Sack, 1987).

It is noteworthy that the deduced systematics bear a remarkable parallel with those established for ferrichromite spinels (e.g. Robbins et al., 1971; Evans and Frost, 1975; Sack and Ghiorso, 1991b). In Fe_3O_4 the preference of Fe^{3+} for tetrahedral sites relative to Fe^{2+} results in inverse cation distributions at low temperatures (i.e. Fe^{3+} is the principal constituent of the single tetrahedral site, and the two octahedral sites are nearly equally occupied by Fe^{3+} and Fe^{2+}). However, with increasing substitution of $FeCr_2O_4$ for Fe_3O_4, Fe^{3+} undergoes a progressive change in site preference with the result that it is primarily octahedral in chromian compositions (e.g. Robbins et al., 1971). The progressive change in site preference is manifest in sigmoid volume–composition relations for Fe_3O_4–$FeCr_2O_4$ spinels (e.g. Francombe, 1957; Robbins et al., 1971) and the distribution coefficients for Fe–Mg exchange between olivines and ferrichromites also display a strongly sigmoid dependence on $Cr^{3+}/(Cr^{3+} + Fe^{3+})$ ratio at a given temperature and Fe/Mg ratio of coexisting olivine (Evans and Frost, 1975; Sack and Ghiorso, 1991b).

The exact mechanism responsible for the parallel behaviour in argentian tetrahedrites is still not unambiguously established. The change of Ag site preference inferred from the observed $\bar{Q}_2^* - X_4$ relations in the context of the present, simple formulation for argentian tetrahedrites (i.e. s changes from positive to negative with increasing substitution of Ag for Cu) has not been substantiated by spectroscopic studies. There are no X-ray structure refinements of relevant tetrahedrites, and extended X-ray absorption fine structure (EXAFS) spectroscopy has been performed on only one natural tetrahedrite with appropriate Ag/Cu and Zn/Fe ratios to have substantial tetrahedral silver, according to the calibration of equation (24). This sample (BM88668, $X_4 \approx 0.55$) exhibits an anomalous Ag-EXAFS spectrum relative to samples in which most of the silver is predicted to be in trigonal coordination (cf Charnock et al., 1989). Charnock and coworkers (1989) conclude that the Ag in this sample is in trigonal coordination, and that the anomalous peak in the Fourier transform of this spectrum is due to Sb scatterers at distances less than those that are consistent with the sum of the Van der Waals radii of Sb atoms and the octahedrally coordinated sulphur atoms, a feature possibly related to the collapsed unit-cell volumes of such tetrahedrites. However, they fail to find evidence for tetrahedrally coordinated Ag even in synthetic tetrahedrites in which it must be present (i.e. $X_4 \geqslant 0.6$) and there is the added possibility that tetrahedrites with intermediate Ag/(Ag+Cu) ratios undergo structural transformation at low temperature.

Despite the lack of definitive constraints on the structural role of Ag at temperatures appropriate to ore deposition, it may be further demonstrated that the present model for the thermodynamic properties of argentian tetrahedrite–tennantites successfully accounts for the remaining known phase-equilibrium features. It does so for the physically plausible assumption that there is no excess *vibrational* entropy of mixing (i.e. the entropies of the reciprocal reactions are zero, and all regular-solution-type parameters are constants independent of temperature). The additional tests of the adequacy of the model are provided by the results of Ag–Cu exchange experiments (Ebel and Sack, 1989) and the composition data from nature (Fig. 7.3).

Ebel and Sack (1989) report experimental constraints relevant to determining the Gibbs energy of the reciprocal reaction expressing the incompatibility between As and Ag in the tetrahedrite–tennantite structure ($\Delta \bar{G}^0_{34}$, eq. 1). They obtained tight reversal brackets for the Ag/Cu ratios of tetrahedrite–tennantites ($0 \leqslant X_3 \leqslant 1$) equilibrated with the assemblage electrum + chalcopyrite + pyrite + sphalerite (400°C; electrums with 20 and 30 mole % Ag). They demonstrate that a tight bound on $\Delta \bar{G}^0_{34}$ (12.25 ± 1.92 kcal/gfw) is obtained by employing the previously derived constraints on the parameters in equations (24) and (28) to evaluate the quantity Y in the condition of Ag–Cu exchange equilibrium for this assemblage written in the form

$$\tfrac{1}{10}Y + RT \ln a^{\text{ELEC}}_{\text{Ag}} = \Delta \bar{G}^0_{\text{Ag(Cu)}-1} + \tfrac{1}{10}\Delta \bar{G}^0_{34}(X_3). \tag{27}$$

In equation (26) the quantity Y is defined by the expression

$$Y = RT \ln \left(\frac{(1-X_4-\tfrac{2}{5}s)^6 (\tfrac{2}{3}[1-X_4]+\tfrac{2}{5}s)^4}{(X_4+\tfrac{2}{5}s)^6 (\tfrac{2}{3}X_4-\tfrac{2}{5}s)^4} \right) - \Delta \bar{G}^0_{24}(X_2)$$
$$- (\Delta \bar{G}^*_{4s} + W^{\text{TET}}_{\text{AgCu}} + W^{\text{TRG}}_{\text{AgCu}})(1-2X_4) - \tfrac{1}{5}(\Delta \bar{G}^*_{4s} + 6W^{\text{TET}}_{\text{AgCu}} - 4W^{\text{TRG}}_{\text{AgCu}})(s), \tag{28}$$

$\Delta \bar{G}^0_{\text{Ag(Cu)}-1}$ is the standard state Gibbs energy of the Ag–Cu exchange reaction

$$\text{Ag} + \tfrac{1}{10}\text{Cu}_{10}\text{Fe}_2\text{Sb}_4\text{S}_{13} + \text{FeS}_2 = \tfrac{1}{10}\text{Ag}_{10}\text{Fe}_2\text{Sb}_4\text{S}_{13} + \text{CuFeS}_2,$$
ELEC TD–TN PYR TD–TN CPY

$a^{\text{ELEC}}_{\text{Ag}}$ is evaluated using the calibration of Hultgren *et al.* (1963), and it is assumed that chalcopyrite and pyrite are unary.

It is noteworthy that the 400°C calibration for equation (27) may be extrapolated to successfully account for Ag/Cu ratios of tetrahedrites in both 300°C experiments (Ebel and Sack, 1991) and in equivalent natural assemblages with up to nearly twice the maximum Ag observed in the 400°C experiments ($\approx 4\tfrac{1}{2}$ Ag atoms/formula unit, Ebel and Sack, 1989). Furthermore, such an extrapolation is achieved by making only the additional assumption that the entropy of the Ag–Cu exchange reaction ($\Delta \bar{S}^0_{\text{Ag(Cu)}-1}$) is identical to that of the reaction between copper metal, pyrite, and chalcopyrite

($\Delta \bar{S}^0 = \approx 10.38$ Gibbs/gfw; Barton and Skinner, 1979),

$$Cu + FeS_2 = CuFeS_2.$$

This assumption may be justified on the basis that the Ag–Cu exchange reaction of interest is the sum of this reaction and the Ag–Cu exchange reaction between tetrahedrite and metal,

$$Ag + \tfrac{1}{10}Cu_{10}Fe_2Sb_4S_{13} = \tfrac{1}{10}Ag_{10}Fe_2Sb_4S_{13} + Cu.$$

Because this latter reaction does not involve the reordering of metals among lattice sites, its entropy should be small to negligible, as observed for many exchange reactions involving only two phases (e.g. Sack and Ghiorso, 1989) such as the Fe–Zn exchange reaction between tetrahedrite–tennantite and sphalerite discussed above. The success of this calibration of equation (27) in predicting both the results of 300°C experiments and Ag/Cu ratios of natural tetrahedrites lends credence to the assertion that the thermodynamic model provides reasonable constraints on activity–composition relations.

A comparison between calculated miscibility gaps and the composition data for natural tetrahedrite–tennantites crystallized over the 200–350°C temperature range (Fig. 7.3) provides another test of the plausibility of the derived activity–composition relations. It is striking that no tetrahedrite–tennantites with intermediate Ag/(Ag+Cu) and As/(As+Sb) ratios are observed in nature, a feature that implies that miscibility gaps are present at temperatures of ore deposition (200–350°C). The calculated gaps of Zn- and Fe-bearing argentian tetrahedrites are in reasonable accord with these constraints. They have maximum critical temperatures of less than 260°C, and have 200–250°C binodal curves that do not enclose compositions of natural tetrahedrite–tennantites crystallized at these temperatures, with the exception of some of the zincian tetrahedrite–tennantites reported by Ixer and Stanley (1983). These tetrahedrite–tennantites are mercurian, and thus are not strictly comparable. Although tetrahedrite–tennantites with intermediate Zn/(Zn+Fe) ratios have critical curves that extend to temperatures 20–30°C higher than those in the end-member Fe and Zn subsystems, they have miscibility gaps that are intermediate between the gaps in the Fe and Zn subsystems at a given temperature. Thus the model for the thermodynamic properties of tetrahedrite–tennantites is consistent with the constraints provided by nature.

7.4 Crystallochemical controls on Metal Zoning

Having placed constraints on the shape of the Gibbs energy surface, it is of interest to ascertain to what extent non-idealities in tetrahedrite–tennantites are reflected in element zoning trends in polymetallic base-metal sulphide deposits of hydrothermal-vein type. Hackbarth and Petersen (1984) have shown that the strong composition dependencies on the distribution coefficients for the exchange the As and Sb, and Ag and Cu between tetrahedrite-tennantite and hydrothermal fluid are required to explain the sense of composition zoning exhibited by tetrahedrite–tennantites where it is a major phase in such deposits. To be consistent with mass balance constraints, the distribution coefficient for As–Sb exchange between tetrahedrite–tennantite and hydrothermal fluid, $[(X_3/1-X_3)(n_{Sb^{3+}}^{FLUID}/n_{As^{3+}}^{FLUID})]$, must be greater than unity for tetrahedrite–tennantites deposited nearest the source of the hydrothermal fluids, but it must be smaller for distal argentian tetrahedrite–tennantites. In contrast, the distribution coefficient for Ag–Cu exchange between tetrahedrite–tennantite and hydrothermal fluid, $[(X_4/1-X_4)(n_{Cu^+}^{FLUID}/n_{Ag^+}^{FLUID})]$, must be less than unity for tetrahedrite–tennantites deposited nearest the source of the hydrothermal fluids, but it must increase for distal argentian tetrahedrite–tennantites. It may be demonstrated that the required composition dependencies to these distribution coefficients are consistent with the deduced non-idealities in tetrahedrite–tennantite by comparing curves of constant $As(Sb)_{-1}$ and $Ag(Cu)_{-1}$ exchange potentials for tetrahedrite–tennantites at a representative temperature of ore deposition with the envelopes on tetrahedrite–tennantite composition paths defined by Hackbarth and Petersen (1984) (Figure 7.4). It is noteworthy that within these envelopes these curves exhibit decreasing $X_3(As/(As+Sb))$ and $X_4(Ag/(Ag+Cu))$ with increasing X_4 and X_3, respectively. These dependencies reflect both the non-idealities associated with the substitutions of As for Sb, and Ag for Cu, and the incompatibility between As and Sb in the tetrahedrite–tennantite structure as expressed by a large positive value for the Gibbs energy of reciprocal reaction given by equation (1). They are consistent with the inferences regarding the changes in distribution coefficients for As–Sb and Ag–Cu exchange required to explain the strong curvature of the tetrahedrite–tennantite composition paths along flow paths of mineralization.

This latter point may be amplified in more direct and practical terms by calculating tetrahedrite-tennantite compositions along flow paths for several simplifying assumptions. To illustrate only the crystallochemical component of such paths, it will be assumed Fe and Zn, As and Sb, and Ag and Cu mix ideally in the hydrothermal fluid. For this assumption we may use the following expressions for the $Zn(Fe)_{-1}$, $As(Sb)_{-1}$, and $Ag(Cu)_{-1}$ exchange potentials in the hydrothermal fluid:

$$\mu_{Zn(Fe)_{-1}}^{FLUID} = \left[\mu_{Zn(Fe)_{-1}}^{0\,FLUID} + RT \ln\left(\frac{\gamma_{Zn^{2+}}^{FLUID}}{\gamma_{Fe^{2+}}^{FLUID}}\right)\right] + RT \ln\left(\frac{n_{Zn^{2+}}^{FLUID}}{n_{Fe^{2+}}^{FLUID}}\right), \quad (29)$$

$$\mu_{As(Sb)_{-1}}^{FLUID} = \left[\mu_{As(Sb)_{-1}}^{0\,FLUID} + RT \ln\left(\frac{\gamma_{As^{3+}}^{FLUID}}{\gamma_{Sb^{3+}}^{FLUID}}\right)\right] + RT \ln\left(\frac{n_{As^{3+}}^{FLUID}}{n_{Sb^{3+}}^{FLUID}}\right), \quad (30)$$

and $\quad \mu_{Ag(Cu)_{-1}}^{FLUID} = \left[\mu_{Ag(Cu)_{-1}}^{0\,FLUID} + RT \ln\left(\frac{\gamma_{Ag^+}^{FLUID}}{\gamma_{Cu^+}^{FLUID}}\right)\right] + RT \ln\left(\frac{n_{Ag^+}^{FLUID}}{n_{Cu^+}^{FLUID}}\right), \quad (31)$

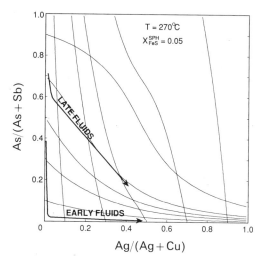

Figure 7.4 Calculated 270°C isotherms of constant $As(Sb)_{-1}$ and $Ag(Cu)_{-1}$ exchange potential for tetrahedrite–tennantites coexisting with sphalerite with $X_{FeS}^{SPH}=0.05$ compared with envelopes of Hackbarth and Petersen (1984). Isotherms are calculated for tetrahedrite–tennantites with the same $As(Sb)_{-1}$ and $Ag(Cu)_{-1}$ exchange potentials as reference Ag-free tetrahedrite–tennantites with $X_3=0.1$, 0.3, 0.5, 0.7, and 0.9 and argentian tetrahedrites with $X_4=0.1$, 0.3, 0.5, 0.7, and 0.9, respectively. Thicker curves labelled EARLY FLUIDS and LATE FLUIDS are from Hackbarth and Petersen (1984). They represent envelopes on the most extreme composition paths exhibited by the first- and last-deposited tetrahedrite–tennantites in crustification bands from the Coeur, Crescent, Galena, and Sunshine mines in the Coeur d'Alene districts, Idaho and from the Casapalca, Julcani, and Orcopampa districts in Peru. Arrows indicate the sense of downstream composition change from the centres of zoned ore deposits outwards.

where the n_{iz+}^{FLUID} and γ_{iz+}^{FLUID} represent the total moles of the appropriate metal or semimetal ions in the hydrothermal fluid and their activity coefficients, respectively, and the quantities in brackets are regarded as constants at a given temperature. These expressions may be combined with the equivalent expressions for tetrahedrite–tennantite to yield to following equations governing the exchange of metals and semi-metals between tetrahedrite–tennantite and hydrothermal fluid:

$$RT \ln\left[\left(\frac{1-X_2}{X_2}\right)\left(\frac{n_{Zn^{2+}}^{FLUID}}{n_{Fe^{2+}}^{FLUID}}\right)\right] = \Delta\bar{G}_{Zn(Fe)_{-1}}^{0\,F-F} + \tfrac{1}{2}\{\Delta\bar{G}_{23}^{0}(X_3) + \Delta\bar{G}_{24}^{0}(X_4) + \Delta\bar{G}_{2s}^{*}(s)\}, \qquad (32)$$

$$RT \ln\left[\left(\frac{1-X_3}{X_3}\right)\left(\frac{n_{As^{3+}}^{FLUID}}{n_{Sb^{3+}}^{FLUID}}\right)\right] = \Delta\bar{G}_{As(Sb)_{-1}}^{0\,F-F} + \tfrac{1}{4}\{\Delta\bar{G}_{23}^{0}(X_2) + \Delta\bar{G}_{34}^{0}(X_4) + \Delta\bar{G}_{3s}^{*}(s)\} + \tfrac{1}{4}W_{AsSb}^{SM}(1-2X_3), \qquad (33)$$

and

$$\tfrac{1}{10}RT\ln\left[\frac{(1-X_4-\tfrac{2}{5}s)^6(\tfrac{2}{3}[1-X_4]+\tfrac{2}{5}s)^4}{(X_4+\tfrac{2}{5}s)^6(\tfrac{2}{3}X_4-\tfrac{2}{5}s)^4}\left(\frac{n^{\text{FLUID}}_{\text{Ag}^+}}{n^{\text{FLUID}}_{\text{Cu}^+}}\right)^{10}\right]=\Delta\bar{G}^{0\,\text{F-F}}_{\text{Ag(Cu)}-1}+\tfrac{1}{10}Z \quad (34)$$

where

$$Z=\Delta\bar{G}^0_{24}(X_2)+\Delta\bar{G}^0_{34}(X_3)+(\Delta\bar{G}^*_{4s}+W^{\text{TET}}_{\text{AgCu}}+W^{\text{TRG}}_{\text{AgCu}})(1-2X_4)$$
$$+\tfrac{1}{5}(\Delta\bar{G}^*_{4s}+6W^{\text{TET}}_{\text{AgCu}}-4W^{\text{TRG}}_{\text{AgCu}})(s)$$

and the terms $\Delta\bar{G}^{0\,\text{F-F}}_{\text{Zn(Fe)}-1}$, $\Delta\bar{G}^{0\,\text{F-F}}_{\text{As(Sb)}-1}$, and $\Delta\bar{G}^{0\,\text{F-F}}_{\text{Ag(Cu)}-1}$ are constants at a given temperature that are defined by the expressions

$$\Delta\bar{G}^{0\,\text{F-F}}_{\text{Zn(Fe)}-1}=\tfrac{1}{2}(\bar{G}^*_{\text{Cu}_{10}\text{Zn}_2\text{Sb}_4\text{S}_{13}}-\bar{G}^*_{\text{Cu}_{10}\text{Fe}_2\text{Sb}_4\text{S}_{13}})-\left[\mu^{0\,\text{FLUID}}_{\text{Zn(Fe)}-1}+RT\ln\left(\frac{\gamma^{\text{FLUID}}_{\text{Zn}^{2+}}}{\gamma^{\text{FLUID}}_{\text{Fe}^{2+}}}\right)\right],$$

$$\Delta\bar{G}^{0\,\text{F-F}}_{\text{As(Sb)}-1}=\tfrac{1}{4}(\bar{G}^*_{\text{Cu}_{10}\text{Fe}_2\text{As}_4\text{S}_{13}}-\bar{G}^*_{\text{Cu}_{10}\text{Fe}_2\text{Sb}_4\text{S}_{13}})-\left[\mu^{0\,\text{FLUID}}_{\text{As(Sb)}-1}+RT\ln\left(\frac{\gamma^{\text{FLUID}}_{\text{As}^{3+}}}{\gamma^{\text{FLUID}}_{\text{Sb}^{3+}}}\right)\right],$$

and

$$\Delta\bar{G}^{0\,\text{F-F}}_{\text{Ag(Cu)}-1}=\tfrac{1}{10}(\bar{G}^*_{\text{Ag}_{10}\text{Fe}_2\text{Sb}_4\text{S}_{13}}-\bar{G}^*_{\text{Cu}_{10}\text{Fe}_2\text{Sb}_4\text{S}_{13}})-\left[\mu^{0\,\text{FLUID}}_{\text{Ag(Cu)}-1}+RT\ln\left(\frac{\gamma^{\text{FLUID}}_{\text{Ag}^+}}{\gamma^{\text{FLUID}}_{\text{Cu}^+}}\right)\right].$$

Tetrahedrite–tennantite composition curves may be readily calculated using equations (32)–(34) for the additional simplifying assumptions of perfect fractionation, constant temperature, and that the metals and semimetals have the same ratios in the hydrothermal fluid as in stoichiometric tetrahedrite–tennantite (Figure 7.5). Such calculated composition paths are in excellent accord with those observed in tetrahedrite–tennantite dominated veins from the Coeur d'Alene district and in Peruvian deposits (cf Wu and Petersen, 1977; Hackbarth and Petersen, 1984) providing $\Delta\bar{G}^{0\,\text{F-F}}_{\text{As(Sb)}-1}$ is negative, and $\Delta\bar{G}^{0\,\text{F-F}}_{\text{Ag(Cu)}-1}$ and $\Delta\bar{G}^{0\,\text{F-F}}_{\text{Zn(Fe)}-1}$ are small in absolute value. $\Delta\bar{G}^{0\,\text{F-F}}_{\text{As(Sb)}-1}$ must be negative for the distribution coefficient $[(X_3/1-X_3)(n^{\text{FLUID}}_{\text{Sb}^{3+}}/n^{\text{FLUID}}_{\text{As}^{3+}})]$ to be greater than unity for silver-poor tetrahedrite–tennantites. $\Delta\bar{G}^{0\,\text{F-F}}_{\text{Ag(Cu)}-1}$ must be small for tetrahedrite–tennantites to have lower Ag/(Ag+Cu) ratios than hydrothermal fluids for relevant bulk compositions. Finally $\Delta\bar{G}^{0\,\text{F-F}}_{\text{Zn(Fe)}-1}$ must be small and positive, if tetrahedrite–tennantites do not exhibit pronounced downstream variations in Zn/(Zn+Fe) ratios (cf Wu and Petersen, 1977; Hackbarth and Petersen, 1984; Sack and Loucks, 1985).

Although some of the approximations employed in constructing Figure 7.5 may not be strictly appropriate for particular fissure-vein deposits, the excellent accord between observed and calculated composition paths for tetrahedrite–tennantite-dominated systems is sufficient to demonstrate that

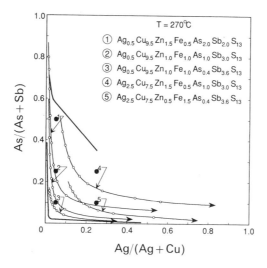

Figure 7.5 270°C fractionation paths calculated using equations (32)–(34) for several representative tetrahedrite–tennantite bulk compositions indicated by filled circles. Composition curves are calculated for $\Delta \bar{G}^{0F-F}_{Zn(Fe)_{-1}} = -0.5$, $\Delta \bar{G}^{0F-F}_{As(Sb)_{-1}} = -1.7$, and $\Delta \bar{G}^{0F-F}_{Ag(Cu)_{-1}} = +0.5$ kcal/gfw assuming perfect fractionation. Open circles on them indicate 0.1 increments of crystallization; tips of arrows indicate 0.99 crystallization. Thick curves are the envelopes on composition paths in tetrahedrite–tennantite-dominated crustification bands defined by Hackbarth and Petersen (1984).

zoning trends in them largely reflect As–Sb fractionation and non-ideality in the thermodynamic mixing properties of tetrahedrite–tennantites. Thus calculated chords of the type displayed in Figure 7.5 may be useful in evaluating Ag reserves in partially developed mining districts. The upstream/downstream grade of ore could be estimated by comparison of assay data with such chords. More generally the restrictive assumption employed may be relaxed and the model for the thermodynamic properties of tetrahedrite–tennantites may be integrated with studies of hydrothermal fluids in ore-forming systems in which tetrahedrite–tennantite is not necessarily the most abundant product of mineralization. A prerequisite to such a development is the establishment of an internally consistent database for the Gibbs energies of tetrahedrite–tennantite end-member components. The latter may be achieved from studies of low-variance assemblages employing the previously summarized constraints on thermodynamic mixing properties and standard-state energies of exchange reactions.

Acknowledgements

Thanks to S. E. Bushnell for introducing me to problems in fahlore thermochemistry; M. Stuve for technical assistance; P. C. Goodell, E. Makovicky, and E. Petersen for constructive criticisms; D. S. Ebel for continuing collaboration; and R.A.D. Pattrick for arranging for my participation in the conference on behalf of the Applied Mineralogy Group of the Mineralogy Society. Material support was provided by the National Science Foundation through grant EAR89-04270.

References

Barton, P. B., Jr., and Skinner, B. J. (1979) Solubilities of ore minerals, in *Geochemistry of Hydrothermal Ore Deposits*, (ed. H. L. Barnes), John Wiley and Sons, pp. 404–60.
Barton, P. B., Jr., and Toulmin P., III (1966) Phase relations involving sphalerite in the Fe–Zn–S system. *Economic Geology*, **61**, 815–49.
Berman, R. G. (1988) Internally-consistent thermodynamic data for minerals in the system $Na_2O-K_2O-CaO-MgO-FeO-Fe_2O_3-Al_2O_3-SiO_2-TiO_2-H_2O-CO_2$: representation, estimation, and high temperature extrapolation. *Journal of Petrology*, **29**, 445–522.
Bishop, A. C., Criddle, A. J., and Clark, A. M. (1977) Plumbian tennantite from Sark, Channel Islands. *Mineralogical Magazine*, **41**, 59–63.

Charnock, J. M., Garner, C. D., Pattrick, R. A. D., *et al.*, (1988) Investigation into the nature of copper and silver sites in argentian tetrahedrites using EXAFS spectroscopy. *Physics and Chemistry of Minerals*, **15**, 296–9.
Charnock, J. M., Garner, C. D., Pattrick, R, A. D., *et al.*, (1989) Coordination sites of metals in tetrahedrite minerals determined by EXAFS. *Journal of Solid State Chemistry*, **82**, 279–89.
Charlat, M, and Levy, C. (1974) Substitutions multiples dans la serie tennantite–tetrahedrite. *Bulletin de la Société française de Minéralogie et de Cristallographie*, **97**, 241–50.

Ebel, D. S. and Sack, R. O. (1989) Ag–Cu and As–Sb exchange energies in tetrahedrite–tennantite fahlores. *Geochimica et Cosmochimica Acta*, **53**, 2301–9.
Ebel, D. S., and Sack, R. O. (1991) Arsenic–silver incompatibility in fahlore. *Mineralogical Magazine*, **55**, 521–8.
Evans, B. W. and Frost, B. R. (1975) Chrome-spinels in progressive metamorphism—a preliminary analysis. *Geochimica et Cosmochimica Acta*, **39**, 959–72.

Francombe, M. H. (1957) Lattice changes in spinel-type iron chromites. *Journal of Physics and Chemistry of Solids*, **3**, 37–43.
Fryklund, V. C., Jr. (1964) Ore deposits of the Coeur d'Alene district, Shoshone country, Idaho. *United States Geological Survey Professional Paper* 445.

Ghiorso, M. S. (1990a) Thermodynamic properties of hematite–ilmenite–geikielite solid solutions. *Contributions to Mineralogy and Petrology*, **104**, 645–67.
Ghiorso, M. S. (1990b) The application of the Darken equation to mineral solid solutions with variable degrees of order–disorder. *American Mineralogist*, **75**, 539–43.
Goodell, P. C. and Petersen, U. (1974) Julcani mining district, Peru: a study of metal ratios. *Economic Geology*, **69**, 347–61.

Hackbarth, C. J. and Petersen, U. (1984) Systematic compositional variations in argentian tetrahedrite. *Economic Geology*, **79**, 448–60.
Hultgren, R., Orr, R. L., Anderson, P. D., *et al.* (1963) *Selected Values of Thermodynamic Properties of Metals and Alloys*. John Wiley and Sons.

Imai, N. and Lee, H. K. (1980) Complex sulfide–sulfosalt ores from Janggun mine, ROK, in *Complex Sulfide Ores*, Institute of Mining and Metallurgy, pp. 248–9.

Indolev, L. N., Nevoysa, I. A., and Bryzgalov, I. A. (1971) New data on the composition of stibnite and the isomorphism of copper and silver. *Doklady Akademii Nauk SSSR*, **199**, 115–18.

Ixer, R. A. and Stanley, C. J. (1983) Silver mineralization at Sark's Hope mine, Channel Islands. *Mineralogical Magazine*, **47**, 539–45.

Johnson, M. L. and Burnham, C. W. (1985) Crystal structure refinement of an arsenic-bearing argentian tetrahedrite. *American Mineralogist*, **70**, 165–70.

Johnson, M. L. and Jeanloz, R. (1983) A Brillouin-zone model for compositional variation in tetrahedrite. *American Mineralogist*, **68**, 220–6.

Johnson, N. E., Craig, J. R., and Rimstidt, J. D. (1986) Compositional trends in tetrahedrite. *Canadian Mineralogist*, **24**, 385–97.

Kalbskopf, R. (1972) Strukturverfeinerung des freibergits. *Tschermaks Mineralogisch und Petrographische Mitteilungen*, **18**, 147–55.

Kvacek, M., Novak, F., and Drabek, M. (1975) Canfieldite and silver-rich tetrahedrite from the Kutna Hora district. *Neues Jahrbuch für Mineralogie Monatschefte*, **4**, 171–9.

Lawson, A. W. (1947) On simple binary solutions. *Journal of Chemical Physics*, **15**, 831–42.

Makovicky, E. and Skinner, B. J. (1978) Studies of the sulfosalts of copper. VI. Low-temperature exsolution in synthetic tetrahedrite solid solution; $Cu_{12+x}Sb_{4+y}S_{13}$. *Canadian Mineralogist*, **16**, 611–23.

Makovicky, E. and Skinner, B. J. (1979) Studies of the sulfosalts of copper VII. Crystal structures of the exsolution products $Cu_{12.3}Sb_4S_{13}$ and $Cu_{13.8}Sb_4S_{13}$ of unsubstituted synthetic tetrahedrite. *Canadian Mineralogist*, **17**, 619–34.

Mishra, B. and Mookherjee, A. (1986) Analytical formulation of phase equilibrium in two observed sulfide–sulfosalt assemblages in the Rajpura–Dariba polymetallic deposit. *Economic Geology*, **81**, 627–39.

O'Leary, M. J. and Sack, R. O. (1987) Fe–Zn exchange reaction between tetrahedrite and sphalerite in natural environments. *Contributions to Mineralogy and Petrology*, **96**, 415–25.

Paar, Von W. H., Chen, T. T., and Cunther, W. (1978) Extrem silberreicher Freibergit in Pb–Zn–Cu–Erzen des Bergbaues 'Knappenstube', Hochtor, Salzburg. *Carinthia II*, **168**, 35–42.

Pattrick, R. A. D. (1978) Microprobe analyses of cadmium-rich tetrahedrites from Tyndrum, Perthshire, Scotland. *Mineralogical Magazine*, **42**, 286–8.

Pattrick, R. A. D. and Hall, A. J. (1983) Silver substitution into synthetic zinc, cadmium, and iron tetrahedrites. *Mineralogical Magazine*, **47**, 441–51.

Pauling, L. and Neuman, E. W. (1934) The crystal structure of binnite, $(Cu,Fe)_{12}As_4S_{13}$, and the chemical composition and structure of minerals of the tetrahedrite group. *Zeitschift für Kristallographie*, **88**, 54–62.

Peterson, R. C. and Miller, I. (1986) Crystal structure refinement and cation distribution in freibergite and tetrahedrite. *Mineralogical Magazine*, **47**, 441–51.

Raabe, K. C. and Sack, R. O. (1984) Growth zoning in tetrahedrite–tennantite from the Hock Hocking mine, Alma, Colorado. *Canadian Mineralogist*, **22**, 577–82.

Riley, J. F. (1974) The tetrahedrite–freibergite series, with reference to the Mount Isa Pb–Zn–Ag ore body. *Minerallium Deposita*, **9**, 117–24.

Robbins, M., Wertheim, G. K., Sherwood, R. C. *et al.* (1971) Magnetic properties and site distributions in the system $FeCr_2O_4$–Fe_3O_4 ($Fe^{2+}Cr_{2-x}Fe_x^{3+}O_4$). *Journal of Physics and Chemistry of Solids*, **32**, 717–29.

Sack, R. O. (1982) Spinels as petrogenetic indicators: activity–composition relations at low pressures. *Contributions to Mineralogy and Petrology*, **79**, 169–82.

Sack, R. O. and Ghiorso, M. S. (1989) Importance of considerations of mixing properties in establishing an internally consistent thermodynamic database: thermochemistry of minerals in the system $Mg_2SiO_4-Fe_2SiO_4-SiO_2$. *Contributions to Mineralogy of Petrology*, **102**, 41–68.

Sack, R. O., and Ghiorso, M. S. (1991a) An internally consistent model for the thermodynamic properties of Fe–Mg–titanomagnetite–aluminate spinels. *Contributions to Mineralogy and Petrology*, **106**, 474–505.

Sack, R. O. and Ghiorso, M. S. (1991b) Chromian spinels as petrogenetic indicators: thermodynamics and petrological applications. *American Mineralogist*, **76**, 827–847.

Sack, R. O. and Loucks, R. R. (1985) Thermodynamic properties of tetrahedrite–tennantites: Constraints on the interdependence of the Ag↔Cu, Fe↔Zn, Cu↔Fe, and As↔Sb exchange reactions. *American Mineralogist*, **70**, 1270–89.

Sack, R. O., Ebel, D. S., and O'Leary, M. J. (1987) Tennahedrite thermochemistry and metal zoning, in *Chemical Transport in Metasomatic Processes*, (ed. H. C. Helgeson), D. Reidel, Dordrecht, pp. 701–31.

Spiridonov, E. M. (1984) Species and varieties of fahlore (tetrahedrite–tennantite) minerals and their rational nomenclature. *Doklady Akademii Nauk SSSR*, **279**, 166–72.

Springer, G. (1969) Electron probe analyses of tetrahedrite. *Neues Jahrbuch für Mineralogie, Monatschefte*, 24–32.

Tatsuka, K. and Morimoto, N. (1977) Tetrahedrite stability relations in the Cu–Fe–Sb–S system. *American Mineralogist*, **62**, 1101–9.

Thompson, J. B., Jr. (1967) Thermodynamics properties of simple solutions, in *Researches in Geochemistry*, ed. P. H. Abelson, vol. 2, New York, John Wiley, pp. 340–61.

Thompson, J. B., Jr. (1969) Chemical reactions in crystals. *American Mineralogist*, **54**, 341–75.

Wu, I. and Petersen, U. (1977) Geochemistry of tetrahedrite–tennantite at Casapalca, Peru. *Economic Geology*, **72**, 993–1016.

Wuensch, B. J. (1964) The crystal structure of tetrahedrite, $Cu_{12}Sb_4S_{13}$. *Zeitschrift für Kristallographie*, **119**, 437–53.

Wuensch, B. J., Takeuchi, Y., and Nowacki, W. (1966) Refinement of the crystal structure of binnite. *Zeitschrift für Kristallographie*, **123**, 1–20.

Yui, S. (1971) Heterogeneity within a single grain of minerals of the tetrahedrite–tennantite series. Society of Mining Geologists of Japan Special Issue, 2, *Proceedings of IMA–IAGOD Meeting '70, Joint Symposium Volume*, 22–9.

CHAPTER EIGHT
Thermodynamic data for minerals: a critical assessment

Martin Engi

8.1 Introduction

Earth scientists in search of thermodynamic data are faced with questions as to the quality and reliability of their finds. Analysis of the currently available thermodynamic data bases for minerals reveals discrepancies of *form* and of *contents*, both of which complicate the quality assessment. Formal differences render comparisons of properties among different data bases cumbersome at least. Where differences of content emerge from such a comparison, even the specialist is often unsure which data are better and how good these might be. Worse, even where a comparison between data bases shows substantial agreement, the user of the data cannot be sure of their quality, because many data-base entries are not independent.

Though the general awareness of users has certainly increased—few people continue to mix independently gathered thermodynamic data—these problems are far from solved. Indeed, despite recent research efforts and progress, there is as yet no generally accepted way of characterizing adequately the quality of refined thermodynamic data.

Owing to these fundamental gaps in our knowledge, users will continue to have to evaluate and choose stability data by themselves for the foreseeable future. Reviews of a general usefulness on this topic do not seem to exist. Some recent textbooks (e.g. Nordstrom and Munoz, 1985) do include chapters on the evaluation of thermodynamic data, but the advice given still remains largely qualitative and does not emphasize *data-base* assessment.

The first part of the present Chapter seeks to provide the basic background users will require to assess a thermodynamic data base. Much emphasis is placed, by necessity, on basic assumptions and on questions of analytical procedure and data representation, because so often these 'details' are at the root of inadequacies within and discrepancies among data bases. Different approaches in data analysis have effects on the quality and, ultimately, the meaning of data-base entries. Hence it is important that users become familiar with the relative merits and pitfalls of least squares regression, linear and

The Stability of Minerals. Edited by G. D. Price and N. L. Ross.
Published in 1992 by Chapman & Hall, London. ISBN 0 412 44150 0

non-linear programming, and other optimization methods currently applied to feed the different thermodynamic data bases.

There are now several data bases, all continually evolving, hopefully improving, and perhaps converging. A comparison of data-base entries at any one time thus provides only a snapshot of that process. Despite these reservations, the second part of this Chapter presents such a comparison among most recently published data-base versions. One hope behind the comparison is that it will help focus attention and research efforts on the problems emerging.

8.1.1 Need for thermodynamic data

The need for reliable thermodynamic data of mineral systems is growing steadily. Much of the demand stems from the availability of increasingly powerful software (e.g. Table 8.1) to aid earth scientists in solving their diverse geochemical, mineralogical, and petrological problems, some of a complexity hitherto deemed prohibitive. Progress has been particularly notable in three areas: in petrogenetic reseach, in the quest to understand the relations between macroscopic stability and microscopic structural features ('Realbau'), and in

Table 8.1 Programs to calculate equilibrium-phase relations or diagrams for chemically complex systems

Program name	Source	Special features
	Group A: Designed primarily for geological systems	
VERTEX	Connolly, 1990	Batch calculation of stable phase relations in a generalized thermodynamic space. Projections are possible onto a variety of phase diagrams
GE0CALC and PTAX	Berman et al., 1987; Perkins et al., 1988; Lieberman, 1991a; b	Interactive calculation and display of phase relations in diagrams (e.g. $P-T$, $T-X_{CO_2}$, $a_i - a_j$); optional inclusion of metastable equilibria
THERMOCALC	Powell and Holland, 1988	Calculation of individual equilibria, incl. the propagation of (correlated) uncertainties
THERIAK	De Capitani and Brown, 1987	Efficient computation of equilibrium phase assemblages, even if complex solutions occur
	Group B: Designed primarily for metallurgical systems	
SOLGASMIX	Eriksson and Hack, 1984	Calculation of complete phase diagrams, incl. multicomponent solutions (alloys)
POLY-II and THERMO-CALC	Jansson, 1984a; Sundman et al., 1985	Calculation of complete phase diagrams and thermodynamic properties, specially for alloys.

the quantitative simulation of irreversible fluid-rock interaction processes and their integrated geochemical effects. Despite their diversity, these three topics share an acute requirement for refined thermodynamic data and models of documented reliability.

8.1.2 *From data to models: a first glance at the problems*

In light of current efforts to construct reliable data bases for thermodynamic properties of minerals, the entire wealth of primary data—especially from experimental petrology and geochemistry, but increasingly also from studies of natural phase relations—serves only one purpose: *to constrain the values of thermodynamic parameters* we seek to feed into a data base.

Two steps of analysis are needed when going from primary observations and measurements to such a data base. Firstly, quantitative models are needed to describe observations in terms of parameters, with due account of errors in the data. In principle, chemical thermodynamics provides the necessary relations to accomplish this first step. Secondly, some algorithm is required to determine the numerical values of thermodynamic parameters such that they collectively satisfy the constraints derived from the primary data as well as possible, ideally within the measurement errors of the data. The algorithm should also yield uncertainties in the parameter values commensurate with the uncertainties of the primary data as a whole.

While it might be expected that controversy would arise over the best algorithm or optimization technique for the problem, one may wonder why the coding of primary data into a numerical model should be controversial. It may be even less clear why phase-equilibrium data, the concepts of which are so widely known and well understood, should be most affected. Do not the combined first and second laws of thermodynamics provide all of the criteria needed for a proper model? They do, of course; the problem has historic roots.

Early experiments in petrology yielded 'synthesis data'; a mineral or assemblage A was synthesized from some highly metastable starting material, such as an oxide mix. It was soon recognized that such data allow no inference on the relative stability of two minerals or assemblages, say A versus B. This led to the concept of an 'experimental reversal' or a pair of 'half-reversals': Assemblage A was directly converted to assemblage B and, at some other conditions, the spontaneous reverse reaction $B \rightarrow A$ was observed. Where possible, such pairs of half-brackets were established in a small $P-T$-range, hence it became natural to look at the constraints pairwise. In principle, however, each half-bracket is an independent experiment, each with its own uncertainties. There is no need to group two experiments as one; indeed the concept of a reversed bracket generally adds ambiguity: How close do two half-brackets have to be in order to constitute a reversal? May a particular half-bracket, if spaced equally close to two opposite half-brackets, be part of two reversals? How do uncertainties in the location of each half-bracket combine to an overall uncertainty in the reversal? These questions are at the root of the coding problem outlined above. A thorough analysis of uncertainties in the primary data and of their effect on the thermodynamic parameters to be determined is required.

Before exploring these points further (Section 8.3), it will be useful to take a first look at currently available thermodynamic data bases and at the nature of primary data used to feed these.

8.1.3 Thermodynamic data bases

Recent progress has been considerable in consolidating our state of knowledge on stability relations amongst minerals and fluids—including mixed (supercritical) fluids, gases and aqueous species—over a large range of temperatures and pressures. One proof of this consolidation is the publication of a series of thermodynamic data bases (Table 8.2) which, for many problems in earth science, supersede to a large extent the more general data compilations, such as the JANAF Tables (Stull and Prophet, 1971) and the CATCH Tables (Pedley and Rylance, 1977). In the present attempt to summarize that progress and to review at least some of the results, the focus will be on *compilations of thermodynamic data* that fulfil four basic conditions, i.e. for which:

1. the stability information for minerals was obtained from consideration of suitably combined thermophysical, thermochemical, volumetric, *and* phase-equilibrium measurements and thus was not confined to purely calorimetric data;
2. the procedures of data analysis were documented;
3. multicomponent systems were examined;
4. the resulting data are available in machine-readable form.

Thermodynamic data bases differ in form and content (Table 8.3). Those denoted '*type K*' contain expressions that yield directly the $\log(K)$ of the hydrolysis[1] for each mineral species. Such data bases are designed to be used primarily in low-temperature applications, specifically in ground-water and hydrothermal systems. They derive their thermodynamic properties principally from solubility experiments and other types of eqilibrium information limited in temperature from near-ambient conditions to the boiling curve of H_2O, approximately. In line with this use at temperatures typically below the critical point of H_2O, the temperature dependence of the thermodynamic properties is parameterized in *type K* data bases by suitable expansions, such as a Taylor series in T of the $\log(K)$ expression for each mineral species. The second kind

---

[1] i.e. the equilibrium constant pertaining to the dissolution of the mineral to a defined set of aqueous species, e.g. for albite: $NaAlSi_3O_8 + 8H_2O = Na^+ + Al^{+++} + 3Si(OH)_4 + 4OH^-$.

Table 8.2 Published thermodynamic data bases designed primarily for petrology and geochemistry

Group	Reference (program)	Format	Special features[‡]
USGS	Robie and Waldbaum, 1968 Robie et al., 1978	data tables	Compilations based on thermochemical and thermophysical data, but do not consider phase equilibrium data
	Haas et al., 1981		Evaluation of CASH data, primarily based on calorimetry
	Robinson et al., 1983		Same for CASCH, MSCH, and FS
Moscow	Karpov et al., 1968 Naumov et al., 1971 Dorogokupetz, 1981	data tables	Extensive compilations incl. soviet calorimetric/volumetric data Evaluation not fully documented
Berkeley	Helgeson et al., 1978 $->$ SUPCRT (DATA0R51*)	data tables and data files	First extensive compilation to refine calorimetric data by means of phase-equilibrium data, using stepwise regression method on some 70 minerals in NKCMFASCH. With updates
	Helgeson et al., 1981	data files	Thermodynamic properties of aqueous species. Several updates to 1989
Cambridge	Powell and Holland, 1985 Holland and Powell, 1985	method data tables	Refinement of calorimetric $\Delta_f H°$ values by phase-equilibrium data, using a global least-squares regression approach. Considers a broad spectrum of published
	$->$ THERMOCALC	data files	data on 43 minerals in NKCMASCH
	Holland and Powell, 1990		Corrections and update to 117 minerals in NKCMmFATSchO by same method
UBC	Berman et al., 1986 Berman, 1988 $->$ GE0CALC[†]	method data tables and data files	Consistency analysis and optimization of $\Delta_f H°$, $S°$, and $V°$ by mathematical programming Considers all of the published thermophysical, thermochemical, and phase-equilibrium data, covering 67 minerals in NKCMFATSCH

*used by EQ3/6 (Wolery, 1979) and by other programs;
[†]used by PTAX (Berman et al., 1989; Lieberman 1991a; b) and THERIAK (deCapitani and Brown, 1987);
[‡]Constituent abbreviations: A: Al_2O_3; C: CO_2; c: C; C: CaO; F: FeO and Fe_2O_3; H: H_2O; h: H_2; K: K_2O; M:MgO; m: MnO; N: Na_2O; O: O_2; S: SiO_2; T: TiO_2.

Table 8.3 Scope of the five thermodynamic data bases compared

Data base	Minerals	Aqueous species	Gases
1. SUPCRT	136[39]	58[103]	8[13]
2. UBC	73	none	5
3. HP-DS	123	none	6
4. EQ3/6	303[322]	153[472]	8[5]
5. MINTEQ	111	71	3

Listed first are the number of species that lie in the composition space e^- – H–C–O–Na–Mg–Al–Si–P–S–Cl–K–Ca–Ti–Mn–Fe only. Numbers in square brackets indicate additional species outside that system, including notably elements such as Pm, U, and Th.

of data base, denoted here 'type S', contains for each mineral species (a sufficient set of) the standard-state properties, such that any and all of the thermodynamic functions for this species may be calculated, typically over an extended interval of physical conditions, and ideally over the entire range of geologically interesting temperatures, 0 to >1500°C, and pressures, 0 to >5 GPa (50 kbar).

Further to these differences in data representation, data bases usually contain a number of specialized representations for certain species and properties. Equations of state are commonly included, e.g. for H_2O and other fluid species, and form an integral part of almost all thermodynamic data bases for minerals. Similarly, the functional dependence of some extensive properties, such as the heat capacity $C_p(T)$ or volume $V(T,P)$ may differ from one data base to the next. Inasmuch as the effects of these differences are not immediately obvious, a word of caution is due already here. Even if the very same experimental data are represented by different parametrizations, the interpolative and especially the extrapolative behaviour may be quite unequal and may lead to substantially different representations. Because many of the measurements, especially of the electrochemical properties (e.g. the enthalpy of formation), are made at temperatures far from the reference state conditions, such differences in representation immediately lead to difficulties. Direct comparison of values (e.g. $\Delta_f H°$) is cumbersome, indeed sometimes impossible; the practice of 'improving' a data base by substituting or importing new values from another data base can be disastrous unless all of the parts fit (Section 8.4.1).

8.1.4 Types and sources of thermodynamic data

Thermodynamic data include thermophysical, volumetric, and thermochemical data. The former class includes the heat capacity (C_p or C_v) of minerals and fluid species, determined as a function of temperature (and sometimes pressure).

Directly related are

the (third law) entropy
$$S°(T') = S^* + \int_{T=0K}^{T'} C_p/T \, dT \qquad (1)$$

and the heat content
$$H°(T_2) - H°(T_1) = \int_{T_1}^{T_2} C_p \, dT. \qquad (2)$$

Mineral data on C_p, $S°$ and $\Delta_T H°$ derive from calorimetric measurements on synthetic or well-characterized natural materials. Recent progress on deriving C_p and $S°$ from spectroscopic data, using quantum-chemical models of solids, is not discussed here; the topic is reviewed by N. L. Ross (Chapter 4, this volume). S^* denotes the zero-point entropy, a non-negligible contribution for some minerals containing 'frozen-in' disorder or imperfections (Section 8.2).

Molar volume data stem primarily from X-ray diffraction on synthetic phases, as do data on the differential properties, i.e.

the thermal expansivity
$$\alpha = \frac{1}{V}\left(\frac{\partial V}{\partial T}\right)_P, \qquad (3)$$

and the compressibility
$$\beta = \frac{-1}{V}\left(\frac{\partial V}{\partial P}\right)_T, \qquad (4)$$

or its inverse, the isothermal bulk modulus $K_T = 1/\beta$.

In addition to the above properties, geophysicists and high-pressure mineralogists use the Grüneisen parameter

$$\gamma = \frac{\alpha K_T}{\rho C_V} \qquad (5)$$

where ρ is the density and the two heat capacities are related by

$$C_P - C_V = \alpha^2 V T K_T. \qquad (6)$$

Thermochemical data result from measurements involving some chemical reaction, i.e. reflecting a change in the arrangement of matter. For example, the enthalpy of formation ($\Delta_f H°$) for a mineral from its (element or oxide) constituents needs to be determined by some some scheme of measurements, such as can be realized in a solution colorimeter. The heat of reaction associated with the dissolution in HF or a fused salt is measured, not only for the mineral species, but for each of the constituents as well. Evidently, the uncertainty in $\Delta_f H°$ contains the combined uncertainty in all of these measurements. Given the complex schemes commonly required for silicates, for example, it is not surprising that the uncertainty of these data can be fairly large and that it is no simple matter to assess it realistically. Furthermore, for refractory substances, complete reaction may not be guaranteed (especially in HF calorimeters); for some systems intermediate states in the reaction sequence may occur or changes in the composition of the solvent may be non-negligible. Such effects add to the uncertainty and obviously increase the size of systematic errors. As a result, enthalpy data are commonly not known to the required accuracy which, depending on the application, may be on the order of a few hundred Joules/mol for each species involved in a chemical equilibrium.

From the users' point of view it is a real blessing, therefore, that thermochemical and thermophysical data for individual mineral species can be checked and refined by means of independent experimental data, especially by phase-equilibrium information (Section 8.2).

8.2 Thermodynamic measurements on chemical species, phases, and equilibria

We may view all of the data pertaining to the (relative) stability of geological substances as serving a single purpose: determining the thermodynamic properties of geological systems. Regardless of the intended use of these thermodynamic data, such a view leads to two questions:

1. Which primary data yield which constraints on thermodynamic properties?
2. How do we best use all of the primary data collectively to refine the thermodynamic properties of geological substances? (Equivalently, we could ask more formally: how can we satisfy optimally the simultaneous constraints imposed by all of the data?)

These questions are addressed in this chapter and are discussed in the light of current practices employed in deriving consistent, thermodynamic data bases for minerals.

8.2.1 *The structure of the problem*

In most applications, thermodynamic data serve to represent or predict the relative stability of a species, phase, or phase assemblage. Here, 'relative' refers to a competing arrangement of matter of identical bulk composition. Practically, heterogeneous chemical equilibria among phases as well as homogeneous equilibria (within phases) are readily described in terms of their energetic behaviour. Particularly convenient descriptions are afforded by either the Gibbs free-energy function (G) or the chemical potential (μ). Though neither function is directly measurable, all of their parts are, at least in principle. In practice, the complete Gibbs energy function is usually not required. It turns out especially simple and sufficient to use the *apparent Gibbs free energy*:

$$\Delta_a G° = \Delta_f H° - T \cdot S° + \int_{T_r}^{T} C_p \, dT - T \cdot \int_{T_r}^{T} C_p/T \, dT + \int_{P_r}^{P} V \, dP + R \cdot T \cdot \ln(a). \tag{7}$$

(Note that this definition follows Berman et al. (1986) and differs slightly from that introduced by Benson (1968) and used by Helgeson et al. (1978).) Though devoid of immediate physical meaning, $\Delta_a G°$ contains just the bare minimum of terms needed to compute chemical equilibrium problems. Except for the enthalpy of formation $\Delta_f H°$ (from elements, oxides, or similar), $\Delta_a G°$ uses only properties of the species or phase itself: its entropy $S°$ and volume V, its heat capacity C_p, and activity a. Of these properties, $\Delta_f H°$, V, $C_p(T)$ and $S°$—by integration—are more or less directly measurable by experiment. V may be measured over a range of T and/or P, data especially crucial for compressible species, including fluids.

Further to these data pertaining to single species or phases, the observation of irreversible chemical changes (reactions and phase transitions) from one set of species or phases to an alternative set, yield information on the change in $\Delta_r G°$ at the conditions (T, P) of the observation or experiment. For any irreversible change from reactants (with stoichiometric coefficients $v_i < 0$) to products ($v_i > 0$), thermodynamics demand that $\Delta_r G° < 0$ and thus

$$0 > \sum_i v_i \Delta_a G_i° \tag{8}$$

Assembling terms, the formal representation of a chemical reaction thus yields an *inequality constraint* amongst thermodynamic properties of the reaction partners, with the sign of the relation depending on the direction of the change:

$$0\{\lessgtr\} \Delta_r H° - T \cdot \Delta_r S° + \int_{T_r}^{T} \Delta_r C_p \, dT - T \cdot \int_{T_r}^{T} \Delta_r C_p/T \, dT$$
$$+ \int_{P_r}^{P} \Delta_r V_T \, dP + R \cdot T \cdot \ln K \tag{9}$$

Each of the reaction properties (indexed r) is again a linear combination now of the properties
for each species i,

$$\Delta_r H° = \sum_i v_i \Delta_f H_i° \tag{10a}$$

$$\Delta_r S° = \sum_i v_i S_i° \tag{10b}$$

$$\Delta_r C_p = \sum_i v_i C_{Pi} \qquad (10c)$$

$$\Delta_r V_T = \sum_i v_i \left(V_i^\circ + \int_{T_r}^{T} \left(\frac{\partial V_i}{\partial T}\right)_{P=P_r} dT \right) \qquad (10d)$$

and the equilibrium constant or activity term,

$$\ln K = \sum_i v_i \ln(a_i). \qquad (10e)$$

Within a constraint set to determine thermodynamic properties of constituent species or phases, the effect of a single observation (of some chemical reaction) is perhaps not obvious and may indeed be modest. The power of such constraints taken collectively and in combination with measurements of the fundamental properties of individual species is, however, quite fantastic! Provided all of the data are compatible, phase-equilibrium information allows substantial refinement of many of the individual species' thermodynamic properties.

8.2.2 Consistency requirements: shape and contents

In a world free of conflict, all of the measurements are mutually compatible within the uncertainties estimated by each experimentalist, and all of the users have unanimously consented to a single reference state ($T_r, P_r, ...$) for all properties. In that world, the power of thermodynamic theory needs to be used only to unify all available data. In our world, users' opinions diverge (and with good reason). Worse, some data are inconsistent with each other, and uncertainty estimates may be rightfully questioned. Worse yet, the procedures of data analysis may lead to thermodynamic properties which are in conflict with some accepted observations, even where these data are not mutually incompatible. What, then, does it mean when 'internal consistency' of a thermodynamic data base is asserted? Practically, a list of increasingly stringent conditions can be formulated:

1. compatibility with basic thermodynamic definitions, e.g. $G = H - TS$; $H_2 - H_1 = \int_{T_1}^{T_2} C_p dT$;
2. adherence to one set of reference values (P_r, T_r, thermodynamic properties of elements or other basic compounds) and constants (gas constant, molecular weights, etc.);
3. representation of (all of the listed) experimental data, in the sense of some 'best compromise' with all of the experimental data considered simultaneously, regardless whether the data themselves are compatible or conflicting;

4. compatibility with all underlying data (within stated uncertainties), except for those data excluded (or error brackets relaxed) on the basis of the experimental procedures reported, in order to achieve consistency with the rest.

For the sake of clarity, it is suggested that a data base be denoted:

1. *formally consistent* if only the first two points are satisfied;
2. *partially consistent* if the third point is also guaranteed;
3. *fully consistent* if points 1, 2, 3 and 4 above are ensured.

Currently available thermodynamic data bases for minerals are readily classifiable according to these criteria (Section 8.7). Additional criteria, such as comprehensiveness (of species and properties), may determine the value of a thermodynamic data base. However, full consistency would seem the most basic quality, owing to the inherent respect of those primary observations and measurements judged best on the basis of reported experimental and analytical procedures. After all, no amount of data analysis should lead to violation of primary data deemed valid!

8.2.3 Nature and magnitude of the uncertainty in the primary data

GENERAL NATURE OF ERRORS

There are two basic kinds of uncertainty or error in the primary data: errors of measurement and uncertainties of identity; both kinds may contain a systematic and a random part.

Errors of measurement attend the chemical composition of the phase(s) in question and the conditions (temperature, pressure, other intensive variables) of experimental environment. By contrast, uncertainties of identity refer principally to the structure of the minerals, not only of the experimental products but, for phase-equilibrium experiments, also of reactant phases. Evidently, errors of measurement are more readily quantified, as are their effects on the free-energy constraints derived from an experiment. In general, these errors are reliably estimated and reported by the experimentalists themselves, based on reproducibility or stability of laboratory measurements. As such they quantify precision. Despite all efforts to eliminate systematic errors in the laboratory, precision may not be a complete estimate of the accuracy of all experimental data. It is a difficult and thankless task to assess the magnitude of suspected systematic error, e.g. of calibration, drift, and operation. However, the results of interlaboratory comparison and experience in the data analysis make it a necessary part of serious data evaluation.

The task is another order of magnitude more difficult yet when we seek to assess the effect of uncertainties of identity. The effect of structural differences between nominally identical minerals is not readily quantified in terms of energy. Where documented, differences in grain size, crystallinity (based on

XRD data), and ordering are becoming a bit more tractable in this respect thanks to recent progress in crystal chemistry and mineral physics (e.g. Salje, 1988). However, with newly acquired analytical techniques—such as high-resolution transmission electron microscopy (HRTEM) and various surface analytical techniques—being applied to experimental products, it is becoming increasingly obvious that the problem of microstructural identity is much more complex and affects more minerals than previously assumed. However, the energetic consequences of, e.g. a known concentration of specific defects, are still poorly known. For the sake of thermodynamic analysis, perhaps the only practical approximation is to regard a mineral thus affected as a solid solution. The chemical activity of the 'desired species' is then computed assuming ideal dilution (Henry's law behaviour) by the 'defect species', the concentration of which is known, e.g. from HRTM. It is obvious that such an approximation is crude in the light of research in defect chemistry, but for the structural complexities common in many silicates, no alternatives appear to have been found.

In many classical studies of experimental petrology, the product phases have not been characterized in detail. For these, the new evidence summarized above may raise questions only. We are left having to assume the magnitude of suspected microstructural problems or denying their presence until the pertinent data are produced. In any case, the effect of structural disorder (chain width, stacking sequence, etc.) on the thermodynamic properties for several mineral groups—amphiboles and clays in particular—cannot be estimated reliably at present.

SPECIFIC ERROR DISTRIBUTION

There appears to be no need to challenge in general the usual assumption of normal error distributions for individual sets of thermophysical or thermochemical data. Despite this, it may be important to recall that 'outliers' do exist and do need elimination and that robust statistics are preferable to *a priori* assumptions about (a Gaussian) error distribution. These remarks apply more obviously to thermochemical than thermophysical data, as will be detailed below.

Adiabatic calorimeters—for low temperature measurements—and drop calorimeters—especially for temperatures above 1000 K—yield heat capacity data for excellent quality (standard error $\sim 0.1\%$) but they are experimentally demanding and require large samples. Modern differential scanning calorimetry, though considerably less precise, is used increasingly because it is technically simpler and allows use of small samples, such as synthesized high pressure minerals. If—and only if—operation is quite careful, the DSC may yield C_p data within 1–2% uncertainty up to fairly high temperature. Above 700–800 K, at least some of the DSC data have standard errors of several percentage points (e.g. Mraw and Naas, 1979). The main difficulty with DSC data lies in the error assessment: Data of low quality are not at all easy to identify!

When representing C_p data of variable uncertainty by means of optimized heat capacity coefficients, weighting factors (proportional to the estimated variance) for each datum should be considered in the least-squares fit. For this purpose, reported calorimetric uncertainties are adequate only if they estimate the accuracy, not just reproducibility. In drop calorimetry at least, systematic errors outweigh the latter and must be included (Douglas and King, 1968). For DSC data a realistic error assessment is very difficult and depends largely on detailed knowledge about the experimental procedures. In the absence of this information, comparison of the measured results with predicted C_p values based on correlation analysis (Berman and Brown, 1985) is one useful though inconclusive test of the quality of new data. Good measurements generally lie within $\pm 2\%$ of such estimates.

Uncertainties in entropy depend not only on the accuracy of available calorimetric data but also on the lower limit of temperature for which C_p measurements exist. This is especially true for minerals containing quenched disorder states or transition metals capable of very low-temperature electronic- or magnetic-phase transitions, most of which lead to λ-anomalies in C_p (Ulbrich and Waldbaum, 1976). Neglect of such phase transitions introduces systematic errors of several percentage points into calorimetric entropy data. As vacancies, substitutions, and disorder account for configurational contributions to $S°$, third-law entropy data for any phase with a zero-point contribution (S^* in equation (1)) due to quenched disorder or imperfections may constitute only a lower bound on the true value of $S°$ (Berman et al., 1986). On the other hand, S^* itself may be smaller than previously thought, since recent studies on several minerals show that short-range ordering occurs at low temperatures even where the (more readily documentable) long-range ordering has ceased due to kinetic barriers (e.g. Putnis and Bish, 1983; Powell and Holland, 1985). Where data to $<10-20$ K exist and S^* is small, modern adiabatic calorimetry measures C_p within better than $\pm 0.1\%$, except at the lowest temperature (perhaps $\pm 1\%$ at $T < 15$ K, Westrum et al., 1968); integration yields $S°$ at 298 K within ± 0.2 to $\pm 3\%$, depending on the uncertainty in S^*. Realistic accuracies are not easy to assess for structurally complex minerals, a difficulty multiplied where corrections for impurities of (natural) samples are required.

Volume data, especially of simpler minerals and at ambient conditions, are fairly complete and generally of sufficient quality (typically 0.1–0.2% standard error) from X-ray cell refinements. However, for many minerals $V(P, T)$ is not well known; especially sparse are data on α at elevated pressure and $\beta(T)$. For some geological applications, e.g. phase relations in the earth's mantle, the available volume data are far from adequate. In addition, the (partial) molal volume of many complex solid solutions still is not adequately measured. (P–V–T properties for fluid species are discussed in Section 8.5). The volume data from several studies on some of the more complex minerals are mutually not compatible within stated uncertainties, presumably owing to real structural differences between samples. Unless of unequal quality, the most practical compromise seems to admit an uncertainty spanning the range of all such volume data.

Data on the enthalpy of reaction are certainly the most difficult to assess, owing to the greater complexity of heat of solution measurements, as compared to thermophysical techniques. Recent advances in HF-solution calorimetry (25–90°C) and in oxide-melt-solution calorimetry (600–1000°C) allow a precision of better than ± 500 J/mol (Navrotsky, 1979) to be attained in measurements of the heat of formation of silicates (from oxides). Many of the older sources of enthalpy data, however, have uncertainties that are definitely larger but are very difficult to estimate reliably. For example, discrepancies among the enthalpy of formation data have been noted even for nominally simple mineral species, such as enstatite (Charlu et al., 1975), anorthite (Charlu et al., 1978), and forsterite (Berman et al., 1986), among many others. Where several data sets of similar reliability exist, it is distressingly common for the uncertainty bands *not* to overlap. This observation is not to belittle the considerable efforts and advances made since the early days of Sahama and Torgeson (1948), but merely to stress that major difficulties remain in our ability to judge the overall uncertainties in heat of formation data.

These problems with enthalpy data have lead some analysts (e.g. Helgeson et al., 1978; Holland and Powell, 1985; 1990) to consider $\Delta_f H°$ data as a special class of thermodynamic parameter, indeed the only one in need of refinement (by means of the 'phase-equilibrium calorimeter', Zen, 1977). It is noted here, however, that this restriction is unnecessary and is a source of systematic errors to the extent that other thermodynamic data, e.g. $S°$, are not optimized and may not be precisely compatible with experimental constraints (see also Section 8.7).

Consideration of uncertainties in phase-equilibrium data is complicated by two factors. Firstly, it is a multivariate problem: Uncertainties of calibration, control, and recording or measurement of T, P, X (phase composition) and possibly other uncertainties yield a combined effect on the freedom of reaction they constrain. Secondly, the uncertainties of individual half-brackets are sometimes combined for the sake of data analysis (Ch. 9), a practice that has been questioned and will be discussed in Section 8.3.2.

Experimentalists take care to minimize and assess the errors of operation in their labs. In general the uncertainties reported with their data appear adequate and should be used in the data analysis. Exceptions to this practice are rare and include systematic errors not suspected by the experimentalist and documented later, such as friction corrections to piston cylinder data and uncertainties of phase identity due to the difficulties of characterizing fine-grained synthetic products. Only where a data set proves incompatible within the uncertainties of other, well-established data sets may it be advisable and indeed indispensible to reassess the errors reported with original data. It may be considered a telling test of any thermodynamic data analysis to see how carefully errors in the experimental data have been evaluated and incorporated!

ESTIMATION OF PROPERTIES

The problem of missing or suspect data has plagued many applications of thermodynamics to mineral systems and has lead to numerous attempts to estimate properties. Prediction algorithms for any thermodynamic property of minerals are based on a model describing their contents in terms of some modules of chemical composition (e.g. oxides) and in some cases of structure (e.g. polyhedral units). Some theoretical or semi-empirical connection to measured properties is typically used to establish the contribution of each such module to the property sought. Most models assume the contributions from each module to be independent to a first approximation; some properties need corrections, such as the well-established volume correction for entropy. Once calibrated, the prediction algorithm may be tested on the data base used to construct the model. The predictive value of any calibration cannot be judged on that basis alone, of course. However, given the state of refinement reached in the past 10 years for many properties, the distribution of residuals may give a good indication of the potential value of a method, especially if the extreme outliers are scrutinized. It might be appropriate also to re-emphasize that correlation analysis is not independent of the units of quantity chosen and that gram-atom units are preferable to molar units for this purpose (Brady and Stout, 1980).

Estimation of C_P has been mentioned above; heat capacity is certainly the property for which correlation analysis (Robinson and Haas, 1983; Berman and Brown, 1985) yields the most reliable of predictions. Entropy estimation, recently refined by Holland (1989), follows closely except for phases with suspected zero-point entropy contributions or other complications. The notorious difficulties in estimating the enthalpy or free energy of formation of a mineral has led to a long sequence of improvements in the approaches or models proposed and, indeed, in the underlying data base used to calibrate them. Recent contributions by Vieillard and Tardy (1988) using detailed crystal chemical information and by Chermak and Rimstidt (1989) using selected polyhedral units constitute hopeful progress. Yet, though the estimation methods have improved in terms of their average deviation, the calibration residuals are still on the order of several kJ/mol, the absolute uncertainties are bound to be considerably higher (owing to the uncertainty in the underlying data base) and there remain distressing outliers, against which users are warned explicitly in both papers! Not many applications come to mind that would warrant the use of thermodynamic data so uncertain.

8.3 Data retrieval and data analysis

Thermodynamic data bases, regardless of their type, are fed by and, at best, representative of the underlying stability data measured. An important aspect of this review concerns the connection between underlying data, the mathematical structure of the constraints they impose, and the procedures of data analysis employed. When discussing methods of data analysis, it is helpful to make a clear distinction between our ideal goal and the practical result we can achieve. In our case, we seek to determine the fundamental thermodynamic properties

$$\Delta_f H_i^\circ \quad S_i^\circ \quad V_i^\circ \quad C_{p,i} \quad \text{(11a)}$$

of every species (i). In practice, we may obtain a set of parameter values

$$h_i \quad s_i \quad v_i \quad c_i \quad \text{(11b)}$$

for each species. The procedures of data analysis need to ensure primarily that the parameters are indeed representative of the properties sought. To that end, the mathematical structure of the constraint matrix must respect basic thermodynamic theory and it should be possible to translate each of the available measurements into a mathematical form so as to constrain one or several of the parameters to be determined.

Since the ultimately accepted set of parameter values is supposed to satisfy all of the mutually compatible experimental constraints, it is essential that all of these constraints be considered simultaneously. In addition, any incompatible data should be identified and removed, so as not to influence the 'best' or optimal parameter set. Only if these conditions are met, can full internal consistency (with the underlying theory *and* data) be guaranteed (Section 8.2.2). The distinction between parameter values and the thermodynamic properties they represent will be important in our discussion of uncertainty in the parameters (Section 8.4).

8.3.1 Goals

In a very general sense, data analysis serves a double purpose: verification and simplification. We shall thus attempt to answer two questions, one concerned with the quality of data, the other with the economy of representation:

1. Which of the (primary) data are *mutually incompatible* or contradictory?
2. What is a *sufficient* (simple yet complete) way to represent all of the compatible data?

More specifically, for the data to be analysed here, all of which constrain the relative stability of different states of matter, both these tasks are helped by thermodynamic theory. It provides a sufficient set of fundamental properties and of relations that must hold between them.

If we had, for a collection of chemical species, a complete set of their fundamental properties, these would contain all that can be known about the stability of a system composed of (nothing but) those species. Within a specified uncertainty, say the standard error, admitted for the value of every property, all of the known stability data would be satisfied and could be reproduced by calculation from the fundamental properties.

In practice, we have a large number of measured constraints, all subject to some uncertainty, and we seek to extract the 'fundamental juice' they collectively contain: our ambition is to determine a set of parameters, such that they reproduce every pertinent measurement within its admitted uncertainty. The parameters are chosen so as to represent the fundamental thermodynamic properties. Whether the values of our parameters turn out to be good representations of these fundamental properties depends both on the data themselves and on our procedure of data analysis.

8.3.2 Methods of data analysis: mathematical programming and regression analysis

Let us start by restating the general problem: For a given chemical system, defined by its mineral and fluid species, we seek parameter values *par*:=($par_1 \cdots par_m$) to represent (equations (10a) and (10b)) the thermodynamic properties of each species. The vector elements of *par* are to satisfy relations corresponding to equations (1)–(4) and lie 'as close as possible' to the primary data, ideally within the tolerance interval corresponding to some estimated confidence level of the data. Phase-equilibrium data are to be satisfied, i.e. phase relations computed using *par* must not conflict with experimentally established half-brackets (Section 8.1.2).

There are two basically different approaches in use to accomplish these goals:

1. linear programming (LIP) as pioneered in this context by Gordon (1973); and
2. regression analysis (REG) as initiated by Helgeson *et al.* (1978).

Both approaches have been extended considerably in recent years: LIP has been generalized to handle non-linear constraints and is now referred to as MAP, mathematical programming (Berman *et al.*, 1986), of which LIP may be thought as a special case. REG has been further developed to consider all of the constraints simultaneously and to include error propagation (Powell and Holland, 1985, Holland and Powell, 1990).

Both approaches, MAP and REG, as well as the thermodynamic data bases developed using them continue to evolve. Therefore, the following section concentrates on the main characteristics of the two approaches; further aspects are considered by Berman *et al.* (1986) and a comparison of the resulting data bases follows in Section 8.7.2.

Certain measurements on individual mineral species—such as their $C_p(T)$ or $V(P, T)$—are commonly considered independent of all other data. If such results are analysed separately, i.e. independent of phase-equilibrium data, all data analysts use linear regression. A weighted least-squares approach is usually employed to fit coefficients k_i of some heat capacity function, preferably one permitting reliable extrapolation, such as

$$c_p(T) = k_0 + k_1 T^{-1/2} + k_2 T^{-2} + k_3 T^{-3} \quad \text{(Berman and Brown, 1985)} \quad (12a)$$

or

$$c_p(T) = 3nR(1 + k_1 T^{-1} + k_2 T^{-2} + k_3 T^{-3}) \\ + k_4 + k_5 T \quad \text{(Fei and Saxena, 1987)} \quad (12b)$$

or, for V, a volume function, such as

$$V(P, T) = V_r(1 + v_1)(P - P_r) + v_2(P - P_r)^2 \\ + v_3(T - T_r) + v_4(T - T_r)^2 \quad \text{(Berman, 1988)} \quad (13)$$

or an integrated version of equations (3) and (4). Here, the REG approach is adequate in most cases, because these properties are represented by equality constraints (equation (15b)). In the absence of phase-equilibrium constraints, optimized parameter values would thus result simply from minimizing the sum of squares of residuals, weighted by the inverse estimated variance. For any property of a species, the weighted mean of individual determinations would represent the best value of the corresponding parameter.

However, it has been found important to refine $C_p(T)$ for a few minerals by considering constraints derived from pertinent phase-equilibrium information. This need arises typically where the thermal stability of a mineral is limited at low pressure but its phase relations were investigated experimentally to high P and T. Where a conflict with the (extrapolated) heat-capacity function occurs, C_p measurements and phase-equilibrium data must be considered simultaneously, so as to constrain the high-temperature portion of the C_p function. Similarly, phase-equilibrium data from high-pressure studies have been used to optimize the volume function (equation (13): $v_1, \ldots, v_4$) of some minerals (Berman, 1988).

As mentioned in Section 8.1.3, the main difference between MAP and REG concerns constraints from phase-equilibrium data. Table 8.4 (from Berman *et al.*, 1986) summarizes the two approaches used and their effects. Whereas MAP translates each half-bracket to an inequality relation (equation (9)), yielding an inequality constraint equation (15a), REG considers the same data pairwise, as a reversal bracket. In this case the data are translated into an equality constraint (equation (15b)) by assuming $\Delta_r G° = 0$ for the mid-point between the half-brackets.

In the language of optimization theory, REG solves the following uncon-

Table 8.4 Comparison of characteristics of the MAP and REG approaches to phase equilibria (Berman et al., 1986)

Mathematical programming analysis (MAP)	Regression analysis (REG)
Treats phase equilibrium data as statements of inequality in $\Delta_r G$	Treats weighted midpoints of brackets as positions where $\Delta_r G = 0$
Analyses individual half-reversals, allowing constraints from unreversed experiments consideration of different effects (solid solution, different starting material) for every experiment	Analyses pairs of half-reversals as brackets
Ensures consistency with all of the data (if consistency is possible)	Minimizes sum of squares of the residuals, but does not ensure consistency with all data
Provides a range of solutions (feasible region) from which a unique solution is obtained with a suitable objective function	Provides a 'unique' solution depending on weighting factors
Uncertainties approximated by the range of parameter values consistent with all data	Uncertainties computed from covariance matrix

strained minimization problem:

Minimize the residual function $RES(q)$

$$RES = \left(\sum_i W_i (Q_i - q_i)^2 \right) \qquad (14)$$

where the parameter values q_i are used as estimators of the data Q_i. These include not only the independently measured thermophysical, volumetric, and thermochemical data for minerals and fluid species, but—at least in the formulation of Powell and Holland (1985)—also the enthalpy of reaction (equation (10a)) data, i.e. the equality constraints derived from the centre of each reversal bracket. The weighting factors W_i scale the influence of every residual by some measure of the uncertainty.

MAP solves an optimization problem also, but in this case it is a constrained minimization problem of the following form (Berman et al., 1986):

Minimize the objective function $OBJ(q')$

$$OBJ = \left(\sum_i W'_i (Q'_i - q'_i)^2 \right) \qquad (15a)$$

$$\text{subject to } f_i = 0, \quad i = 1, 2, \ldots, m \qquad (15b)$$

$$g_i > 0, \quad i = m+1, \ldots, n \qquad (15c)$$

The primed symbols in equation (15a) correspond to those in (14), but here they do not include the phase-equilibrium constraints; these are added as inequality constraints to equation (15c) and complement any explicit equality relations in equation (15b) that have to be satisfied (without error) among the parameter values q'.

MAP thus operates directly on *individual* half-brackets; REG operates on *pairs* of half-brackets and makes an assumption about the location of the equilibrium point and, more generally, about the probability distribution between half-brackets (Figure 8.1). Indeed, the least-squares approach advocated by Powell and Holland (1985; 1990) assumes a normally distributed (Gaussian) probability distribution, an assumption questioned by Berman *et al.* (1986) but defended by Holland and Powell (1990). Kolassa (1991) analysed

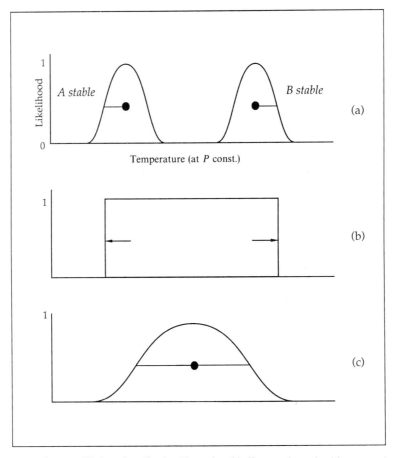

Figure 8.1 Phase-equilibrium data (for A⇔B) consist of half-reversals, each with an experimental uncertainty. Schematically, (a) depicts the inferred likelihood of A→B and B→A being 'true'; the approximate likelihood assumed by MAP is shown in (b), the one assumed by REG in (c).

such probability distributions—derived directly from individual experimental constraints—and found them to be far from normal (see also Section 8.4.2). He emphasized that REG as it has been used to date (unconstrained regression) is very likely to yield optimal thermodynamic parameters that are not consistent with some of the phase-equilibrium constraints. This is due to the fact that the width of a reversal bracket is commonly two to four times greater than the estimated experimental standard error in the location of each half-bracket; for such cases at least, the likelihood surface for the locus satisfying $\Delta_r G° = 0$ is flat (Kolassa, 1991) and the central tendency imposed by assuming a normal likelihood distribution is undesirable. Recognizing this, Holland and Powell (1990) introduced a weighting scheme to reduce the effect of particularly wide reversal brackets on their least-squares fit. Weighting factors are given the careful attention they need by these authors who adjust them iteratively with the aid of regression diagnostics that test for undesirable effects in the fit. In a general sense, this process changes the meaning of the W_i which are no longer proportional to the inverse estimated uncertainty of each datum.

MAP, by contrast, cannot produce inconsistencies in the process of minimizing the objective function. In fact, MAP points directly to any inconsistencies that might be inherent in the data themselves. The need to resolve such inconsistencies may well be the most powerful aspect of the approach. Note that REG has no rigorous way to discern whether an inconsistency is inherent in the data or due solely to the optimization. However, obvious discrepancies in the experimental data can be detected by inspection of the REG results, of course, and are generally resolved by ignoring the suspect data (Holland and Powell, 1990, p.102). As for weighting factors in MAP, the values of W_i' are usually taken as the inverse estimated variance of each experimental datum. Phase-equilibrium half-brackets asymmetrically distributed about the equilibrium locus—as they commonly are owing to unequal kinetics of forward and backward reactions—are not given an artificially small weight (large uncertainty) in the fit.

Constraints derived from phase equilibria require incorporation of the uncertainties in the location of each half-bracket (MAP) or of the width of the reversal bracket (REG). In the former approach, each constraint is usually relaxed by the experimental error estimates in temperature, pressure, and, if appropriate, the activity of participating species or composition of phases in the equilibrium (see Berman et al., 1986). In REG the experimental brackets are converted to an average enthalpy of reaction constraint; for improved robustness, Holland and Powell (1990) advocate use of the mean of two, median enthalpy values derived for example, from the low- and the high-temperature extremes of brackets collected over a certain pressure range. The standard error in the derived $\Delta_r H$ is estimated as one quarter of the difference between the high- and low-bracket ends. This may be rather optimistic, given that the data required in the derivation of $\Delta_r H$—notably the $\Delta_r S, \Delta_r V$ and $\Delta_r C_P$ data—can hardly be assumed to be known free of error.

Before pursuing further the effects of uncertainty in the data and in the weighting, a brief account is given here of the matrix formulation useful in setting up the general optimization problem outlined. Emphasis here is not on the method used to solve such problems but on the formal simplicity of what might by now appear a hopelessly complicated data analysis.

MATRIX FORMULATION

The diverse primary data have been shown above to yield a set of equally diverse constraints on the thermodynamic properties of minerals. Simultaneous consideration of all of these constraints is readily accomplished by means of matrix analytical methods. It is practical to divide the constraint matrix, consisting simply of an ordered array of constraints written in matrix format, into a submatrix assembling inequality constraints and another consisting of the equality constraints. In both parts, expressions ($\exp_1 \ldots \exp_m$) containing the parameters to be determined or refined—the unknowns in the data analysis—are assembled on the left hand side (*LHS*), whereas the properties assumed to be known or fixed in the data analysis are combined into constants ($con_1 \ldots con_n$) and form the *RHS* of each constraint expression. The form of each constraint thus becomes:

$$LHS\ (\exp_1 \ldots \exp_m) \{\lessgtr\} RHS\ (con_1 \ldots con_n) \tag{16a}$$

or

$$LHS(\exp_1 \ldots \exp_m) = RHS\ (con_1 \ldots con_n) \tag{16b}$$

The *LHS* is readily split into a vector product by assembling the real coefficients ($coe_1 \ldots coe_m$) associated with each parameter expression in a row vector; the parameter values ($par_1 \ldots par_m$) themselves form a column vector:

$$LHS = \left([coe_1 \ldots coe_m] \begin{bmatrix} par_1 \\ \ldots \\ par_m \end{bmatrix} \right) \tag{17}$$

By adopting a unique ordering of the parameters, the complete set of simultaneous constraints may be represented by a coefficient matrix **COE**, the parameter vector **par**, and a vector of constants **con**:

$$\boldsymbol{COE} \cdot \boldsymbol{par}\ \{\lessgtr\}\ \boldsymbol{con} \tag{18}$$

For practical purposes, each row entry representing an inequality constraint may be transformed to an equality constraint simply by adding to its associated con_i-value a 'slack' variable sla_i that is inherently positive for '<'

constraints and negative for '>' constraints. It does not matter that the numerical values of these slack variables are not known *a priori*; practical optimization techniques exist (Gill et al., 1981; Schittkowski, 1985a; b) to take advantage of this general matrix formulation, which thus takes the simple form:

$$COE * par = c \qquad (19a)$$

where

$$c = con + sla \qquad (19b)$$

and *sla* represents the vector of slack variables (zero for equality constraints).

The procedure of formalizing the data-analysis problem and an example of applying it to an extensively researched three-component system are described in some detail by Berman et al. (1986).

8.4 Quality of optimized thermodynamic properties

The quality of thermodynamic data has traditionally been a very central theme of those parts of science producing the data. With the widespread use in increasingly complex predictive models, many of undeniably practical importance (e.g. hazardous waste disposal), the questions of reliability are becoming crucial to many 'users' of the data and indeed affect us all. This development has had obvious reflections in the earth science literature. For example, uncertainties in the parameters derived by REG and MAP methods (Section 8.4) have been the focus of a recent debate between protagonists of either method (Halbach and Chatterjee, 1982; Powell and Holland, 1985; Berman et al., 1986; Berman, 1988; Holland and Powell, 1990; Kolassa, 1991). Apart from algorithmic particularities—which might be best left to the specialist, were it not for some obvious effects on the results—the two main goals of error analysis do not seem to be disputed:

1. Uncertainties in the final optimized parameters are to reflect adequately the collective uncertainty in the data;
2. contributions introduced by the optimization procedure must be negligible.

Both of these conditions are difficult to fulfil and to ascertain convincingly.

8.4.1 Error propagation analysis and internal consistency

The central goal of data analysis was defined above as a *consistent and optimal representation of those data which are mutually compatible within the framework of thermodynamics*. The term optimal is understood in the sense defined by the objective function (equation (15a)); consistency requires satisfaction of every experimental constraint (equations (15b) and (15c)). The first question, then, is how uncertain the constraints are owing to the experimental errors. Error-propagation theory provides the answer rigorously, at least in principle, since a functional relation exists linking the experimental variables $(Y_1, Y_2, \ldots, Y_n)$ to each variable Λ in which the constraints are cast (equations (1) to (10)). Because the experimental errors are nearly uncorrelated, we may express the error in Λ as

$$\varepsilon^2(\Lambda) = \sum_j \varepsilon^2(Y_j)(\partial \Lambda / \partial Y_j)^2 \qquad (20)$$

where $\varepsilon(Y_j)$ represents the standard error in the jth experimental variable. The requisite partial differentials may be evaluated explicitly (e.g. $\partial \Delta_r H^\circ / \partial T = \Delta_r C_P$) or by a finite-difference approximation. For phase-equilibrium constraints, it is not rigorously possible to associate the resulting estimate of uncertainty (in the constraint derived from a half-bracket) with a particular confidence level, let alone to infer a specific probability distribution. However, since parameter optimization generally uses a maximum likelihood criterion, some explicit or implicit formulation of the likelihood function is needed. Generally, a normal or Gaussian distribution has been assumed. The problem of combining two opposing half-brackets has already been addressed; the question here is how to combine the uncertainties of each half-bracket constraint—Gaussian or otherwise—into the uncertainty of the entire feasible region. Recent work on this by Kolassa (1990) extends an earlier study by Demarest and Haselton (1981) and shows that the likelihood function describing the location of an equilibrium as constrained by several half-brackets is regular and generally yields a flat likelihood surface. Because the experimental errors in the location of each half-bracket are relativey small, so are the errors in the location of each constraint small compared to the size of the feasible region they define collectively. Kolassa concludes that parameter values derived using maximum likelihood methods—such as REG analysis assumes—are 'most likely not consistent, let alone asymptotically normal' (Kolassa, 1990, p. 5).

Anderson (1977) and Powell and Holland (1985) emphasized the need to consider covariance, i.e. the correlation between parameters in error propagation. Fortunately, REG would seem to lend itself to explicit error propagation analysis because standard least-squares regression packages can compute a covariance matrix for the optimized parameters. Kolassa's results pose a number of questions, however: How representative of the uncertainty in the data are these covariances? If the probability distribution in an important part of the data is not normal, are the uncertainties in the fit parameters lower or upper bounds? How robust an estimate of the combined uncertainty in the data is the covariance matrix, given the importance of a weighting scheme? These questions await further analysis but cast serious doubts on the validity of regression-derived parameter uncertainties. As discussed previously, Holland and Powell (1990) advocate use of weighting fators in the least-squares fit as a way to minimize the danger or effect of inconsistencies introduced by REG. However, reducing the weighting factor of a wide reversal bracket introduces uncertainty that is not inherent in the experiments, at least in cases where one half-bracket of a reversal lies close to the equilibrium. Consequently, uncertainties in parameters optimized by REG may be larger than commensurate with the data. On the other hand, REG tolerates incompatibilities within the data, which translates to an *underestimation* of the actual error in derived parameters. Certainly, weighting factors make it difficult to interpret the uncertainties in optimized thermodynamic parameters.

Finally, the probabilistic character of uncertainty in refined parameters emphasizes a conceptual difficulty that is exemplified by the following question: is it permissible that an internally consistent data base leads to predictions which are (nominally) in conflict with underlying data and consistent with these *only* if uncertainties are considered? If yes, what does internal consistency mean—50%, or 90%, or 99% confidence? The question is non-trivial because the primary data themselves, i.e. experimental half brackets, are composed of two parts, of which only one is probabilistic (e.g. the location of a P–T half reversal); the other part is free of error (e.g. whether it is a low- or a high-T bracket). Kolassa (1991) addresses this problem and his approach combines the two parts successfully. His results would permit quantitative criteria, based on likelihood functions, to define internal consistency.

The promise of explicit error propagation, such as REG suggests, has great attraction. However, uncertainty estimates provided by propagation of least-squares–derived covariance data do not appear to be realistic in distribution or magnitude; indeed they represent neither upper nor lower bounds. Unless or until more reliable estimates of the desired covariance matrix appear, error-propagation calculations lack a sound basis. Two methods based on MAP were proposed recently and promise improvements in the near future: an extended application of Kolassa's (1991) explicit formulation of confidence intervals and/or a stochastic approach, such as a Monte Carlo method (Engi *et al.*, 1990).

It may also be appropriate to stress again that, even once the errors of measurement in the primary data are adequately propagated into the optimized thermodynamic parameters, uncertainties of phase-identity and other systematic errors remain to be addressed. Some of these problems are the topic of the following two Sections.

8.5 Equations of state for fluids and aqueous species

Geologically interesting processes commonly involve at least one volatile phase or fluid, whether gaseous, liquid, or supercritical. In some cases fluids act as solvents in that they may contain significant concentrations of solute species, especially at high temperatures. In order to simplify computations, thermodynamic properties of fluids are normally parameterized as equations of state that express volume or density or free energy as a function of pressure and temperature. The need for this stems from the relatively rapid variation of V (and hence G) for fluids with P and T. Beside the computational advantages of compact representation and interpolation of data, at least some of the equations of state form a useful basis for extrapolation beyond the P and T covered by experiment. Confidence in such extrapolations depends much on theoretically correct behaviour, especially in the limits of high and low P and/or T (e.g. Haar et al., 1984); these conditions are fulfilled to unequal degrees by various formulations. For example, most versions of the popular (modified) Redlich–Kwong equation of state (Holloway, 1977; 1987) represent data inadequately in some parts of the P–T-range interesting to geologists; this defect is shared by most virial equations. Much effort has gone into the development of equations of state that do satisfy both the available data—especially P–V–T measurements—and the theoretical limits. A single equation suitable for all (geological) purposes has not been found, except perhaps for H_2O.

It has been emphasized earlier that optimized thermodynamic properties of mineral species are tightly linked to those of fluid species. Hence it is crucial that mineral data be used with the *same* equation of state that was employed in the refinement procedures. Failure to observe this rule will destroy whatever consistency there may be in a data base and will thus violate underlying experimental data!

The following review emphasizes equations of state that are tied to existing thermodynamic data bases for minerals and concentrates specifically on models for volatiles of the COH system. Equations of state for the species in aqueous solutions and electrolytes are not reviewed in detail, but references are included that attest to the rapid progress being made in this field.

FORM OF CURRENTLY USED EQUATIONS OF STATE FOR FLUIDS

Third-power equations (in the molal volume $\bar{V}$) dominate in the recent literature, either based on the Van der Waals equation or, more commonly, on a Redlich–Kwong form (Holloway, 1977)

$$P = \frac{RT}{\bar{V}-b} - \frac{a}{T^{1/2}\bar{V}(\bar{V}+b)} \tag{21a}$$

The coefficients a and b are specific to each fluid species and, for polar fluid species, a is usually represented by a power series in temperature,

$$a = a^\circ + (L + MT + NT^2 + \cdots) \tag{21b}$$

$a^\circ, L, M, N, \ldots$ are treated as fit coefficients and optimized using P–V–T data

(Ferry and Baumgartner, 1987). Modification of these equations by inclusion of further parameters lacks a sound theoretical basis; it may improve agreement with experimental data, though sometimes at the expense of extrapolative performance or greater computational effort ($\int V \, dP$).

Some virial and virial-like equations are also in use (e.g. Saxena and Fei, 1987a; b) as are simple polynomial equations expressing the fugacity ($\ln f$) as polynomials in P and T (e.g. Powell and Holland, 1985, Holland and Powell, 1990). Depending on the calibration, these representations are more or less adequate for interpolative use; in general, they do not perform well in the limits and demand care anywhere outside the calibrated range.

EQUATIONS OF STATE FOR H_2O, CO_2, AND OTHER MOLECULAR SPECIES

The state-of-the-art representation of thermodynamic properties for water remains the one known as the Haar equation (Haar *et al.*, 1984). This equation of state is superior to others because it relies on first principles as much as possible, introducing empirical terms only where its theoretical basis is weak—notably near the critical point—and thereby ensuring proper limiting behaviour, especially for high pressures and/or temperatures. The Haar equation parameterizes density $\rho = \rho(P, T)$ and provides all of the relevant thermodynamic functions, though at some computational expense. Data representation and extrapolation are excellent, at least for $1 \, \text{bar} < P < 10 \, \text{kbar}$ and $25° < T < 1200°C$. In this range it may be fair to compare predictions (Table 8.5 and Figure 8.2) of other equations of state (Helgeson and Kirkham, 1974a; Delany and Helgeson, 1978; Kerrick and Jacobs, 1981; Holland and Powell, 1990; Saxena and Fei, 1987a; b) to the Haar equation as if the latter represented 'truth'. Such a comparison points out the scale and distribution of discrepancies in a P–T-window of geological interest. Clearly, information on the average deviation of an equation from the data characterizes its performance inadequately; the distribution of the residuals matters. Both the Kerrick and Jacobs (1981) and the Holland and Powell (1990) equations show systematic deviations from the data. If used in connection with phase-equilibrium data, such deviations demand corresponding differences in the standard-state thermodynamic properties of hydrous minerals. Neither the mineral properties nor the fluid properties are interchangeable or directly comparable otherwise!

For $P > 10 \, \text{kbar}$ useful equations of state were developed by Delany and Helgeson (1978), Halbach and Chatterjee (1982), Saxena and Fei (1987a), and Holland and Powell (1990). These were derived by incorporating shock-wave data and static P–V–T data, principally those of Burnham *et al.* (1969). The high-pressure data and equations of state derived from them are considerably less reliable and difficult to compare meaningfully, because systematic uncertainties in the shock-wave data may outweigh the discrepancies between equations of state. Saxena and Fei (1987b) summarize the performance of their own and earlier equations; the simple polynomial of Holland and Powell (1990) shows relatively minor deviations.

THERMODYNAMIC DATA FOR MINERALS: A CRITICAL ASSESSMENT

Table 8.5 Hierarchy of aqueous species selection

Primary species (*PS*)	A defined set of independent (aqueous) chemical species. Taken as a set of basis vectors they span the chemical space of the derived species
Derived species	Include all non-primary species (i.e. secondary aqueous species, minerals, etc.) whose energetic properties are referred to the *PS* (e.g. as log *K*)
Extended *PS*	A defined set of derived species which substitute for specific *PS*, conserving linear independence of the combined basis

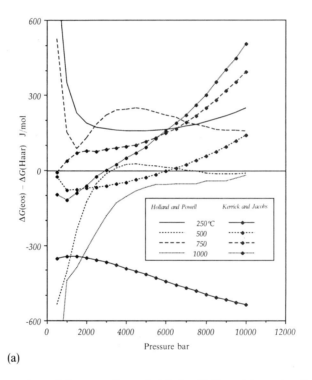

(a)

Figure 8.2 Comparison of three equations of state for H_2O used in conjunction with thermodynamic data bases for minerals. The apparent molal Gibbs free energy calculated by the equations of Kerrick and Jacobs (1981) and Holland and Powell (1990) is shown as a difference to values from the equations of Haar *et al.* (1984). (a) Isotherms between 250 and 1000°C are compared for $0.5 < P < 10$ kbar for these equations of state; (b) contours at constant pressure of the discrepancies of Holland and Powell's polynomial equation; (c) contours at constant pressure of the discrepancies of Kerrick and Jacobs' modified Redlich–Kwong equation. Within graphical resolution, free-energy values from Helgeson and Kirkham's (1974a) equations agree with those from the Haar equation and are not shown here.

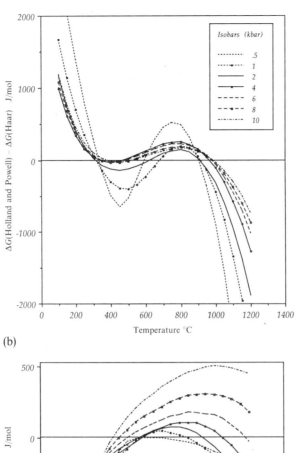

(b)

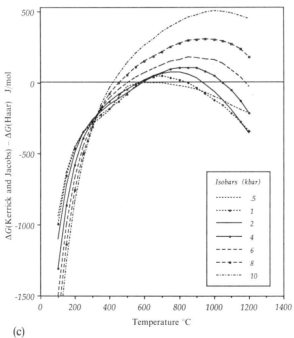

(c)

Figure 8.2—*contd.*

Compared with H_2O, the situation is less satisfactory for carbon dioxide, in that none of the equations of state used in conjunction with thermodynamic data bases does complete justice to the combined experimental data on CO_2. For example, a long-known discrepancy between high-pressure phase-equilibrium data and extrapolations from equations of state calibrated by means of $P-V-T$ data has only recently been resolved by Mäder and Berman (1991). They developed a Van der Waals type formulation:

$$P = \frac{RT}{V-(B_1+B_2T-B_3(V^3+C)^{-1})} - \frac{A_1}{TV^2} + \frac{A_2}{V^4} \quad (22)$$

in which $C = B_3/(B_1+B_2T)$. The five coefficients A_1, A_2, B_1, B_2, and B_3 were optimized simultaneously to $P-V-T$ data (to 8 kbar) and phase equilibrium constraints to 42 kbar. This new calibration yields much improved thermodynamic properties for CO_2. Unfortunately, these are not consistent with published thermodynamic data bases and thus cannot be used yet. Mäder and Berman (1991) also include a comparison of their equation of state with those of Holloway (1977), Bottinga and Richet (1981), Kerrick and Jacobs (1981), Powell and Holland (1985), and Saxena and Fei (1987a; b). Discrepancies among these are somewhat larger yet than for water.

For mixtures of H_2O and CO_2 the MRK-equation of Kerrick and Jacobs (1981) is still widely in use, despite known, inadequate data representation (at $P<500$ bar for $400° < T < 500°C$ and at $P>6000$ bar) and a formal insufficiency leading to divergence (at $T<325°C$ and $T>1050°C$). Over a large range of $P-T$ conditions, the performance of the Kerrick and Jacobs equation is satisfactory, however, as documented by Ferry and Baumgartner (1987) in a series of comparisons with phase-equilibrium data and with predictions afforded by some alternative equations of state (see also Mäder (1991) for a recent interview).

Ferry and Baumgartner (1987) used the Redlich–Kwong equation also for speciation calculations in fluid mixtures in C–O–H ($\pm S \pm N$). The calculations are straightforward but their results largely defy evaluation, for no data exist at present to test the computed distributions of species for these systems.

8.5.1 Aqueous species

Thermodynamic data of minerals in data bases designed for low-temperature applications depend on the activity models chosen for aqueous species. For neutral species, a simple dependence of the activity coefficient λ_i with the ionic strength I of the solution is commonly assumed: $\log \lambda_i = 0.1\,I$ (Helgeson, 1969). Electrolyte or ionic species are treated either by the Debye–Hückel law, the Pitzer equation, or a modified version of either (Helgeson and Kirkham, 1974b; Pitzer, 1973; 1987; Weare, 1987). These models differ conceptually in their account of changes in aqueous speciation with temperature and density. In Pitzer's approach, activities of aqueous complexes are formulated in terms of double- and triple-ion interaction. By contrast, explicit speciation models, such as those developed by Helgeson and coworkers, adopt specific activity expressions for each known (or inferred) aqueous species. In either case, the range of partial association remains poorly understood primarily because data at supercritical conditions are relatively sparse. Recent advances in the theory and data analysis on aqueous species are given in Oelkers and Helgeson (1988; 1990), Shock and Helgeson, (1990), Tanger and Helgeson (1988), Tanger and Pitzer (1989), Shock et al. (1990), and Sverjensky et al. (1990); these papers also include compilations of the thermodynamic properties to be used in preference over earlier versions for low-temperature applications.

8.6 Complications due to solid solution, ordering, and phase transitions in minerals

Most mineral phases are not completely stoichiometric; their structure tolerates more-or-less compositional variability. The thermodynamically relevant activity of a species (fixed composition) in a solid phase of variable composition is usually computed on the basis of a solution model that is built on the concept of mixing of ions on one or more sites; the substitutions may be independent or coupled (e.g. Cohen, 1986). In addition, within a phase of fixed composition, site occupancies by the ions may show more-or-less ordered patterns and/or clustering. These phenomena are usually temperature-dependent and may be relatively sluggish, at least at low temperatures. Finally, comparatively rapid structural adjustments occur in some minerals in response to changes in environment (P, T); these may lead to phase transitions of first, second or higher order, causing discontinuities in the thermodynamic properties of the respective order (first: G, H, or S; second: V or C_P, etc.).

In minerals where they occur, the above complications have important thermodynamic consequences, the description of which requires approximation by models. A large number of formulations and calibrations of these exist in the literature. Even where based on the same data, models are generally non-equivalent. It is obvious that whatever consistency a set of mineral data may have, needs preserving by use of the appropriate models describing solid solution, ordering, and other special effects.

8.6.1 *Solution models (activity models)*

An excellent introduction to models (Navrotsky, 1987) stresses the proper formulation of the ideal or configurational part of the solution model and gives a brief overview of common formulations of the excess properties that link the ideal to the real solution behaviour. Formulations of the excess functions vary in form; the requisite empirical calibrations, scattered in numerous papers, are all too often based on very limited, experimental data sets. The resulting models are not generally consistent with any of the current thermodynamic data bases, because the standard-state thermodynamic properties (SSP) in these were derived independently. Unless solution models were derived on the basis of a previously established set of internally consistent SSP (e.g. Engi and Wersin, 1987; Sack and Ghiorso, 1989; Ghiorso, 1990b) or, better yet, *simultaneously* optimized with these SSP, the activity models are likely to be inconsistent with the SSP and these should not be used together. Much remains to be done to fill the gap of analysing experimental-phase equilibria involving solid solutions. Furthermore, the experimental data base for such activity calibrations is not sufficient for many important mineral solutions. Use of select natural data is advocated (Holland and Powell 1990), especially distribution coefficient (K_D) data, as they may be our only means to construct a comprehensive petrologic data base. It is obvious that particular care is needed in selecting suitable natural equilibria, whose dependence on the P and T of (re-)equilibration is small (e.g. exchange couples with small absolute values of $\ln K_D$). Precise K_D-data, whether derived from experiments or rocks, yield precise differences in the chemical potentials (hence ratios of the activities) of coexisting phases. Unfortunatey, separation into individual phase properties invariably results in highly correlated functions. In part, this problem may be reduced by considering data simultaneously for a large number of phases; furthermore, at least one phase per exchange component is ideally tied to 'anchor data' (e.g. e.m.f. measurements, heat of mixing data) that are not correlated with the others. Careful error propagation analysis is particularly crucial in the analysis of such data sets, as the dangers of overfitting are real.

From the point of view of applications, the form of solution models is not crucial; however Ghiorso's (1987) plea for theoretically justifiable formulations is worth reiterating. A common problem for many minerals regards the coupled phenomena of ordering and mixing (Navrotsky, 1987; Ghiorso, 1990a). Though conceptually independent, their effects are not readily separable, e.g. on the basis of phase-equilibrium data alone. Where data exist, such as for feldspars, recent models based on Landau theory (Salje, 1987) are a major advance, as they keep the (independently measurable) structural parameters separate from the thermochemical ones.

8.6.2 Order–disorder and (rapid) phase transitions

The basic property affected is C_P—typically resulting in the λ-shape—and, by integration, all others ($S, H, G, \ldots$). Descriptions are based either on Landau theory (Salje, 1987; Holland and Powell, 1990) or on some purely empirical polynomial fit (Berman and Brown, 1985). For the reasons given above, a Landau formulation is clearly preferable. In both cases, a base line (or lattice) contribution that varies slowly with temperature is parameterized separately from the addition due to the transition itself.

The thermodynamic effects of phase transitions are crucial for some phases, such as quartz and cannot be ignored in applications of consistent data bases.

8.7 A comparison of widely used data bases

As emphasized in Section 8.1, any comparison of scientific results—as much in a state of flux as the development of a comprehensive, consistent data base of optimized thermodynamic data for mineral systems—is highly arbitrary in its scope and timing. Progress has never been faster than right now—or so it always seems—and it must seem grossly unfair, especially to those making the progress (by regression or otherwise!) to have selected exactly this time for an account of the status quo. These criticisms are justified and do limit the value of the comparison presented here.

8.7.1 Goals and procedures

Improvement of any data base should be based on a comparison of existing candidate data sets. Differences in the data format require special procedures for such a comparison.

Two classes of data bases are in use, tabulating, for each species, either its standard state properties (*Type S*) or log K, a pertinent equilibrium constant (*Type K*), or some parametrization thereof. Comparisons between *Type S* data bases may be based, in principle, on direct comparison of parameter values. On account of the high correlation between these parameters as well as their intended use, it may be more meaningful to compare the performance of each data base in a series of typical calculations. Phase diagrams are an excellent choice for such comparison because:

1. the free-energy maps they represent are very sensitive to the underlying data; and
2. the computations may be directly compared to data, both those used and those not used in in the derivation of thermodynamic data bases.

Comparison between data bases of *Type S* and *Type K* is possible by the computation of corresponding log K, i.e. the solubility constants of minerals, from standard-state properties. Owing to differences in the reference set of aqueous species (usually primary species, Table 8.6), comparison between two sets of log K data from different data bases necessitates special procedures, equivalent to a change of basis. Comparisons are then readily performed for 25°C and 1 bar. In addition, either the gradient $(\partial \log K/\partial T)_{P=1\,\text{bar}}$ or comparison of log K values at some other temperature is useful. Limitations of some of the low-temperature data and of equation (25) would suggest a conservative upper limit of perhaps 100°C for this.

The method of comparison used here is to convert Gibbs free energies $\Delta_f G^\circ$ and enthalpies $\Delta_f H^\circ$ of formation to equivalent log K values and their derivatives according to the equations:

$$\log K = -\Delta_f G^\circ/(R \cdot T_r) \log e \qquad (23a)$$

THERMODYNAMIC DATA FOR MINERALS: A CRITICAL ASSESSMENT

Table 8.6 Difference of log K data expressed as $\delta \log K$ (MINTEQ–EQ3/6)

Mineral species*	log K EQ3/6	log K MINTEQ	$\delta \log K$	$\Delta_r H°$ EQ3/6	$\Delta_r H°$ MINTEQ	$\delta \Delta_r H°$ kJ/mol
AKERMANITE	45.19	47.47	2.28	−297.94	−319.87	−21.93
ALBITE-LOW	3.10	2.59	−0.51	−60.08	−72.80	−12.72
ALUNITE	0.67	−1.35	−2.02			
ANALCIME	7.28	6.72	−0.56	−87.28	−95.56	−8.28
ANHYDRITE	−4.27	−4.64	−0.37	−24.81	−15.77	9.04
ANNITE	29.75	23.29	−6.46	−266.56	−274.97	−8.41
ANORTHITE	27.06	25.43	−1.63	−296.06	−295.64	0.42
ARAGONITE	−8.47	−8.36	0.11	−20.04	−10.96	9.08
ARTINITE	9.60	9.60	0.00	−123.93	−120.25	3.68
BOEHMITE	9.60	8.58	−1.02	−117.49	−117.70	−0.21
BRUCITE	16.44	16.79	0.35	−111.46	−108.11	3.35
CA3SIO5	74.01	73.87	−0.14	−442.67	−444.93	−2.26
CALCITE	−8.63	−8.48	0.16	−19.83	−10.84	9.00
CHALCEDONY	−3.73	−3.52	0.21	25.44	19.33	−6.11
CHRYSOTILE	31.55	32.19	0.64	−228.82	−219.62	9.20
CRISTOBALITE	−3.45	−3.59	−0.14	23.22	23.01	−0.21
DIASPORE	8.75	6.87	−1.88	−108.74	−103.05	5.69
DIOPSIDE	20.97	19.89	−1.08	−144.26	−135.06	9.20
DOLOMITE	−18.17	−17.00	1.17	−49.54	−34.69	14.85
EPSOMITE	−1.80	−2.14	−0.34	14.43	11.80	−2.64
FE(OH)3(PPD)	4.90	4.89	0.00	−77.66		
FE2(SO4)3(C)	0.81	3.58	2.77	−262.38	−247.36	15.02
FERRITE-MG	21.12	16.77	−4.36	−277.52	−278.82	−1.30
FORSTERITE	28.15	28.30	0.15	−211.67	−202.97	8.70
GIBBSITE	7.96	8.77	0.81	−92.97	−95.40	−2.43
GOETHITE	0.50	0.50	0.00	−60.17	−60.58	−0.42
GREENALITE	22.59	20.81	−1.78	−174.64		
GYPSUM	−4.44	−4.85	−0.41	−7.15	1.09	8.24
HALITE	1.59	1.58	−0.01	1.76	3.85	2.09
HAUSMANNITE	61.33	61.54	0.21	−415.47	−335.31	80.17
HEMATITE	0.05	−4.01	−4.06	−126.90	−129.08	−2.18
HERCYNITE	29.45	27.16	−2.29	−327.73	−327.86	−0.13
HUNTITE	−30.62	−29.97	0.65	−152.55	−107.78	44.77
HYDROMAGNESITE	−9.80	−8.77	1.04	−269.11	−218.45	50.67
JAROSITE-K	−8.88	−14.80	−5.92	−194.89	−130.88	64.02
JAROSITE-NA	−4.86	−11.20	−6.34		−151.38	
KALSILITE	11.26	12.84	1.58	−105.44	−121.00	−15.56
KAOLINITE	7.43	5.73	−1.70	−144.81	−147.61	−2.80
LARNITE	38.98	39.14	0.17	−234.97	−239.49	−4.52
LAUMONTITE	14.15	14.46	0.31	−187.74	−211.08	−23.35
LIME	32.70	32.80	0.10	−193.47	−193.59	−0.13
MAGNESITE	−7.91	-8.03	−0.12	−39.96	−25.82	14.14
MAGNETITE	10.30	3.74	−6.57	−213.26	−211.12	2.13
MAXIMUM–MICROCLINE	0.08	0.62	0.53	−32.68	−51.51	−18.83
MELANTERITE	−2.36	−2.47	−0.11	9.25	11.97	2.72
MERWINITE	68.25	68.54	0.30	−438.06	−448.15	−10.08
MIRABILITE	−1.07	−1.11	−0.05	88.70	79.45	−9.25
MNCL2.4H2O	2.72	2.71	−0.01	−9.29	72.72	82.01

Mineral species*	log K EQ3/6	log K MINTEQ	δ log K	$\Delta_r H°$ EQ3/6	$\Delta_r H°$ MINTEQ	$\delta \Delta_r H°$ kJ/mol
MNSO4(C)	2.67	2.67	0.00	−70.79	−64.77	6.02
MONTICELLITE	29.75	30.27	0.52	−201.29	−206.77	−5.48
MUSCOVITE	14.56	12.99	−1.57	−233.72	−248.28	−14.56
NATRON	−0.77	−1.31	−0.54	61.63	65.90	4.27
NEPHELINE	14.13	14.22	0.08	−131.50	−138.91	−7.41
NESQUEHONITE	−5.20	−5.62	−0.42	−26.48	−24.23	2.26
PHLOGOPITE	38.22	66.30	28.08	−318.78	−361.33	−42.55
PORTLANDITE	22.58	22.68	0.09	−127.86	−128.41	−0.54
PSEUDOWOLLASTONITE	14.02	13.85	−0.17	−85.06	−88.16	−3.10
PYRITE	−16.29	−18.48	−2.19	37.24	47.28	10.04
PYROPHYLLITE	1.06	−1.60	−2.66	−107.86		
QUARTZ	−4.00	−4.01	−0.01	26.99	26.02	−0.96
RHODOCHROSITE	−10.58	−10.41	0.17	−15.61	−8.70	6.90
SANIDINE-HIGH	1.28	1.06	−0.22	−43.76	−59.62	−15.86
SEPIOLITE	31.00	15.91	−15.09	−189.20	−114.10	75.10
SIDERITE	−10.57	−10.55	0.02	−27.24	−22.30	4.94
SPINEL	38.39	36.33	−2.06	−380.53	−372.75	7.78
SULFUR-RHMB	−2.11	−2.11	0.00	−21.76	−17.57	4.18
TALC	21.56	23.06	1.49	−172.21	−146.48	25.73
THENARDITE	−0.24	−0.18	0.06	−7.61	−2.38	5.23
THERMONATRITE	0.65	0.13	−0.52	−20.75	−11.72	9.04
TREMOLITE	61.67	56.55	−5.13	−450.87	−404.26	46.61
WAIRAKITE	18.56	18.87	0.31	−241.79	−264.22	−22.43
WOLLASTONITE	13.62	13.00	−0.63	−81.21	−81.59	−0.38
WUSTITE	12.38	11.69	−0.69	−101.88	−103.97	−2.09
Aqueous species*						
$AL(OH)2^+$	10.10	10.10	0.00	−93.47		
$AL(OH)3$	16.17	16.00	−0.17	−156.69		
$AL(OH)4^-$	22.16	23.00	0.84	−168.78	−184.35	−15.56
$AL(SO4)2^-$	−4.90	−4.92	−0.02	−10.13	−11.88	−1.76
$ALOH^{++}$	4.93	4.99	0.06	−45.48	−49.79	−4.31
$ALSO4^+$	−3.01	−3.02	−0.01	−15.86	−9.00	6.86
$CACO3$	−3.22	−3.15	0.07	−14.10	−16.86	−2.76
$CAH2PO4^+$	−20.91	−20.96	−0.05	−4.69	4.69	9.37
$CAHCO3^+$	−11.57	−11.33	0.24	−0.54	−7.49	−6.95
$CAHPO4$	−15.05	−15.09	−0.04	−7.24	0.96	8.20
$CAOH^+$	−1.30	12.60	13.90	−11.63	−60.84	−49.20
$CAPO4^-$	−6.46	−6.46	0.00	−12.18	−12.97	−0.79
$CASO4$	−2.32	−2.31	0.01	−10.67	−6.15	4.52
$FE(OH)2^+$	5.67	5.67	0.00	−69.41		
$FE(OH)3^-$	34.23	31.00	−3.23	−164.39	−126.78	37.61
$FE(OH)4^-$	21.66	21.60	−0.06	−119.12		
$FE(SO4)2^-$	−5.39	−5.42	−0.03	−34.23	−19.25	14.98
$FE2(OH)2^{++++}$	2.95	2.95	0.00	−52.51	−56.48	−3.97
$FE3(OH)4(5^+)$	6.31	6.30	−0.01	−39.25	−59.83	−20.59
$FECL^{++}$	−1.48	−1.48	0.00	−35.15	−23.43	11.72

THERMODYNAMIC DATA FOR MINERALS: A CRITICAL ASSESSMENT

Mineral species*	log K EQ3/6	log K MINTEQ	δ log K	$\Delta_r H°$ EQ3/6	$\Delta_r H°$ MINTEQ	$\delta\Delta_r H°$ kJ/mol
FECL2+	−2.13	−2.13	0.00	−38.45		
FECL3	−1.13	−1.13	0.00	−46.61		
FEH2PO4++	−23.68	−24.98	−1.30			
FEHPO4+	−22.23	−17.78	4.45		30.54	
FEOH++	2.19	2.19	0.00	−42.55	−43.51	−0.96
FESO4+	−4.11	−3.92	0.19	−31.80	−16.36	15.44
H2PO4−	−19.51	−19.55	−0.04	9.46	18.91	9.46
H2SIO4−−	22.91	21.62	−1.29	−65.56	−124.31	−58.74
H3SIO4−	9.81	9.93	0.12	−19.54	−37.40	−17.87
HCO3−	−10.34	−10.33	0.01	7.28	15.15	7.87
HPO4−−	−12.31	−12.35	−0.04	9.58	14.77	5.19
HSO4−	−1.99	−1.99	0.01	−28.07	−20.54	7.53
KHPO4−	−13.35	−12.64	0.71	−24.73	0.00	24.73
KSO4−	−0.85	−0.85	0.00	−6.23	−9.41	−3.18
MGCO3	−13.26	−2.98	10.28	−2.59	−8.45	−5.86
MGH2PO4+	−21.01	−21.07	−0.05	−4.52	4.69	9.20
MGHCO3+	−11.35	−11.40	−0.05	1.42	10.17	8.74
MGHPO4	−15.22	−15.22	0.00	−5.06	0.96	6.02
MGOH+	−2.1962	11.79	13.986	−14.08	−66.69	−52.61
MGPO4−	−6.4591	−6.589	−0.13	−9.31	−12.97	−3.66
MGSO4	−2.2278	−2.25	−0.022	−8.73	−5.85	2.87
MN(OH)3−	34.2212	34.8	0.5788			
MNCL+	0.1977	−0.607	−0.805			
MNCL3−	0.3148	0.305	−0.01			
MNHCO3+	−11.6106	−11.6	0.0106	2.06		
MNO4−−	118.417	118.44	0.023	−705.35	−627.60	77.75
MNOH+	10.591	10.59	−0.001	−55.53	−60.25	−4.72
NACO3−	−10.8481	−1.268	9.5801	28.39	−37.28	−65.67
NAHCO3	−10.4729	−10.08	0.3929	21.19		
NAHPO4−	−13.4591	−12.636	0.8231	−33.03		
NASO4−	−0.694	−0.7	−0.006	−5.22	−4.69	0.54
O2(AQ)	86.0018	85.98	−0.022	−564.33	−559.82	4.51
S−−	13.9032	12.918	−0.985	−52.25	−50.63	1.62
Gas species*						
CH4(G)	−41.0596	−40.1	0.9596	251.18	255.22	4.04
CO2(G)	−18.1622	−18.16	0.0022	−5.98	2.22	8.20
O2(G)	83.1028	83.12	0.0172	−572.97	−571.53	1.44

*Species names as they appear in the data bases.

and

$$(d \log K/dT)_{T_r} = \Delta_f H°/(R \cdot T_r^2) \log e \qquad (23b)$$

where T_r denotes the reference temperature (25°C). The latter equation is referred to as the van't Hoff relation. The standard-state entropy of formation from the elements $\Delta_f S°$ then follows from the equation

$$\Delta_f S°/(R \cdot \ln 10) = (\Delta_f H° - \Delta_f G°)/(R \cdot T_r \cdot \ln 10) \qquad (24a)$$
$$= \log K + T_r (\mathrm{d} \log K/\mathrm{d}T)_{T_r}. \qquad (24b)$$

Differences in $\log K$ between different data sets, denoted by a prefix δ, are defined according to

$$\delta \log K = \log K - \log K' = \log(K/K') \qquad (25a)$$

and

$$\delta \mathrm{d} \log K/\mathrm{d}T = \mathrm{d} \log K/\mathrm{d}T - \mathrm{d} \log K'/\mathrm{d}T = \mathrm{d}/\mathrm{d}T \log(K/K') \qquad (25b)$$

where the primed and unprimed quantities refer to distinct data bases. These equations give the true difference in equilibrium constant or its derivative for any reaction involving the species in question, provided all other species in the reaction have identical free energies or enthalpies of formation in the data sets being compared and thus cancel. However, it should be kept in mind that for reactions which include several species which have different values in each data set, the true log K difference reflects the combined differences in all such species.

Analysis of the overlap in content among the candidate data bases and of the relative and absolute differences in thermodynamic data they contain serves as a basis for further analysis: which data or models are trustworthy, which of them suspect, which of them (demonstrably) faulty? These questions will be addressed only to a very limited extent here, however. The possibilities of improving a current data base are limited by several factors, including fundamental differences in the models of aqueous species. Elimination of values that are clearly wrong is straightforward; but nearly every other step towards an improved data base would necessitate extensive further analysis, if not further experimental data. Thus, on the basis of the current comparison, only a few recommendations can be made. These are based as much on considerations of principle as on the actual data-base comparison, since the latter cannot establish the accuracy of the data.

PRACTICAL ASPECTS OF THE COMPARISON

For the log K comparison the various data sets were analysed with the help of text-processing utilities belonging to the UNIX operating system. These included the utilities *awk*, *cat*, *grep*, *join*, *lex*, *tee*, *sed*, and *sort*. Shell scripts written for the *C-Shell* extracted and manipulated the necessary information from each data set and combined the results with other data sets. Results are presented as tables in Section 8.7.3.

Comparisons among *type S* data bases make use of equilibrium phase relations computed by means of Ge0Calc software (Table 8.1). These phase diagrams afford direct comparison with experimental data.

8.7.2 Data bases selected for comparison

Experience in data analysis by MAP techniques stresses the need for a comprehensive approach. To ascertain internal consistency of some data in isolation—without considering related data—turns out to be not a very stringent test of quality of a data set. (Probably, a similar statement could be made on the basis of REG analysis.) Indeed, one can almost always obtain a set of 'fit parameters' or at least upper and lower bounds for the parameters h, s, v, c, ... (equation (11b)) so as to satisfy the experimental phase equilibrium constraints (equation (9)) of only a few equilibria. Only when a large number of experimentally independent constraints are considered simultaneously can we hope that the (optimized) parameter values obtained for each participating species are indeed reliable estimates of their true thermodynamic properties $\Delta_f H°, S°, V°, C_p, \ldots$.

For these reasons only five of the larger, current mineral data bases are compared here. More restricted thermodynamic data collections—derived from analysis of smaller sets of data—exist in the recent literature. Though of use in some applications, they do not lend themselves to a comprehensive comparison. Included in the present comparison are the following:

1. SUPCRT: the Berkeley data base of Helgeson *et al.* (1978, with updates);
2. UBC: the Vancouver data base in the version of Berman (1988);
3. HP-DS: the Cambridge data base in the edition of Holland and Powell (1990);
4. EQ3/6: the Livermore data set associated with the codes of Wolery (1979; 1983) and Wolery *et al.* (1984);
5. MINTEQ: The USGS data set associated with the programs of Felmy *et al.* (1984) and Nordstrom *et al.* (1984).

The first three data bases are of *Type S* (Section 8.1.3); they result from comprehensive refinement of experimental data. The latter two are more appropriately termed data compilations; they are of *Type K* and are included here chiefly because they are widely used in low-temperature applications.

The data sets SUPCRT, UBC and HP-DS were constructed primarily by analysing calorimetric data together with high-temperature mineral equilibria and extrapolating the results to 25°C and 1 bar to obtain standard-state Gibbs free energies and enthalpies and of formation, entropies, etc. Each of the data sets SUPCRT, UBC and HP-DS is at least partially consistent. By contrast, the data sets EQ3/6 and MINTEQ are compilations of log K values from different sources, and no attempt has been made to ensure that these data sets are consistent with all the available calorimetric and experimental phase equilibrium data. Consequently, these data sets are only formally consistent.

In addition to this basic distinction, the above five data sets differ substantially in contents (Table 8.3) because the data bases were designed for different purposes and at different times; their authors employed fundamentally different techniques of data analysis and relied, to some extent, on different experimental data.

The following brief introduction purposely aims to highlight differences in the content and form of these data bases. Similarly, the subsequent section stresses differences in the aim and approach of the authors who contributed to the five data bases. These differences notwithstanding, it is essential to recognize that the data sets hinge largely on the same experimental data. One would thus expect to find agreement between some parts of these data bases and it would be optimistic or naive to infer high accuracy from such agreement. The emphasis on discrepancies should be seen as a first step to resolve them.

THE SUPCRT DATA BASE

The SUPCRT data base for minerals represents the first attempt to construct and fully document a comprehensive, internally consistent data set (Helgeson et al., 1978) that considers phase-equilibrium data in addition to the calorimetry. In the strict sense defined here, this data base is only formally consistent, however. An incremental REG strategy was employed in the data analysis: starting with the limited data sets deemed most trustworthy, thermodynamic parameters for some well-constrained species were derived. Subsequently, the phase-equilibrium data for larger systems were incorporated, and properties for further species were derived, holding the previously determined ones constant. The resulting SUPCRT data set for minerals is separated and nearly independent from that for aqueous species. The latter is more recent (Helgeson et al., 1981) and has been updated repeatedly (Tanger and Helgeson, 1988; Oelkers and Helgeson, 1988; 1990; Shock and Helgeson, 1990; Shock et al., 1990; Sverjensky et al., 1990).

THE UBC DATA BASE

The UBC data set is an internally consistent data set for minerals constructed at the University of British Columbia, Vancouver. Berman (1988) reported fully documented results for the 11-component system Na_2O–K_2O–CaO–MgO–FeO–Fe_2O_3–Al_2O_3–SiO_2–TiO_2–H_2O–CO_2 based on an analysis of all available experimental data. These were optimized by means of the non-linear mathematical programming techniques of Berman et al. (1986). Refined properties include all of the standard-state properties for each mineral with provision for ordering and phase transitions where needed; so far, only a few nonideal solution models have been incorporated. No aqueous species are included in this data base but, to the small extent that solubility data considered provide a link, the internal consistency of the mineral data is preserved with the data and formulation for aqueous species in SUPCRT.

THE HP–DS DATA BASE

Holland and Powell's (1990) data set, incorporating most recent experimental phase-equilibrium data as well as some constraints from metamorphic rocks, represents one of the best documented and largest among current, mineral data bases. Enthalpy data in the 13-component system K_2O–Na_2O–CaO–MgO–MnO–FeO–Fe_2O_3–Al_2O_3–TiO_2–SiO_2–C–H_2–O_2 were refined by specially tailored REG methods. Thermodynamic parameters include provision for ordering; solid solutions are treated by an ideal mixing-on-sites model. Regression uncertainties and their correlations are given for all parameter values. No solubility data were considered, hence aqueous species are not included. Strictly speaking, this data base is partially consistent.

A number of data collections that were based on similar REG methods are largely superseded by the HP–DS data base and are not included in the current comparison for this reason. Prominent among these are the USGS data bases (Robie *et al.*, 1978; Haas *et al.*, 1981; Hemingway *et al.*, 1982; Robinson *et al.*, 1983; Hemingway and Sposito, 1989) which are at least partially consistent. Note, however, that data included on aqueous species in these data bases are inconsistent with the mineral data.

THE EQ3/6 DATA BASE

The EQ3/6 data base compiles $\log K$ values from 0 to 300°C along the saturation curve of water (Wolery, 1983). Included in the EQ3/6 data base is a quality factor for each entry: excellent, good, fair, poor, bad, uncertain, speculative, unspecified, error, unknown, or blank (=unspecified). For a selected set of aqueous species, the data base also includes the data required for Harvie–Møller–Weare models designed to handle concentrated solutions accurately (Weare, 1987).

In the current analysis, version DATA0R51 of the EQ3/6 data base has been used; it is but formally consistent.

THE MINTEQ DATA BASE

Compiled in this data base are values (at $T_r = 25°C$) of $\log K$ and, for many species, $\Delta_r H^0$. Enthalpy data serve to approximate the temperature dependence of the equilibrium constant by the van't Hoff equation which yields—for small intervals in temperature:

$$\Delta_T \log K_{T_r} \approx (T - T_r) \Delta_r H^\circ \log(e)/(RT_r^2). \tag{26}$$

The data stem from a number of authors, principally those associated with WATEQ (Truesdell and Jones, 1974; Ball *et al.*, 1980; 1981; Krupka and Jenne, 1982). No documentation is included with this data base. Instead, reference is made to the documentation of its predecessors (WATEQ, REDEQL, MINEQL), whose authors appear to have selected with care which data to include. However, the methods of analysis are not detailed, nor are the choices or the improvements made by others since 1981. Nordstrom *et al.* (1990) give sources and some justification for many of these updates but their documentation includes some more recent changes not yet considered in the present comparison. The version of the data base used here belongs to the program MINTEQ Version A1.0 (December 1986, Center for Water Quality Modeling, US Environmental Protection Agency, Environmental Research Lab, Athens, Georgia).

This data base was principally developed for low-temperature applications and is currently in use for a variety of purposes. Similar in scope and use to the MINTEQ and EQ3/6 data bases are a number of data collections associated with specific geochemical programs, such as GEOCHEM (Sposito and Mattigod, 1980), PHREEQE (Parkhurst *et al.*, 1980), and SOLMINEQ.88 (Kharaka *et al.*, 1988). Nordstrom *et al.* (1990) present the most recent assessments and improvements on these.

AQUEOUS SPECIES

For low-temperature models, the various programs and data bases are linked with the activity models chosen for aqueous species, i.e. for ions the simple Debye–Hückel law, the Pitzer equation, or a modified version of either. Despite the considerable overlap in the primary experimental data considered—solubility measurements figure prominently among these—different log K values for minerals resulted due to differences in the activity relations used. Obviously, comparisons among mineral data are meaningful only if the activity coefficients are computed using the appropriate model for each data base. The same precautions were observed here for neutral species, though differences in their activity coefficients are minor.

Some parts of the data bases can be regarded as complementary: while the data for minerals in SUPCRT are, to the extent of overlap, superseded by those in UBC and HP–DS, the latter data bases do not contain data of their own for aqueous species. Insread, the data for minerals from UBC are combined with aqueous species from SUPCRT. Similar combinations might be sensible for MINTEQ and for EQ3/6, the mineral data of which derive in part from SUPCRT.

8.7.3 Results

The results are presented and briefly discussed in two sections. First, the two petrologically relevant data bases HP–DS and UBC are compared, then those data sets currently in use in low-temperature applications are examined, i.e. SUPCRT, EQ3/6, and MINTEQ. For these three, the analysis is presented separately for minerals and aqueous species because, in general, the values of the thermodynamic parameters for aqueous species are almost independent of those for minerals. (However, where standard-state properties were refined for a mineral by means of equilibria involving aqueous species, the mineral data do depend on the models employed for aqueous species.)

HP–DS VERSUS UBC

To the extent of overlap (Table 8.2), the two data bases show substantial agreement for a majority of minerals. In view of the different methods of data analysis used and despite the largely overlapping primary data considered, this agreement attests a certain maturity to both data bases. These encouraging statements notwithstanding, there are noticeable and non-trivial differences. For reasons detailed earlier (Sections 8.5 and 8.6), value by value comparison of parameters yields little interpretable information. A better basis of comparison for the thermodynamic data is shown in Figures 8.3–8.5, in which computed phase equilibria for a series of chemical systems are tested against all available experimental half-brackets. Nearly all of the experimental data shown had in fact been employed in the data analyses that yielded the two data bases, hence one should expect excellent agreement.

In these graphs every arrow depicts one experimentally established half-reversal; these are simply listed in the abscissa. For the ordinate, each arrow's position was calculated as the difference between the equilibrium temperature computed (for the nominal pressure of the experiment) and the reported half-reversal. The length of the arrow marks the experimental uncertainty in T; its orientation gives the sense of the half-bracket: experiments producing the high-temperature assemblage (from low-T reactants) point downward and vice versa. The desired consistency between data and calculated equilibria is thus simply tested by checking for arrows that point 'the wrong way', i.e. away from the zero line. (Note that the distance from the zero line is immaterial if the arrow points the 'right way', i.e. as long as it indicates no inconsistency.) Inconsistencies may occur for different reasons, of course. Some are readily explained as experimental problems, e.g. where high-quality data contradict suspect ones; some experiments may be doubted owing to poor phase characterization. Other conflicts may have been produced because certain parameters $(S°, C_P)$ were not included in the refinement. Indeed, owing to recent progress in data analysis, very few inconsistencies remain that defy explanation. However, a full discussion of each inconsistency is beyond the scope of this comparison; the interested reader is referred to Holland and Powell (1990) and especially to Berman (1988) who address these questions in more detail.

The following discussion of Figures 8.3–8.5 simply counts inconsistencies and categorizes them arbitrarily as minor ($<5°$), serious (5–20°), and major ($>20°$). Only where they occur grouped will an explanation be attempted. To avoid confusion, it should be pointed out that inconsistencies which could not be remedied, were—indeed they had to be—omitted in the MAP analysis leading to the UBC data base. Similarly, Holland and Powell (1990) reported that their initial analysis led them to reject some data. However, all of these data are included in the present comparison.

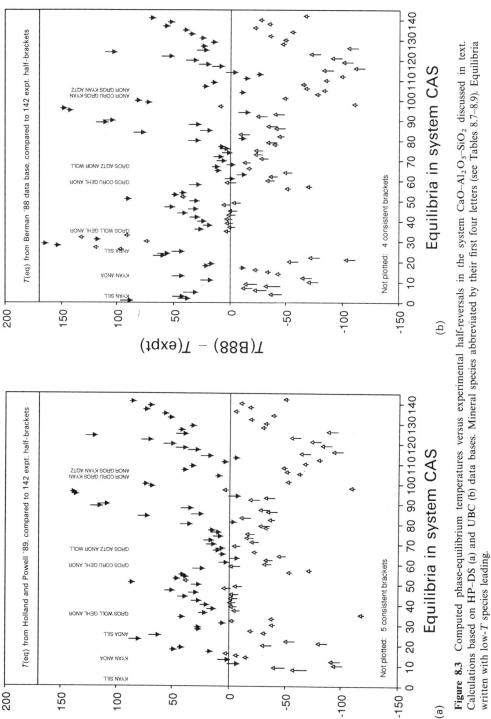

Figure 8.3 Computed phase-equilibrium temperatures versus experimental half-reversals in the system $CaO-Al_2O_3-SiO_2$ discussed in text. Calculations based on HP-DS (a) and UBC (b) data bases. Mineral species abbreviated by their first four letters (see Tables 8.7–8.9). Equilibria written with low-T species leading.

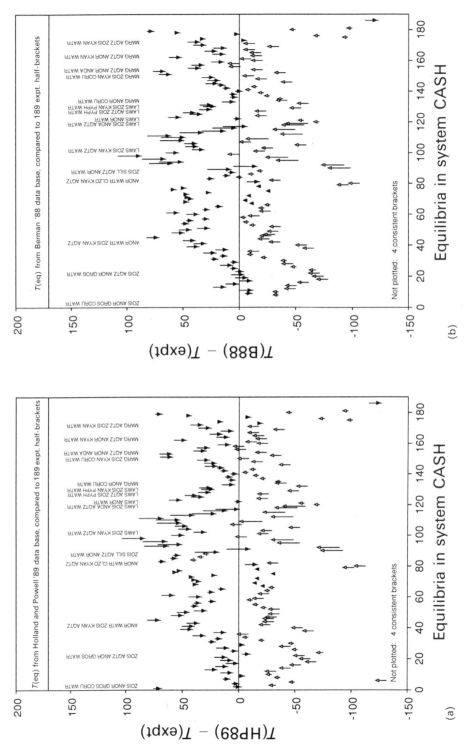

Figure 8.4 As for Figure 8.3, but for system CaO–Al$_2$O$_3$–SiO$_2$–H$_2$O.

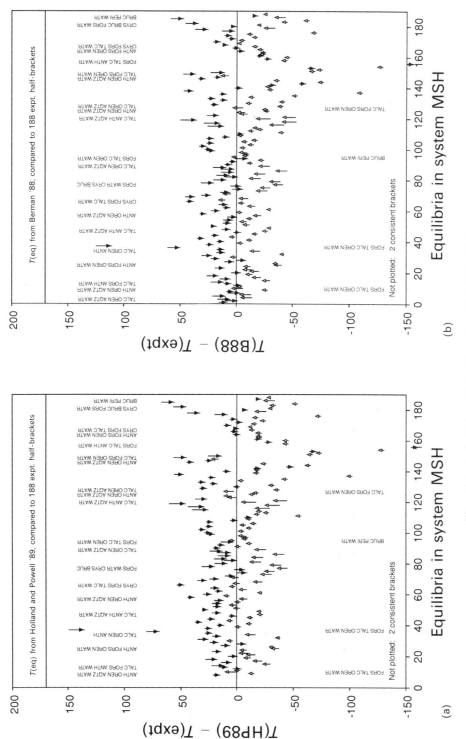

Figure 8.5 As for Figure 8.3, but for system $MgO-SiO_2-H_2O$.

In the system $CaO-Al_2O_3-SiO_2$ (Fig. 8.3), some 140 half-reversals are tested. The UBC data base is consistent with all but two of these; one inconsistency is very minor, the other is major. The HP–DS data base predicts equilibrium temperatures inconsistent with 14 half-reversals; eight are serious inconsistencies, six are major. Five of the latter lie between 70 and 140° from Holdaway's (1971) sillimanite→andalusite brackets; it should be emphasized that the data of Richardson et al. (1969) are not shown as they used fibrolite and thus observed a metastable reaction.

In the system $CaO-Al_2O_3-SiO_2-H_2O$ (Fig. 5.4), some 180 half-reversals are shown, none from the CAS subsystem. The UBC data base produces three inconsistencies with these data, one major, one serious, and one minor. Two of these pertain to a margarite breakdown reaction, for which newer data by Jenkins (1984) supersede those of Storre and Nitsch (1974); the latter cause the conflict documented here which may be due to andalusite instead of kyanite having been involved. The HP–DS data base predicts seven incompatibilities, four of them serious. Of these, three pertain to the magarite experiments just discussed.

As a final example, the system $MgO-SiO_2-H_2O$ (Fig. 8.5) with over 180 experiments is shown. The UBC data set produces nine inconsistencies, of which two are serious and seven are major. Eight of them are conflicts with high-pressure experiments of very short duration (Kitihara et al., 1966) in which product phases were in part not well crystallized. HP–DS finds 16 inconsistencies, though six of these are minor. The seven major ones occur for the same data and presumably the same reasons discussed above.

Of course, the three systems considered briefly here do not do justice to the entire data-base contents of the HP–DS and UBC data bases, but the examples given are typical of their quality and remaining problems in them. Certainly, a most central shortcoming of both of them is the limited availability of consistent and optimized solution models for rock-forming minerals.

LOW-TEMPERATURE DATA SETS
The following discussion relies largely on Lichtner and Engi (1989). Tables 8.7 and 8.8 contain the listings of the comparisons of log K values; the same results are displayed in Figures 8.6–8.8. Data-base pairs most akin to each other or of particular concern in low-T applications are presented. Discussion of the results has to be minimal, not only for the sake of space, but because most of the discrepancies—and many very major ones are among these—cannot presently be explained, much less reconciled.

MINERALS
General remarks: Since the data sets SUPCRT, UBC, and HP–DS refer to standard-state thermodynamic properties for which no chemical reaction is specified, differences in log K and d log K/dT values give true log K differences for any reaction for which all other species in the reaction have identical values

Table 8.7 Difference of log K data expressed as $\delta \log K$ (SUPCRT–MINTEQ)

Mineral species*	log K MINTEQ	log K SUPCRT	$\delta \log K$	$\Delta_r H°$ MINTEQ	$\Delta_r H°$ SUPCRT	$\delta \Delta_r H°$ kJ/mol
AKERMANITE	47.47	44.99	−2.48	−319.85	−294.46	25.39
ALUNITE	−1.35	−2.92	−1.57	16.39		
ANALCIME	6.72	6.09	−0.63	−95.56	−86.19	9.38
ANHYDRITE	−4.64	−4.31	0.33	−15.77	−18.58	−2.81
ANNITE	23.29	28.61	5.32	−274.97	−264.56	10.41
ANORTHITE	25.43	24.70	−0.73	−295.64	−295.87	−0.23
ARAGONITE	−8.36	−8.48	−0.12	−10.94	−10.30	0.64
ARTINITE	9.60	9.32	−0.28	−120.26	−115.70	4.55
BOEHMITE	8.58	8.43	−0.15	−117.70	−119.52	−1.82
BRUCITE	16.79	16.30	−0.51	−108.11	−111.33	−3.21
CALCITE	−8.48	−8.64	−0.17	−10.82	−10.10	0.72
CASO4	−2.31	−2.11	0.20			
CHALCEDONY	−3.52	−3.73	−0.21	119.31	27.54	8.23
CHRYSOTILE	32.19	31.12	−1.07	−219.60	−225.75	−6.15
CRISTOBALITE	−3.59	−3.45	0.14	23.01	25.33	2.32
DIASPORE	6.87	7.59	0.71	−103.05	−110.76	−7.71
DIOPSIDE	19.89	20.80	0.92	−135.06	−140.58	−5.52
DOLOMITE	−17.00	−18.30	−1.30	−34.69	−29.65	5.04
FORSTERITE	28.30	27.86	−0.44	−202.97	−209.47	−6.51
GIBBSITE	8.77	6.79	−1.98	−95.40	−95.79	−0.39
GREENALITE	20.81					
HALITE	1.58	1.59	0.01	3.84	3.74	−0.10
HALLOYSITE	8.99			−166.23		
HEMATITE	−4.01	0.11	4.11	−129.06	−129.40	−0.34
HIGH∼SANIDINE	1.06	0.07	−0.99	−59.63	−39.64	19.98
HUNTITE	−29.97	−31.01	−1.05	−107.78	−112.30	−4.52
HYDROMAGNESITE	−8.77	−10.47	−1.70	−218.45	−230.87	−12.42
K-FELDSPAR	0.62	−1.13	−1.75	−51.50	−28.56	22.94
KALSILITE	12.84	10.04	−2.80	−121.00	−105.46	15.54
KAOLINITE	5.73	5.10	−0.63	−147.61	−145.52	2.08
LARNITE	39.14			−239.48		
LAUMONTITE	14.46	11.79	−2.67	−211.08	−185.20	25.87
LEONHARDITE	16.49			−357.15		
LIME	32.80	32.57	−0.23	−193.57	−193.82	−0.25
LOW-ALBITE	2.59	1.91	−0.68	−72.80	−56.47	16.33
MAGNESITE	−8.03	−8.03	−0.00	−25.81	−29.80	−3.99
MAGNETITE	3.74	10.47	6.73	−211.12	−216.57	−5.45
MERWINITE	68.54	68.03	−0.52	−448.15	−435.03	13.13
MONTICELLITE	30.27	29.58	−0.69	−206.78	−199.57	7.20
MUSCOVITE	12.99	11.02	−1.97	−248.28	−233.85	14.42
NEPHELINE	14.22	12.94	−1.27	−138.93	−131.94	6.98
NESQUEHONITE	−5.62	−5.34	0.29	−24.22	−21.43	2.78
PERICLASE	21.51	21.33	−0.18	−151.19	−150.13	1.05
PHLOGOPITE	66.30	36.58	−29.72	−361.33	−315.10	46.23
PYRITE	−18.48	−16.23	2.25	47.28	47.07	−0.21
PYROPHYLLITE	−1.60	−1.27	0.33		−103.65	
QUARTZ	−4.01	−4.00	0.01	26.02	29.08	3.05
RHODOCHROSITE	−10.41	−10.52	−0.11	−8.70	−6.64	2.05

Mineral species*	log K MINTEQ	log K SUPCRT	δ log K	$\Delta_r H°$ MINTEQ	$\Delta_r H°$ SUPCRT	$\delta\Delta_r H°$ kJ/mol
SEPIOLITE	15.91			−114.09		
SIDERITE	−10.55	−10.52	0.03	−22.29	−17.83	4.46
SPINEL	36.33	35.92	−0.42	−372.75	−384.12	−11.37
TALC	23.06	21.14	−1.92	−146.46	−164.19	−17.73
TREMOLITE	56.55	60.91	4.36	−404.24	−435.48	−31.25
WAIRAKITE	18.87	16.20	−2.67	−264.22	−238.34	25.87
WOLLASTONITE	13.00	13.60	0.60	−81.58	−79.52	2.06
Aqueous species*						
AL(OH)$^{++}$	4.99	3.83	−1.16	−49.79		
H2PO4$^-$	−19.55	−26.14	−6.59	18.91	18.91	0.00
H2S-(AQ)	−6.99	−6.99	0.00	22.18	21.55	−0.62
HCO3$^-$	−10.33	−10.33	0.00	15.13	14.70	−0.44
HPO4$^-$2	−12.35	−18.93	−6.59	14.77	14.71	−0.06
HSO4$^-$	−1.99	−1.98	0.01	−20.54	−20.50	0.04
KSO4$^-$	−0.85	−0.88	−0.03	−9.41		
MGOH$^+$	11.79	11.83	0.04	−66.67		
MNO4$^-$2	118.44	118.41	−0.03	−627.68	−711.23	−83.54
O2-(AQ)	85.98	86.00	0.02	−559.94	−559.53	0.41
OH$^-$	14.00	13.99	−0.01	−55.84	−55.81	0.03
Gas species*						
O2(G)	83.12	83.10	−0.02	−571.65	−571.66	−0.01

*Species names as they appear in the data bases.

for each pair of data sets. Thus, provided the $\Delta_f G°$ and $\Delta_f H°$ values for aqueous species are identical in each data base for a reaction involving a single mineral, the differences in log K and d log K/dT values between the two data bases can be directly computed from differences in $\Delta_f G°$ and $\Delta_f H°$ for that mineral.

For many minerals at 25°C considerable differences, several log K units, are observed between some of the data sets. Since most of the data in these data sets are obtained by extrapolation from high-temperature experiments, it is noteworthy that for more than half of the minerals the sign of Δlog K is opposite to the sign of the difference in the log K temperature derivative. This implies that when plotted as a function of temperature, the log K curves of such a mineral tend to converge at higher temperature. Therefore, at higher temperatures better agreement can be expected between the three data bases. Indeed, many of the minerals included in these data bases are only stable at high temperatures and pressures, and it is from such temperatures and pressures that experimental and field data were used to extrapolate to 25°C to obtain values for standard state properties.

Table 8.8 Difference of log K data expressed as $\delta \log K$ (UBC−MINTEQ)

Mineral species*	log K MINTEQ	log K UBC	$\delta \log K$	$\Delta_r H°$ MINTEQ	$\Delta_r H°$ UBC	$\delta \Delta_r H°$ kJ/mol
AKERMANITE	47.472	48.024	0.552	−319.85	−312.32	7.53
ANORTHITE	25.43	22.875	−2.554	−295.64	−283.66	11.98
BRUCITE	16.792	16.375	−0.416	−108.11	−111.69	−3.57
CALCITE	−8.475	−8.325	0.149	−10.82	−11.50	−0.68
CASO4	−2.309	−2.107	0.201	−6.15		
CHRYSOTILE	32.188	31.409	−0.778	−219.60	−226.82	−7.23
CRISTOBALITE	−3.587	−3.592	−0.005	23.01	26.18	3.17
DIASPORE	6.873	6.356	−0.516	−103.05	−103.70	−0.65
DIOPSIDE	19.886	21.329	1.443	−135.06	−143.26	−8.20
DOLOMITE	−17	−17.45	−0.45	−34.69	−34.26	0.42
FORSTERITE	28.298	28.154	−0.143	−202.97	−210.73	−7.77
HEMATITE	−4.008	0.407	4.415	−129.06	−131.03	−1.97
HIGH-SANIDINE	1.062	0.172	−0.889	−59.63	−40.25	19.38
K-FELDSPAR	0.616	−0.985	−1.601	−51.50	−29.17	22.33
KAOLINITE	5.726	3.225	−2.5	−147.61	−134.81	12.80
LIME	32.797	32.693	−0.103	−193.57	−193.82	−0.25
LOW-ALBITE	2.592	1.312	−1.279	−72.80	−52.99	19.82
MAGNESITE	−8.029	−8.392	−0.363	−25.81	−27.56	−1.74
MAGNETITE	3.737	10.591	6.854	−211.12	−217.35	−6.22
MERWINITE	68.543	73.26	4.717	−448.15	−464.18	−16.03
MONTICELLITE	30.272	31.94	1.668	−206.78	−212.25	−5.47
MUSCOVITE	12.99	10.029	−2.96	−248.28	−229.39	18.89
PERICLASE	21.51	21.364	−0.145	−151.19	−150.29	0.90
PHLOGOPITE	66.3	39.106	−27.193	−361.33	−333.82	27.51
PYROPHYLLITE	−1.598	−3.334	−1.736			
QUARTZ	−4.006	−4.007	−0.001	26.02	29.13	3.10
TALC	23.055	22.187	−0.867	−146.46	−170.10	−23.64
TREMOLITE	56.546	63.359	6.813	−404.24	−449.60	−45.36
WOLLASTONITE	12.996	13.534	0.538	−81.58	−78.98	2.60

*Species names as they appear in the data bases.
Gases/Aqueous Species: UBC uses SUPCRT data; compare Table 8.7 (MINTEQ−SUPCRT).

MINTEQ VERSUS EQ3/6

Quite a large number of serious discrepancies emerge for the mineral data. In addition to the Fe-phases (oxides, pyrite, annite, ferrisulfate), several important Mg- and Ca–Mg-silicates and carbonates (talc, tremolite, phlogopite, chrysotile, sepiolite; dolomite, hydromagnesite, and others) are discrepant, as well as a variety of Mn- and Al-minerals. While for any of these, the disagreement decreases with temperature, there are notable exceptions to that rule: for pyrophyllite and talc the discrepancies are worse at 100° than at 25°C. Since the documentation available is sparse regarding the procedures and criteria used in updating these two data bases, discrepancies (and, for that matter, cases of agreement) remain difficult to interpret.

THERMODYNAMIC DATA FOR MINERALS: A CRITICAL ASSESSMENT

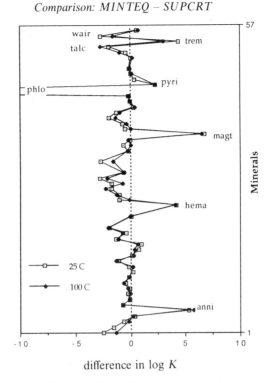

Figure 8.6 Discrepancies in log K (at 25 and 100°C) of minerals common to the SUPCRT and the MINTEQ data bases. Abbreviated names (compare Table 8.6) are given only for extreme outliers.

Since many of the mineral data in the EQ3/6 data base were adopted from SUPCRT, several problems and limitations inherent in the latter also apply to the former. They will be discussed specifically below (UBC–SUPCRT).

MINTEQ VERSUS SUPCRT
Discrepancies are substantial for the Fe-oxides, pyrite, biotites, talc, and wairakite. Notably, agreement is relatively close for aluminum-bearing phases (pyrophyllite, diaspore, boehmite, etc.). As will be discussed below, an error in the SUPCRT data base should have prevented such agreement, were the data for these minerals not faulty in both these data bases (see SUPCRT versus UBC).

MINTEQ VERSUS UBC
In addition to the problems with haematite and magnetite, large discrepancies exist for the feldspars, especially K-feldspar and anorthite, for diopside and tremolite, as well as several sheet silicates.

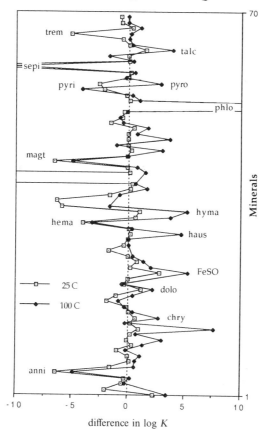

Figure 8.7 Discrepancies in log K (at 25 and 100°C) of minerals common to the EQ3/6 and the MINTEQ data bases. Abbreviated names (compare Table 8.7) are given only for extreme outliers.

It seems noteworthy that the properties for tremolite, kaolinite, and the micas are based, in the UBC data base, on recent phase-equilibrium data and should be quite reliable (Berman, 1988). For potassium feldspar, the UBC data base employs a special model to account for Al–Si-ordering; hence the standard-state data for sanidine and mirocline should be superior to any data base that does not consider this effect. The same is consequently true for muscovite, because its properties depend directly on those of K-feldspar, owing to the well-constrained loci of its breakdown reactions muscovite + quartz = Kspar + Al_2SiO_5 + H_2O and muscovite = Kspar + corundum + H_2O.

Anorthite properties in many of the earlier data bases were adopted from Helgeson *et al.* (1978), who appear to have overestimated its entropy. While the data in MINTEQ are likely to be affected, these shortcomings have been

THERMODYNAMIC DATA FOR MINERALS: A CRITICAL ASSESSMENT

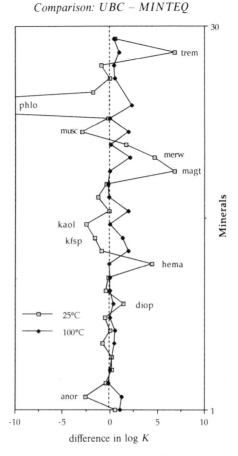

Figure 8.8 Discrepancies in log K (at 25 and 100°C) of minerals common to the UBC and the MINTEQ data bases. Abbreviated names (compare Table 8.8) are given only for extreme outliers.

eliminated in the UBC data base, though several difficulties led Berman (1988) to caution that the low-temperature properties of anorthite are still uncertain.

SUPCRT VERSUS UBC
Substantial differences exist between the SUPCRT and UBC data sets at 25°C for the Gibbs free energy, enthalpy, and entropy of formation. Heat capacities were compared for a selected set of minerals and were found to be essentially equivalent at low temperatures, although for certain minerals large differences exist, especially at higher temperatures (Berman, 1988). The heat-capacity function used in the UBC data set is based on Berman and Brown (1985) and can be extrapolated to very high temperatures, in contrast to the Maier–Kelley function used in SUPCRT.

Some general statements can be made regarding the differences found between these two data sets. Firstly, large differences are found for the aluminum-bearing minerals: pyrophyllite, diaspore, corundum, andalusite, kyanite, and sillimanite. This discrepancy for Al-bearing minerals in SUPCRT was previously noted by Hemingway et al. (1982) and can be traced to an incorrect $\Delta_f G°$ value used for corundum in the SUPCRT data base. A correction for this error can be made by adding the difference in Gibbs free energy for Al_2O_3 (weighted by the appropriate stoichiometric factor for each mineral) between the SUPCRT data base and any other data bases. The log K values thus corrected result in closer agreement with the other data bases.

Secondly, many of the remaining discrepancies may be attributed to uncertainties in the experimental data for enthalpies and entropies of formation from which the data are derived. For example, large discrepancies in log K values are found for åkermanite, monticellite, and merwinite, discrepancies attributable to the lack of accurate data for enthalpies and, in the case of merwinite, the incorrect molar volume used in the SUPCRT data base (Berman, 1988). Recent calorimetric redeterminations of the heats of solution for these minerals are 10–13 kJ/mol less negative than earlier measurements (Brousse et al., 1984; Neuvonen, 1952a; b). In addition the molar volume of merwinite is approximately 6 cm^3/mole smaller than that tabulated by Robie et al. (1967), the value used by Helgeson et al. (1978). Berman (1988) gives further details.

UBC VERSUS HP–DS

For slightly less than half of the minerals compared the equilibrium constants differ by more than a factor of two between the two data sets. Six of these differ by an order of magnitude or greater. However the differences are generally much smaller than the differences between the SUPCRT and UBC data sets. Again a difference in sign between $\delta \log K$ and its temperature derivative is noted, indicating better agreement at higher temperatures than at 25°C.

EQ3/6 VERSUS SUPCRT

Most minerals common to EQ3/6 and SUPCRT have very similar properties. The similarity is not surprising since the EQ3/6 data set was originally based on SUPCRT. Major differences occur for the minerals clinochlore (7Å and 14Å), pargasite, phlogopite, pyrrhotite, sepiolite, strontianite, and tremolite.

AQUEOUS SPECIES

Comparison of aqueous species was made between the data sets SUPCRT, EQ3/6, and MINTEQ. By and large, there is considerably better agreement between the various data sets for thermodynamic data of aqueous species than for minerals.

MINTEQ VERSUS SUPCRT

Comparison of the MINTEQ data set with SUPCRT for aqueous species shown in Table 8.7 yields essentially identical results except for $Al(OH)^{2+}$, $H_2PO_4^-$, and HPO_4^{2-}. The data for phosphate species in MINTEQ agrees with those in the EQ3/6 data base; both differ from SUPCRT, with a difference of more than six log units! (The independence of the latter two is not guaranteed, however.) The species $Al(OH)^{2+}$ has a significantly different enthalpy of formation between the two data sets.

MINTEQ VERSUS EQ3/6

Table 8.6 documents relatively close agreement with a few very notable outliers for Mg-species ($MgCO_3$ and $MgOH^+$, discrepancies of >10 log-units!), the two Fe-phosphates, two hydroxides ($Ca(OH)^+$, $Fe(OH)_3^-$), and $NaCO_3^-$. For $H_2SiO_4^{2-}$, a notable discrepancy also exists and, though not terrible at 25°C, gets worse with increasing temperature.

Note, however, that most of the species in the current version of the EQ3/6 *db* are not present in MINTEQ (or in SUPCRT, below).

SUPCRT VERSUS EQ3/6

The EQ3/6 and SUPCRT data sets are essentially identical for those species they have in common, indicating again that the EQ3/6 data set was originally based on SUPCRT. Major differences occur for the species $CaCl^+$, $CsCl^+$, $MgCl^+$, NaOH, PO_4^{3-}, S_2^{2-}, S_3^{2-}, and S_4^{2-}.

GASES

Several errors occur in MINTEQ's $\log K$ listings for $O_2(g)$. These were corrected prior to the analysis reported here. A substantial discrepancy occurs for CH_4 between EQ3/6 and MINTEQ.

8.8 General assessment and outlook

The critical comparison in Section 8.7 makes plain that there remain substantial differences, gaps and inconsistencies among most of the widely used thermodynamic data bases for mineral systems. At present, the situation seems far worse for data bases at low temperature, where discrepancies are abundant and no completely consistent set of thermodynamic data exists—recent efforts (e.g. Nordstrom *et al.*, 1990) notwithstanding. Nevertheless, progress in recent years has been quite considerable, and the state of the art has certainly improved very rapidly since the days when raw calorimetric data (albeit carefully selected, e.g. by Robie *et al.*, 1978) were thought to be the best available. To the earth scientists who needs to perform stability calculations today, there is no real alternative to the use of refined thermodynamic data from one of the internally consistent data bases. And indeed, at least the best of these data bases are being widely used, especially by petrologists. Unless a particular problem at hand leaves no choices (e.g. because some necessary properties are available only in one data base) it is clearly essential to decide which data base to use. It may thus be appropriate to summarize and compare the main achievements and limits of the two most widely used mineral data bases, HP–DS and UBC. By necessity, some of the statements below are qualitative, and arguably some of the conclusions are subjective.

The primary scope behind HP–DS has been to provide a comprehensive set of thermodynamic data for petrologic applications, an ambitious task that the Holland and Powell (1990) version of the data base largely meets. Inconsistencies in the primary stability data were minimized—using regression methods—by adjusting enthalpy data for all species (except for one 'anchor value' per element) on the basis of a broad selection of phase equilibrium data. Where experimental data do not constrain the enthalpies of some relevant species accurately, data from natural assemblages have been used to provide estimates. A laudable attempt was made by Holland and Powell to assess the uncertainties remaining after their data analysis, but it remains a non-trivial task to verify how realistic these error estimates are, especially where natural parageneses served as constraints. A few weaknesses affect the current version of HP–DS:

1. The entropy, volume and heat capacity functions are not optimized.
2. An overly simplistic equation of state is adopted for fluids.
3. The compressibility and thermal expansivity of each mineral are assumed to be constant.
4. Ideal site-mixing models are assumed for mineral activities.
5. Some polymorphic transitions, notably for quartz, are not properly accounted for.

Future improvements of this data base will certainly seek to remedy at least some of these shortcomings, especially since today's computers (even PCs) make the simplifying assumptions—such as those behind the last four of the listed problems—quite unnecessary.

The main goal of the UBC data base has been to produce refined thermodynamic data for minerals by careful scrutiny of *all* pertinent experimental stability data. Where some of the primary data proved inconsistent, data analysis—by mathematical programming methods—necessitated some reinterpretation and in some cases rejection of suspect data sets (Berman, 1988). The scope of data has not, up to now, included natural data in some systematic fashion. As a result, where experimental data are lacking, some species of considerable mineralogical or petrological interest are missing. Current efforts towards an extension of the UBC data base can be expected to fill some of these gaps in the near future.

Neither the HP–DS nor the UBC data base thus far incorporate the available low-temperature phase (dis-)equilibrium data. Consideration of these, especially solubility data, will require simultaneous refinement of the properties for mineral and (major) aqueous species, a testing task! The endeavour is, however, sorely needed in view of the requirements for hydro- and geothermal problems and for some pressing technical applications. Similarly, the list of mineral solution phases may be expected to grow, for which realistic activity models are developed that are consistent with the end-member thermodynamic properties. By necessity these additions will in part rely on natural mineral pairs (and larger metamorphic assemblages) deemed to have equilibrated at well constrained $P-T$-conditions. Provided that the pertinent uncertainties are correctly estimated and propagated, there is no reason why natural assemblages should not be combined with the experimental data to refine thermodynamic data, particularly solution models. The importance of adding such activity models for further rock-forming minerals cannot be overestimated, since many applications depend on them. Notably, some modern methods of thermobarometry (Berman, 1991; Lieberman and Petrakakis, 1991, Powell and Holland, 1988) make full use of optimized thermodynamic models, especially those relying on *all* simultaneous equilibria represented by a phase assemblage. Feedback from selected natural samples investigated by such methods is likely to yield the next major step of improvement in consistent thermodynamic data bases for minerals.

Acknowledgements

Thanks to the Mineralogical Society and especially David Price, Nancy Ross and Ross Angel for organizing a most stimulating meeting. To them and Paul Henderson I am grateful for their kind invitation to participate. David Price and Nancy Ross also took on the task of editing the present book, and I thank them for their courage and endless patience. For this paper I have depended on help from several colleagues. Rob Berman, Tim Holland, Peter Lichtner, and Urs Mäder contributed results from ongoing research prior to publication. Assistance in the data analysis, with various programs, and with the manuscript was provided by Helmut Horn, Peter Lichtner, Josh Lieberman, Rob Berman, and Tom Brown. I am most grateful for the time and effort they have contributed so generously. Funding from the Schweizerischer Nationalfonds (2.752.87/5459) and, for a smaller part of the research summarized here, from the International Energy Agency (Annex VI) is gratefully acknowledged.

References

Anderson, G. M. (1977) The accuracy and precision of calculated mineral dehydration equilibria, in *Thermodynamics in Geology*, (ed. D. G. Fraser), Reidel, Dordrecht, pp. 115–36.

Anderson, G. M. (1977) Uncertainties in calculations involving thermodynamic data, in *Application of Thermodynamics to Petrology and Ore Deposits*, (ed. H. J. Greenwood), Mineralogical Association of Canada, Ottawa, pp. 199–215.

Ball, F. W., Jenne, E. A., and Cantrell, M. W. (1981) *WATEQ3: a Geochemical Model with Uranium Added*, US Geol. Survey, Open-File Report, pp. 81–1183.

Ball, F. W., Nordstrom, D. K. and Jenne, E. A. (1980) *Additional and Revised Thermochemical Data and Computer Code for WATEQ2—A Computerized Chemical Model for Trace Element Speciation and Mineral Equilibria of Natural Waters*, U.S. Geol. Survey, Water Resources Investigation, pp. 78–116.

Berman, R. G. (1988) Internally-consistent thermodynamic data for minerals in the system $Na_2O-K_2O-CaO-MgO-FeO-Fe_2O_3-Al_2O_3-SiO_2-TiO_2-H_2O-CO_2$. *Journal of Petrology*, **29** (2), 445–522.

Berman, R. G. and Brown, T. H. (1985) Heat capacity of minerals in the system $Na_2O-K_2O-CaO-MgO-FeO-Fe_2O_3-Al_2O_3-SiO_2-TiO_2-H_2O-CO_2$: representation, estimation, and high temperature extrapolation. *Contributions to Mineralogy and Petrology*, **89**, 168–83.

Berman, R. G., Engi, M., Greenwood, H.J., et al. (1986) Derivation of internally-consistent thermodynamic properties by the technique of mathematical programming: a review with application to the system $MgO-SiO_2-H_2O$. *Journal of Petrology*, **27** (6), 1331–64.

Bottinga, Y. and Richet, P. (1981) High pressure and temperature equation of state and calculation of thermodynamic properties of gaseous carbon dioxide. *American Journal of Science*, **281**, 615–60.

Brady, J. B. and Stout, J. H. (1980) Normalization of thermodynamic properties and some implications for graphical and analytical problems in petrology. *American Journal of Science*, **280**, 173–89.

Brousse, C., Newton, R. C., and Kleppa, O. J. (1984) Enthalpy of formation of forsterite, enstatite, åkermanite, montecellite and merwinite at 1073 K determined by alkali borate solute ion calorimetry. *Geochimica et Cosmochimica Acta*, **48**, 1081–8.

Burnham, C. W., Holloway, J. R., and Davis, N. E. (1969) The specific volume of water in the range 10–8900 bar, 20°–900°C. *American Journal of Science*, **267A**, 70–95.

Charlu, T. V., Newton, R. C., and Kleppa, O. J. (1975) Enthalpies of formation at 970 K of compounds in the system $MgO-Al_2O_3-SiO_2$ from high temperature solution calorimetry. *Geochimica et Cosmochimica Acta* **39**, 1487–97.

Charlu, T. V., Newton, R. C., and Kleppa, O. J. (1978) Enthalpy of formation of some lime silicates by high-temperature solution calorimetry, with discussion of high pressure phase equilibria. *Geochimica et Cosmochimica Acta*, **42**, 367–75.

Chermak, J. A. and Rimstidt, J. D. (1989) Estimating the thermodynamic properties (ΔG_f° and ΔH_f°) of silicate minerals at 298 K from the sum of polyhedral contributions. *American Mineralogist*, **74**, 1023–31.

Cohen, R. E. (1986) Statistical mechanics of coupled solid solutions in the dilute limit. *Physics and Chemistry of Minerals*, **13**, 174–82.

Connolly, J. A. D. (1990) Multivariate phase diagrams: an algorithm based on generalized thermodynamics. *American Journal of Science*, **290**, 666–718.

CODATA recommended key values for thermodynamics. (1978) Report of the CODATA Task Group on key values for thermodynamics, 1977. *Journal of Chemical Thermodynamics*, **10**, 902–6.

De Capitani, C. and Brown, T. H. (1987) The computation of chemical equilibrium in complex systems containing non-ideal solutions. *Geochimica et Cosmochimica Acta*, **51**, 2639–52.

Delany, J. M. and Helgeson, H. C. (1978) Calculation of the thermodynamic consequences of dehydration in subducting oceanic crust to 100 kbar and >800°C. *American Journal of Science*, **278**, 638–86.

Demarest, H. H., Jr. and Haselton, H. T., Jr. (1981) Error analysis for bracketed phase equilibrium data. *Geochimica et Cosmochimica Acta*, **45**, 217–24.

Douglas, T. B. and King, E. K. (1968) High-temperature drop calorimetry, in *Experimental*

Thermodynamics, (eds J. P. McCulloch and D. W. Scott), vol. 1: *Calorimetry of Non-reacting Systems*, Plenum Press, New York, pp. 293–331.

Engi, M. and Wersin, P. (1987) Derivation and application of a solution model for calcic garnet. *Schweizerische Mineralogische und Petrographische Mitteilungen*, **67**, 53–73.

Engi, M., Lieberman, J., and Berman, R. G. (1990) Uncertainties in thermodynamic data for mineral systems. *Geological/Mineralogical Association of Canada Abstracts 15 (Vancouver)*, **A36**.

Felmy, A. R., Girvin, D. C., and Jenne, E. A. (1984) *MINTEQ: A Computer Program for Calculating Aqueous Geochemical Equilibria*, EPA-600/3-84-032, U. S. Environmental Protection Agency, Atlanta, Georgia. (NTIS PB84–157148, National Technical Information Service, Springfield, Virginia).

Ferry, J. M. and Baumgartner, L. (1987) Thermodynamic models of molecular fluids at the elevated pressures and temperatures of crustal metamorphism. In *Thermodynamic Modeling of Geological Materials: Minerals, Fluids and Melts*, (eds I. S. E. Carmichael and H. P. Eugster), American Mineralogical Society, Reviews in Mineralogy 17, pp. 323–65.

Ghiorso, M. S. (1987) Thermodynamics of minerals and melts. *Reviews of Geophysics*, **25**, 1054–64.

Ghiorso, M. S. (1990a) Application of the Darken equation to mineral solid solutions with variable degrees of order-disorder. *American Mineralogist*, **75**, 539–43.

Ghiorso, M. S. (1990b) A solution model for ilmenite–hematite–geikielite. *Contributions to Mineralogy and Petrology*.

Gill, P. E., Murray, W., and Wright, M. H. (1981) *Practical Optimization*, Academic Press, London.

Gordon, T. M. (1973) Determination of internally consistent thermodynamic data from phase equilibrium experiments. *Journal of Geology*, **81**, 199–208.

Haar, L., Gallagher, J. S., and Kell, G. S. (1984) *NBS/NCR Steam Table*. Hemisphere Publishing, Washington.

Haas, J. L., Jr., Robinson, G. R., and Hemingway, B. S. (1981) Thermodynamic tabulations for selected phases in the system $CaO–Al_2O_3–SiO_2–H_2O$ at 101.325 kPa (1 atm) between 273.15 and 1800 K. *Journal of Physical and Chemical Reference Data*, **10**, 575–669.

Halbach, H. and Chatterjee, N. D. (1982) The use of linear parametric programming for determining internally consistent thermodynamic data. In *High-pressure Researches in Geoscience*, (ed. W. Schreyer), Schweizerbart, Stuttgart, pp. 475–91.

Helgeson, H. C., Delany, J. M., Nesbitt, H. W., et al. (1978) Summary and critique of the thermodynamic properties of rock-forming minerals. *American Journal of Science*, **278A**, 1–229.

Helgeson, H. C. and Kirkham, D. H. (1974a) Theoretical prediction of the thermodynamic behaviour of aqueous electrolytes at high pressures and temperatures: I. Summary of the thermodynamic/electrostatic properties of the solvent. *American Journal of Science*, **274**, 1089–1198.

Helgeson, H. C. and Kirkham, D. H. (1974b) Theoretical predication of the thermodynamic behaviour of aqueous electrolytes at high pressures and temperatures: II. Debye–Hückel parameters for activity coefficients and relative partial molal properties. *American Journal of Science*, **274**, 1199–261.

Helgeson, H. C., Kirkham, D. H., and Flowers, G. C. (1981) Theoretical prediction of the thermodynamic behavior of aqueous electrolytes at high pressures and temperatures: IV. Calculation of activity coefficients, osmotic coefficients, and apparent molal and standard and relative partial molal properties to 600°C and 5 kb. *American Journal of Science*, **281**, 1249–1516.

Hemingway, B. S., Haas, J. L., Jr. and Robinson, G. R., Jr. (1982) *Thermodynamic properties of selected minerals in the system $Al_2O_3–CaO–SiO_2–H_2O$ at 298.15 K and 1 bar (10^5 pascals) pressure and at higher temperatures*, US Geol. Surv. Bull. 1544.

Holdaway, M. J. (1971) Stability of andalusite and the aluminum silicate phase diagram. *American Journal of Science*, **271**, 97–131.

Holland, T. J. B. (1989) Dependence of entropy on volume for silicate and oxide minerals: a review and a predictive model. *American Mineralogist*, **74**, 5–13.

Holland, T. J. B. and Powell, R. (1985) An internally consistent thermodynamic dataset with uncertainties and correlations: 2. Data and results. *Journal of Metamorphic Geology*, **3**, 343–70.

Holland, T. J. B. and Powell, R. (1990) An enlarged and updated internally consistent thermodynamic dataset with uncertainties and correlations: The system $K_2O-Na_2O-CaO-MgO-MnO-FeO-Fe_2O_3-Al_2O_3-TiO_2-SiO_2-C-H_2-O_2$. *Journal of Metamorphic Geology*, **8** (1), 89–124.

Holloway, J. R. (1977) Fugacity and activity of molecular species in supercritical fluids, in *Thermodynamics in Geology*, (ed. D. G. Fraser), Reidel, Dordrecht, pp. 161–81.

Holloway, J. R. (1987) Igneous fluids, in *Thermodynamic Modeling of Geological Materials: Minerals, Fluids and Melts*, (ed. I. S. E. Carmichael and H. P. Eugster), American Mineralogical Society, Reviews in Mineralogy 17, pp. 211–33.

Jansson, B. (1984a) *A General Method for Calculating Phase Equilibria Under Different Types of Conditions*. TRITA-MAC-0233, Materials Center, Royal Institute of Technology, Stockholm.

Jenkins, D. M. (1984) Upper pressure stability of synthetic margarite+quartz. *Contributions to Mineralogy and Petrology*, **88**, 332–9.

Kerrick, D. M. and Jacobs, G. K. (1981) A modified Redlich–Kwong equation or H_2O, CO_2, and H_2O–mixtures at elevated pressures and temperatures. *American Journal of Science*, **281**, 735–67.

Kharaka, Y. J., Gunter, W. D., Aggarwal, P. K., et al. (1989) *SOLMINEQ.88: A Computer Program for Geochemical Modeling of Water-rock Interactions*. US Geol. Survey, Water Resources Investigations Report 88–4227.

Kitihara, S., Takenouchi, S., and Kennedy, G. C. (1966) Phase relations in the system $MgO-SiO_2-H_2O$ at high temperatures and pressures. *American Journal of Science*, **264**, 223–33.

Kolassa J. E. (1990) *Confidence Intervals for Thermodynamic constants*, IBM Reseach Report 15769 and *Geochimica et Cosmochimica Acta*, **55**, 3543–52.

Krupka, K. M. and Jenne, E. A. (1982) *WATEQ3 Geochemical Model: Thermodynamic Data for Several Additional Solids*. PNL–4276, Pacific Northwest Laboratory, Richland, Washington.

Lichtner, P. C. and Engi, M. (1989) Data bases of thermodynamic properties used to model thermal energy storage in aquifers. *International Energy Agency* Annex VI, Subtask A, Doc. 26–A.

Lieberman, J. (1991a) GridLoc: Macintosh graphics tools for phase diagrams (in preparation).

Lieberman, J. (1991b) PTAX: Calculation of mineral fluid stability relations and software for Macintosh and Unix computers (in preparation).

Lieberman, J. and Petrakakis, K. (1991) TWEEQU thermobarometry: analysis of analytical uncertainties and applications to granulites from Western Alaska and Austria. *Canadian Mineralogist*, **29**, 857–88.

Mäder, U.K. (1991) H_2O-CO_2 mixtures: a review of P-V-T-X data and an assessment from a phase equilibrium point of view. *Canadian Mineralogist*, **29**, 767–90.

Mäder, U. K. and Berman, R. G. (1991) An equation of state for carbon dioxide to high pressure and temperature. *American Mineralogist*, **76**, 1547–59.

Mraw, S. C. and Naas, D. F. (1979) The measurement of accurate heat capacities by differential scanning calorimetry. Comparison of d.s.c. results on pyrite (100 to 800 K) with literature values from precision adiabatic calorimetry. *Journal of Chemical Thermodynamics*, **11**, 567–84.

Navrotsky, A. (1979) Calorimetry: its application to petrology. *Annual Reviews of Earth and Planetary Science*, **7**, 93–115.

Navrotsky, A. (1987) Models for crystalline solutions, in *Thermodynamic Modeling of Geological Materials: Minerals, Fluids and Melts*, (ed. I. S. E. Carmichael and H. P. Eugster), American Mineralogical Society, Reviews in Mineralogy 17, pp. 35–69.

Neuvonen, K. J. (1952) Heat of formation of merwinite and monticellite. *American Journal of Science, Bowen*, **2**, 373–80.

Newton, R. C., Navrotsky, A., and Wood, B. J. (eds) (1981) *Thermodynamics of Minerals and Melts*, Springer, New York.

Newton, R. C., Wood, B. J., and Kleppa, O. J. (1981) Thermochemistry of silicate solid solutions. *Bulletin minéralogique*, **104**, 162–72.

Nordstrom, D. K. and Munoz, J. L. (1985) *Geochemical Thermodynamics*, Benjamin/Cummings, Menlo Park, California.

Nordstrom, D. K., Plummer, L. N., Langmuir, D., et al. (1990) Revised chemical equilibrium data for major water–mineral reactions and their limitations; in *Chemical Modeling of Aqueous Systems II*, (eds D. C. Melchior and R. L. Bassett), 298–413, American Chemical Society, Wasington DC, ACS Symposium Series 416.

Nordstrom, D. K., Valentine, S. D., Ball, J. W., et al. (1984) Partial compilation and revision of basic data in the WATEQ programs. *US Geol. Survey Water-Resources Investigation Report* 84-4186, 10-4

Oelkers, E. H. and Helgeson, H. C. (1988) Calculation of the thermodynamic and transport properties of aqueous species at temperatures from 400 to 800°C and pressures from 500 to 4000 bars. *Journal of Physical Chemistry*, **92**, 1631–9.

Oelkers, E. H. and Helgeson H. C. (1990) Triple-anions and polynuclear complexing in supercritical electrolyte solutions. *Geochimica et Cosmochimica Acta*, **54**, 727–38.

Parkhurst, D. L., Thorsteson, D. C., and Plummer, L. N. (1980) PHREEQE – a computer program for geochemical calculations. *US Geol. Survey Water-Resources Investigation Report* 80–96.

Pedley, J. B. and Rylance, J. (1977) Computer analysis of thermochemical data, in *Proceedings, Fifth Biennial CODATA Conference*, (ed. B. Dreyfus), Pergamon Press, Oxford, pp. 557–80.

Pitzer, K. S. (1987) A thermodynamic model for aqueous solutions of liquid-like density, in *Thermodynamic Modeling of Geological Materials: Minerals, Fluids and Melts*, (eds I. S. E. Carmichael and H. P. Eugster), American Mineralogical Society, Reviews in Mineralogy 17, pp. 97–142.

Pitzer, K. S. (1973) Thermodynamics of electrolytes. I. Theoretical basis and general equations. *Journal of Physical Chemistry*, **77**, 268–77.

Powell, R. and Holland, T. J. B. (1985) An internally consistent thermodynamic dataset with uncertainties and correlations: 1. Methods and a worked example. *Journal of Metamorphic Geology*, **3**, 327–42.

Powell, R. and Holland, T. J. B. (1988) An internally consistent thermodynamic dataset with uncertainties and correlations: 3. applications to geobarometry, worked examples and a computer program. *Journal of Metamorphic Geology*, **6**, 173–204.

Putnis, A. and Bish, D. H. (1983) The mechanism and kinetics of Al/Si ordering in Mg-cordierite. *American Mineralogist*, **68**, 60–5.

Richardson, S. W., Gilbert, M. C., and Bell, P. M. (1969) Experimental determination of kyanite–andalusite and andalusite–sillimanite equilibria; the aluminum silicate triple point. *American Journal of Science*, **267**, 259–72.

Robie, R. A., Hemingway, B. S., and Fisher, J. R. (1979) *Thermodynamic Properties of Minerals and Related Substances at 298.15 K and 1 bar (10^5 Pascals) Pressure and at Higher Temperatures*, US Geol. Surv. Bull. 1452.

Robinson, G. R., Haas, J. L., Schafer, C. M., et al. (1983) *Thermodynamic and Thermophysical Properties of Selected Phases in the $MgO-SiO_2-H_2O-CO_2$, $CaO-Al_2O_3-SiO_2-H_2O-CO_2$, and $Fe-FeO-Fe_2O_3-SiO_2$ Chemical Systems, with Special Emphasis on the Properties of Basalts and Their Mineral Components*, US Geol. Survey, Open File Report 83–79.

Sack, R. D. and Ghiorso, M. S. (1989) Importance of considerations of mixing properties in establishing an internally consistent thermodynamic data base: Thermochemistry of minerals in the system Mg_2SiO_4–Fe_2SiO_4–SiO_2. *Contributions to Mineralogy and Petrology*, **102**, 41–68.

Salje, E. K. H. (ed.) (1988) *Physical Properties and Thermodynamic Behaviour of Minerals*, NATO ASI Series C, vol. 225. Reidel, Dordrecht.

Saxena, S. K. and Fei, Y. (1987a) Fluids at crustal pressures and temperatures. I. Pure species. *Contributions to Mineralogy and Petrology*, **95**, 370–5.

Saxena, S. K. and Fei, Y. (1987b) High pressure and high temperature fluid fugacities. *Geochimica et Cosmochimica Acta*, **51**, 783–91.

Schittkowski, K. (1985a) NLPQL: A Fortran-subroutine solving constrained nonlinear programming problems. *Annals of Operations Research*, **5**, 485–500.

Schittkowski, K. (1985b) *Solving Constrained Least Squares Problems by a General Purpose SQP-method*, Report, Institut für Informatik, Stuttgart GFR, Universität Stuttgart.

Shock, E. L. and Helgeson, H. C. (1990) Calculation of the thermodynamic and transport properties of aqueous species at high pressures and temperatures: standard partial molal properties of organic species. *Geochimica et Cosmochimica Acta*, **54**, 915–45.

Shock, E. L., Helgeson, H. C., and Sverjensky, D. A. (1990) Calculation of the thermodynamic and transport properties of aqueous species at high pressures and temperatures: standard partial molal properties of inorganic neutral species. *Geochimica et Cosmochimica Acta*, **53**, 2157–83.

Sposito, G. and Mattigod, S. V. (1980) *GEOCHEM: a computer program for the calculation of chemical equilibria in soil solutions and other water systems*, Kearny Foundation, University of California, Riverside, California.

Storre, B. and Nitsch, K.-H. (1974) Zur Stabilität von Margarit im System CaO–Al_2O_3–SiO_2–H_2O. *Contributions to Mineralogy and Petrology*, **43**, 1–24.

Stull, D. R. and Prophet, H. (1971) *JANAF Thermochemical Tables*, 2nd edn, National Bureau of Standards NSRDS 37. (Supplements in the *Journal of Physical and Chemical Reference Data* **3**, 311–480, 1974; **4**, 1–185, 1975; **7**, 793–940, 1978.)

Sundman, B., Jansson, B., and Andersson, J.-O. (1985) The Thermo-Calc databank system. *CALPHAD*, **9**, 153–86.

Sverjensky, D. A., Shock, E. L., and Helgeson, H. C. (1990) Prediction of the thermodynamic properties of inorganic aqueous metal complexes to 1000°C and 5 kbar (in preparation).

Tanger, J. C. and Helgeson, H. C. (1988) Calculation of the thermodynamic and transport properties of aqueous species at high pressures and temperatures: Revised equations of state for the standard partial molal properties of ions and electrolytes. *American Journal of Science*, **288**, 19–98.

Tanger, J. C. and Pitzer, K. S. (1989) Calculation of the thermodynamic properties of aqueous electrolytes to 1000°C and 5000 bar from a semicontinuum model for ion hydration. *Journal of Physical Chemistry*, **93**, 4941–51.

Torgeson, D. R. and Sahama, T. G. (1948) A hydrofluoric acid solution calorimeter and the determination of the heats of formation of Mg_2SiO_4, $MgSiO_3$, and $CaSiO_3$. *Journal of the American Chemical Society*, **70**, 2156–60.

Truesdell, A. H. and Jones, B. F. (1974) WATEQ – a computer program for calculating chemical equilibria of natural water, US Geol. Survey, *Journal of Research* **2**, 233–48.

Ulbrich, H. H. and Waldbaum, D. R. (1976) Structural and other contributions to the third-law entropy of silicates. *Geochimica et Cosmochimica Acta*, **40**, 1–24.

Vieillard, P. and Tardy, Y. (1988) Estimation on enthalpies of formation of minerals based on their refined crystal structures. *American Journal of Science*, **288**, 997–1040.

Weare, J. H. (1987) Models of mineral solubility in concentrated brines with application to field observations, in *Thermodynamic Modeling of Geological Materials: Minerals, Fluids and Melts*, I. (eds S. E. Carmichael and H. P. Eugster), American Mineralogical Society, Reviews in Mineralogy 17, pp. 143–74.

Westrum, E. F., Jr., Furukawa, G. T., and McCulloch, J. P. (1968) Adiabatic low-temperature calorimetry, in *Experimental Thermodynamics*, (eds J. P. McCulloch and D. W. Scott), vol. 1, *Calorimetry of Non-reacting Systems*, Plenum Press, New York, pp. 133–214.

Wolery, T. J. (1979) *Calculation of Chemical Equilibrium between Aqueous Solutions and Minerals: the EQ3/EQ6 Software Package*, Lawrence Livermore Laboratory, UCRL-52658, Livermore, California.

Wolery, T. J. (1983) *EQ3NR. A Computer Program for Geochemical Aqueous Speciation-solubility Calculations: User's Guide and Documentation*, Lawrence Livermore Laboratory, UCRL-53414, Livermore, California.

Wolery, T. J., Sherwood, D. J., Jackson, K. J., *et al.* (1984) *EQ3/6: Status and Applications*, Lawrence Livermore Laboratory, UCRL-91884, Livermore, California.

Zen, E-an (1977) The phase-equilibrium calorimeter, the petrogenetic grid, and a tyranny of numbers. *American Mineralogist*, **62**, 189–204.

CHAPTER NINE
The stability of clays
Bruce Velde

9.1 Introduction

Some 30 years ago it was observed that the type of clay-mineral assemblage found in an argillaceous rock changed as a function of burial depth in sedimentary basins. This observation leads eventually to a direct comparison with metamorphic reactions which are a function of temperature, pressure, composition, and time variables. In recent years, numerous studies have shown that the clay mineralogy of a sedimentary rock can be a function of its bulk composition (sedimentary facies) or its position in the stratigraphic series of the basin in which it was buried (age and thermal history). A good synopsis of this information can be found in Weaver (1989). This reinforces the idea that the changes in clay minerals are governed by normal parameters of mineral stability (temperature, T and composition, x). Up until now, there is little evidence that pressure (lithostatic) influences the stability of clay minerals in natural environments.

However, the change of clay-mineral facies or assemblages often occurs over significant distances of geologic space, thousands of metres. By contrast, metamorphic mineral assemblages change, usually, over distances of several metres. In this respect the study of clay-mineral stabilities is different from those normally practised on metamorphic systems (temperatures above 300°C or so) where the problems of reaction kinetics (time) are generally ignored and attention is for the most part paid to temperature and composition variables. Most metamorphic systems are assumed to have attained thermodynamic equilibrium whereupon the classical phase rule can be employed.

For those who doubt that the clay systems in nature are to be considered a part of the normal silicate world, one must demonstrate that time, as well as temperature and composition effect phase changes in a regular and consistent manner according to our ideas of kinetics. The task of the present paper is to demonstrate the relations between classic thermodynamic parameters of metamorphism (temperature, pressure, and composition) and those of clay mineralogy (temperaure, composition, pressure, and time).

A selection of different clay-mineral reactions will be used to demonstrate the action of the different kinetic and chemical controls on clay mineral

The Stability of Minerals. Edited by G. D. Price and N. L. Ross.
Published in 1992 by Chapman & Hall, London. ISBN 0 412 44150 0

stability. In order to do this, one must give strong arguments that temperature is important in clay-mineral change, but a more important aspect is that time is a consistently present factor as well as that of chemical composition. The major problem in establishing these parameters is that the clay facies is one of low temperature, between 4 and about 250°C. In this range the reaction rate is always low due to the rate-controlling role of temperature. The second problem is that clays occur in very different chemical environments. Clay diagenesis is most often accomplished during burial of muds which tend to form closed chemical systems, i.e. those which allow little transfer of material into or out of the chemical system. In contrast, the origin of clays, at the surface of the earth is effected under conditions of extreme chemical disequilibrium, those of weathering. Here the chemistry is changed radically, with a large part of the solids present entering into the liquid, aqueous phase and being eventually transported out of the zone where the clays are found to form. Weathering is an environment of an open chemical system where the phases formed respond first to chemical forces and only more slowly to those of thermal constraints. Thus, the genesis of clays and the evolution of clays are often accomplished in quite different thermodynamic settings.

In order to determie the limits of stability of clays, one must consider the *temperatures* to which they were subjected as a function of *time* and the *chemistry* of their surroundings.

Initially, we will look at a demonstration of the effect of time and temperature on the amount of clay-mineral transformation and at the effect of chemistry on the rate of mineral transformation. This is the realm of diagenesis or transformation of clays under conditions of burial in sedimentary basins. This concerns the kinetic aspect of clay transformation. We will also look at the effect of chemistry as it affects the stability of clays. This is the realm of weathering.

Finally, we will look at the chemical homogeneity of clay grains compared to the appearance of the species in a clay assemblage. This concerns the concept of phase equilibrium and chemical composition.

9.2 Time–temperature control of clay stability

Kinetics classically deals with the time and temperature relations of reactions. Initially we will look at several aspects of kinetics which directly affect the configuration of clays and clay assemblages in nature.

9.2.1 Time–temperature space

The following is a very brief summary of certain aspects of phase equilibria and chemical reactions which are important to understand the arguments of the kinetic relations of clay phase 'equilibria'. A more thorough description of the few notions expressed in this section can be found in the various chapters of the review book of Lasaga and Kirkpatrick (1981) among other works.

First, we will represent phase equilibria in classical, 'static' space, i.e. that without the dimension of time.

According to equilibrium thermodynamics (i.e. static systems) applicable to geologic problems, at some point in pressure–temperature space for a given chemical system two or several phases or assemblages of phases can coexist. When pressure or temperature are changed independently, the number of phases will be reduced. This is shown in Figure 9.1 in a very simple case where we consider a system with only one chemical component and where just two phases or minerals are present in the range of pressure (P) – temperature (T) conditions considered. The pressure–temperature space can be divided into two zones where either phase X or phase Y will be present and a line in P–T space where both phases can be present stably together.

If we assume that the reaction X to Y is very simple, not being dependent upon the concentration of the reactants as should be the case in a one-component system, the rate of change of X will follow the relations

$$dX/dt = k(Y^n) \qquad (1)$$

where n is the order of the reaction concerning the presence of Y.

The effect of temperature is seen in

$$k = A \exp(-E/RT). \qquad (2)$$

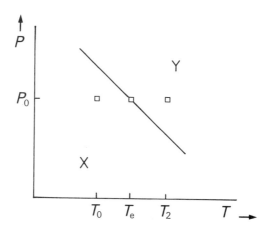

Figure 9.1 Representation of a phase boundary in pressure (P)–temperature (T) space for a single component system where two phases are present, X and Y. The change of phase investigated is effected at constant pressure (P_0) between temperatures T_0, T_e, and T_2. T_e is the temperature, at pressure P_0, where the two phases can coexist.

THE STABILITY OF CLAYS

Here T indicates a temperature greater than that at which the new phase is at equilibrium with the old phase, which will be designated as T_e, the equilibrium temperature. At constant pressure, when the temprature increases from that where X is stable to that where Y is stable (where $T > T_e$), the rate of change of X to Y will depend upon the difference in temperature between T_e and the new temperature. In Figure 9.1, this relation is shown by a change of temperature from T_0 to T_2 at constant pressure P_0.

At point P_0, T_0 only phase X will be present, at point P_0, T_e (that of coexistence of phases X and Y) both phases will be present, and at point P_0, T_2 only phase Y will be stable. In changing from point P_0, T_e to a point P_0, T_2, the greater the temperature difference between T_e and the new point, the greater will be the rate at which X will be transformed to Y, following the reaction relation (equation (2)).

To illustrate this idea, let us start from point P_0, T_0 in Figure 9.1 where only phase, X, exists. When we change the temperature to T_e, nothing will happen regardless of the rate at which the temperature is changed. At point T_e at constant pressure, phase X will be converted in part to phase Y. Even if the system is left at T_e for an infinite period of time, both phases will remain present. They are stable together under these physical conditions. If we increase the temperature from T_e (the equilibrium temperature) to T_2, after a certain time, all of phase X will be converted to Y. However, the amount of conversion wll depend upon the time given for the reaction to take place. Figure 9.2 shows the relation between the time taken to change the temperature from T_e to T_2 and the proportion of conversion of X to Y. The amount of change (distance of the point in Figure 9.2 between X and Y at temperature T_2) is a function of the time (t) over which the temperature change was effected. When the time of temperature change t, is great, the proportion of X and Y will be seen to change

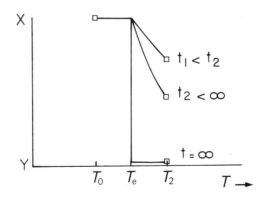

Figure 9.2 Representation of the quantity of phases X and Y present at temperatures T_0, T_e, and T_2 for three different rates of thermal change, infinitely slow (t), slow (t_2), and rapid (t_1). The proportion of reaction effected is indicated by the coordinates of the square between X and Y.

abruptly at T_e. When the time of temperature change is infinitely small the reaction will not take place to any significant extent and the amount of change of X to Y will be inconsequential. If the time over which the change from T_e to T_2 takes place is intermediate in duration, there will be a mixture of X and Y present in the sample at point P_0, T_2 after time t_2 ($t_2 < t$). If the change of temperature is effected over a shorter period of time ($t_1 < t_2 < t$), there will be still less of Y present in the sample. In this example, when temperature changes at a finite rate the result is a mixed-phase system. This means that at $0 < t_2 < t$, one should find a certain portion of a phase which has not reacted and is metastable.

By contrast, it is known that in metamorphic reactions at high temperature ($> 300°C$), the problems of the amount of conversion, relations of time and temperature, have been observed to be unimportant. The change from one mineral assemblage or facies to another in rocks of the same or similar composition is seen to occur over short physical distances and thus temperature and time considerations are not important. Metastable phases tend to be uncommon.

If we consider geological space where a thermal gradient has been applied to a sequence of rocks, for reactions at low temperature, the line separating phases X and Y will appear in nature as a band whose width depends upon the time span over which the temperature gradient was maintained. For clay minerals subjected to burial at low temperatures, time will be an important factor.

In order to test this hypothesis we must look carefully at the clay minerals in a simple, continuously changing burial environment. Here we can be sure of the time factor by dating the different layers of the sediment but we must make some assumptions about the geothermal gradient in the sedimentary column.

9.2.2 Observations on the kinetics of clay reactions

In order to determine the thermal effect on clay stability, it will be necessary to consider clay-mineral assemblages as a function of their age and burial history in order to determine the parameters which gave them the mineralogic configuration that they now have. In order to appreciate the temporal effect in clay-mineral stability, we will look at the smectite-to-illite conversion which is possibly the most commonly found in terriginous rocks. A general summary of information on these minerals is given in Velde (1985) and Weaver (1989).

Intermediate steps in the smectite-to-illite conversion are expressed as illite/smectite (I/S) mixed-layer minerals. X-ray diffraction (XRD) of oriented clay aggregates indicates that the clay crystallites diffracting are composed of different proportions of illite and smectite layers (For a discussion of these ideas see Reynolds and Hower, 1970; Reynolds, 1980; and Moore and Reynolds, 1989). The exact numbers of different layers, their composition and position in the crystallite are at present under sufficient debate so that no unequivocal statement currently can be made concerning the identification of I/S clays using XRD methods alone as underlined by Jeans (1989). However, in series of sediments having experienced regular deep burial it has been found systematically that the aggregate I/S mineral compositions (proportions of smectite and illite) change in a regular manner varying rather regulary as a function of depth, and hence maximum temperature attained. Upper levels of sediments in deep drill holes in sedimentary basins contain clays with only smectite layers and the deepest levels contain illite and no smectite. These changes have been taken to establish the effects of burial diagenesis as promoting the smectite-to-illite conversion. In some way the total of smectitic layers decreases and is replaced by illite. We will look at this apparent reaction at first using XRD data alone.

THE STARTING POINT

The first term of the series, smectite, can be found repeatedly to form from a wide variety of materials (glass, micas, mixed layer minerals, feldspars, etc.) in the temperature range of 7–25°C at or near the earth's surface. Smectite is a common component of soils formed from aluminous composition rocks and volcanic materials. It is common in the sediments of lakes and ocean basins. New smectite has been observed to form on detrital grains of smectite in ocean-bottom sediments of ages ranging from Recent to Cretaceous (Steinberg *et al.*, 1988), which indicates that it is a stable mineral forming at low-temperature, low-pressure conditions. Since the initial term of the smectite-to-illite reaction series seems to form systematically from many different materials, it would thus appear to be stable at earth-surface conditions. We can then seriously consider the possibility of a normal reaction relationship from smectite to the other phase illite as one being initiated by a change in physical conditions (temperature and perhaps pressure).

REACTION PROGRESS

In order to estimate the conditions which effect the reaction or transformation of smectite to illite, one can investigate the compositions of the I/S clays in basins which have been subjected to continuous, uninterrupted burial diagenesis up to the present time. In doing this, one eliminates lapses of time which will produce irregular reaction progress. It is necessary to make some assumptions concerning the thermal gradient in these basins. For lack of better data, it will be assumed here that the present thermal gradient will have been similar to that in the past. This is subject to discussion, but we will see that it is useful as a rough approximation for our demonstration. Therefore, if the sediments have experienced similar temperatures at similar depths, the major difference in their history will be the length of time which they were maintained at each depth, i.e. the sedimentation rate will determine the differences in the stage of advancement of the reaction.

Clay-mineral studies by Velde and Iijima (1989) and Velde and Espitalié (1989) are used to investigate this effect. The present-day thermal gradient in all of these basins is near 30°C/km; it is slightly higher in the older basins than in younger ones. The information necessary for our study of clay minerals will be a determination of the composition of the I/S mineral assemblage at a given depth, the temperature at that depth and the age of the sediment. With this data one can determine the range of conditions of temperature and time which were necessary to obtain a given amount of apparent mineral conversion.

The initiation of the smectite-to-illite reaction beginning with a pure smectite starting material seems to occur between 50°C and 80°C in rocks of less than 15 Ma age (see data in Velde and Iijima, 1988) whereas the reaction seems to have begun at lower temperatures in rocks of over 50 Ma age (see data in Velde and Espitalié, 1989). The difference in starting point in the reaction suggests that the formation of a pure smectite starting material could itself be influenced by reaction kinetics and therefore that not all smectite has its origin at surface conditions.

The reaction appears to terminate at near 180°C in recent or young series of rocks (<50 Ma) and can be found to have progressed to near completion at about 90°C after 200 Ma (Figure 9.3).

These data indicate the range of conditions (Temperature, T, and time, t) where we can expect to find intermediate members of the reaction series. The

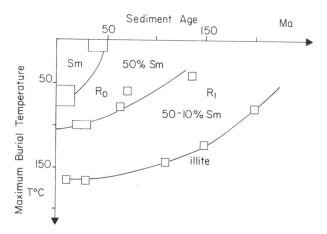

Figure 9.3 Indication of illite/smectite (I/S) mineral type as a function of maximum burial temperature and age of sedimentary layer for wells taken from Velde and Iijima (1988) and Velde and Espitalie (1989). Sm = full smectite mineral present; R_0 indicates the presence of a disordered structure interlayered mineral having a flake-shaped form. The R_0 minerals contain 100 to 50% smectite layers. R_1 is an ordered structure, lath-shaped crystal with from 50 to <10% smectite layers. There is a distinct relation between the maximum temperature attained and the age of the sediment which controls the I/S mineral composition and structural type. Squares represent the zones where the mineral types were identified. Ma = 10^6 years.

THE STABILITY OF CLAYS

present-day temperatures give a maximum value over which the I/S minerals are found in sedimentary basins undergoing continuous deepening. Each sample has, of course, experienced lower temperatures for various periods of time.

This information, then, gives only an approximate view of clay diagenesis because each sedimentary basin has a different burial history due to a variable burial rate. Four wells from the studies of Velde and Iijima (1988) and Velde and Espitalié (1989) given in Figure 9.4 show significant differences in burial history which will give different clay-mineral-reaction progress profiles in a depth–I/S composition plot. The wells are from the Paris basin, two from the Texas Gulf Coast area (Peeler and Carter) and a well from the Los Angeles, CA Basin. However, a plot of depth and I/S composition will indicate the general profile of I/S change. Each basin has the same geothermal gradient and therefore the same temperature at the same depth. Figure 9.5 gives the I/S composition vs depth plot where it is obvious that the shape of the curve is different for each type of diagenesis situation. The youngest is slow to react. In fact, one finds clays of highly varying composition from one sample to another which are taken to be unreacted clays in the upper 2 km of the profile. As the reaction begins from the pure smectite composition produced at some depth, it presents a concave-upwards shape. The oldest sedimentary series, from the Paris Basin, shows exactly the opposite reaction profile. The reaction is already advanced at the uppermost levels and the plot is convex-upwards. The Texas

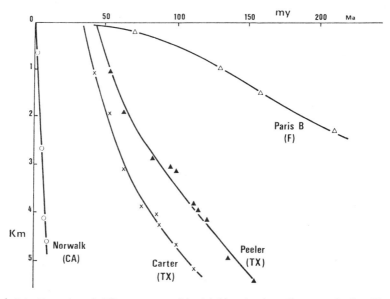

Figure 9.4 Examples of different types of burial histories in sedimentary basins. The clay mineralogy of these wells was studied in Velde and Iijima (1988) and Velde and Espitalié (1989). Names are those of specific wells in California (CA) and Texas (TX) USA, and France (Paris Basin).

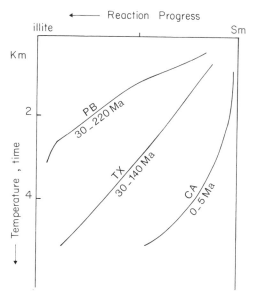

Figure 9.5 Depth – I/S composition plot for samples in three wells of different sedimentation rate. Depth in the well indicates an increase of total time and temperature while a decrease in smectite content of the aggregate sample composition indicates an increase in reaction progress. PB = Paris Basin well, TX = Texas Gulf Coast Peeler well and CA a well in the Los Angeles Basin of California.

well (Peeler), of intermediate age, shows a rather linear composition depth plot. Figure 9.5 clearly shows that the distance, burial depth, over which the I/S minerals are present is greatly reduced with longer periods of burial. As the reaction progresses, the depth interval over which it occurs is shortened.

Using a single-reaction kinetic formulation of first-order reaction type ($n = 1$ in equation (1)), and assuming a single, unchanging pre-exponential factor (see equation (2)) and a constant thermal gradient during burial, by varying the burial rate, the shape of the I/S smectite content in the clays will form the different curve shapes similar to the natural examples (Figure 9.5).

The changes in smectite content with depth calculated using the very simplified system indicate that the effect of burial rate produces different shaped curves similar to those seen in the natural mineral series of Figure 9.5. Thus, it appears that the initial conditions needed to establish whether or not a clay-mineral reaction is taking place are met and it also appears that this reaction follows the general rules of kinetics which could be expected in a low-temperature environment.

REACTION MECHANISM BY PHASE CHANGE
At this point it might be useful to observe the mineralogy during diagenesis in order to see just how the smectite-to-illite reaction could occur.

X-ray diffraction (XRD) Using XRD identification means, it has been well established, that the sequence of I/S minerals found in burial diagenetic environments can be divided into two structural types and possibly three (see Velde (1985) for a review of natural occurrence and Reynolds (1980) for a detailed description of the X-ray interpretation). The first structural type, with a random disposition of illite and smectite layers in the crystallites called R_0, is found in the compositional range of 50 to 100% smectite; the second type with a regular sequence of illite and smectite layering called R_1, is found in the range 50 to near 0% smectite. The illite crystallites, the end member of the series, may possibly have some expandable layers associated with them. In Figure 9.3 the two symbols for the I/S mineral types, R_0 and R_1, are shown according to the range of time – temperature conditions in which they occur.

Transmission electron microscopy (TEM) The difference in response to XRD (structure type) is echoed in the shape of the clay particles as seen under the transmission electron microscope (TEM). The R_0 types are rather irregular and flake-like while the R_1 types are regular, lath-shaped particles. The last term of the sequence, illite, is found to be based upon a hexagonal shape such as is common for higher-temperature micas (see Inoue *et al.*, 1988; Champion and Lanson, 1991). This information indicates that the smectite-to-illite transformation occurs, at best in four steps:

1. smectite to smectite 50% R0 structure;
2. conversion of R0 to R1 structure;
3. R1 50% smectite to R1 10%; and
4. conversion of R1 10% smectite to illite.

It is obvious that the smectite-to-illite conversion is not a homogeneous, continuous reaction.

We will investigate a portion of the reaction series, the R_1-to-illite transition, in more detail in order to establish how at least a part of the conversion series is effected.

X-ray fluorescence chemical analysis (XRF) Champion and Lanson (1991) report an electron microscope study of $<2\,\mu$ clay fraction grains in samples coming from a sequence of diagenetically altered Upper Paleozoic Paris Basin arenaceous sediments. Electron microscope X-ray fluorescence determinations of the composition of individual grains showed that the hexagonal-shaped grains have an illite composition and the laths contain varying amounts of smectite. R_1 and illite clay minerals were identified by XRD methods. Thus it is clear that the particle shape and mineral structure (stacking order) reflect a different mineralogy or composition of smectite content.

Image analysis, grain abundance Image analysis methods were used by Lanson and Champion to describe the relative abundance of the grain shapes in TEM photographs. Both size and grain shape were determined for several hundred grains at several depths in a diagenetic series. Laths could be distinguished from hexagons by aspect ratios (length/width) where the ratio of 3 could delimit the two populations.

In Figure 9.6 where grain type (lath-hexagon) abundance is plotted against the surface area of a grain, expressed as an equivalent diameter of a circle, it can be seen that samples from 1000 m depth have a greater abundance of laths compared to hexagons (aspect ratio <3). This is corroborated by XRD patterns of the bulk clay ($<2\,\mu$ fraction) extracted from the samples which show the presence of two I/S mineral grain populations in each sample, an R_1 type and a phase very close to illite (<10% smectite). TEM shows that at greater depth, 2130 m, the relative abundance of the two grain shapes is inverted and the more abundant illite hexagons are found to be of smaller grain size on the average than the laths. XRD of the bulk sample indicates that the illite-type mineral is more abundant and that the I/S smectite-containing phase has fewer smectite layers present. These two observations indicate that the I/S lath population has lost the less stable, small grains (dissolved) and small hexagonal-shaped illite grains replace them. This is what one would expect to find in a system where a mineral (I/S) has become unstable and has dissolved to form another, more stable one (illite).

However, it should be noted that the overall population of the laths tends to have larger grains at greater depth. The less stable phase has continued to grow and its bulk composition is more illitic. This was seen by bulk XRD also. The mineral which precipitates on both lath shaped I/S and hexagonal grains is of an illite composition. That is, the solution is oversaturated with illite and not smectite. Again this is what is to be expected when a phase (smectite-containing I/S) becomes unstable and another (illite) is stable. The continued growth of the unstable I/S grains is possible when the growth layer is illitic in character. As long as the system dissolves the smaller crystallites, the fluids will be saturated by this material and the stable illite layers will accumulate on the laths as well as the hexagons.

The apparent increase in illite content of R_1 I/S minerals is not a transformation of smectite flakes into illite flakes, but one where the smaller I/S grains dissolve and the material in solution precipitates as illite layers on the surface of the old grains which contain some smectite. This is a problem, in part at least, of growth kinetics as well as one of dissolution and precipitation. Small grains of unstable phases (I/S) dissolve and small grains of stable phases (illite) are created and grow. However, the medium-sized I/S grains are more stable than small I/S grains and they are as stable as small-to-large illite under these conditions of grain size. The I/S grains continue to grow until they become the smallest of the unstable grains present. Then they will eventually dissolve even though they might be larger than the average illite grains. Two factors are important to the ultimate stability of a mineral grain: grain size, and bulk grain composition. Any I/S grain will eventually dissolve (become unstable) regardless of its size when illite is the stable phase.

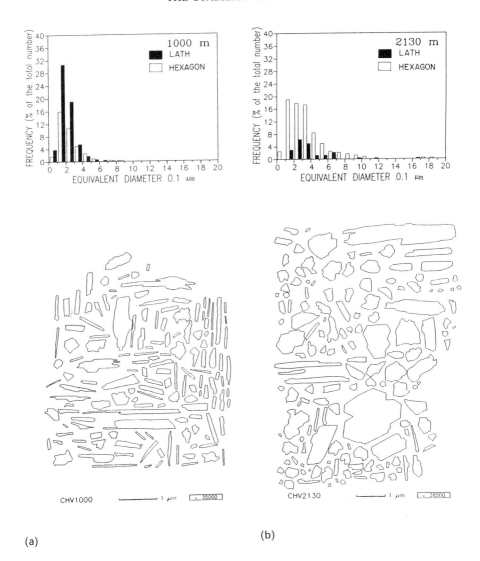

Figure 9.6 Representations of TEM observations for two populations of illite/smectite (I/S) clay particles from different levels (1000 m and 2130 m) from a well in the Paris Basin (Champion and Lanson, 1991). Laths were distinguished from hexagons using an aspect ratio of 3 for length/width. The relative abundance of the particles in the different grain-size categories (shown as an equivalent diameter) is shown in histogram form. At 1000 m laths are more abundant (a) and at 2130 m (b) hexagons are more abundant. More small hexagonal grains are present at 2300 m depth than at 1000 m. This indicates that new, seed-like crystals of hexagons are created at 2130 m at the expense of lath-shaped particles. Nevertheless the lath grains show a greater relative abundance in the larger grain sizes indicating continued, metastable grain growth despite a loss of numbers of grains (Lanson and Champion, in press).

KINETICS OF PHASE CHANGES

What are the consequences of these observations? If there are three major grain or mineral types present in the I/S series, how does this affect the kinetics of the overall smectite-to-illite reaction? Figure 9.7, a plot of the temperature of change in structural type as a function of the age of the specimen, shows that the points in the reaction series where the R_0 mineral changes to R_1 and R_1 transforms to illite occur at about the same or similar rates. A plot of the values $1/T$ and age gives two parallel limits to the mineral types. Thus even though the reaction series of smectite to illite is broken into several portions, the overall progress of the transformation seems to follow similar kinetic functions.

In summary it is possible to demonstrate that the overall change from smectite to illite is determined by the variables of temperature and time and that the I/S-to-illite transformation seems to be one of destabilization of the I/S grains which are replaced by illite even though the overall reaction, smectite to illite, is complex.

The major difference between this complex clay-mineral reaction and those observed in the realm of metamorphism is undoubtedly one of temperature related to the energy of activation necessary to effect the reactions. It is probable that the energies of activation of reaction between similar silicate minerals, smectites, illite, and micas, for example, will not be enormously different. The structures are similar, their compositions are similar, at least compared to alternative mineral species possible given the general bulk composition of the rocks. If the kinetic constants are similar for clays and metamorphic phyllosilicates, the reactions observed at 100°C or below where clay minerals are stable will occur over a larger range of a thermal gradient or

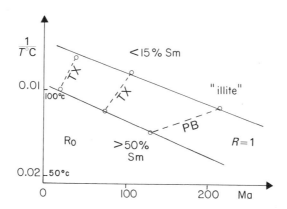

Figure 9.7 Sediment age-present day temperature relation of transition points where R_0 changes to R_1 structure I/S minerals and R_1 I/S minerals to illite on time $- 1/T$ coordinates. PB = Paris Basin well, and TX = the two Texas wells. The two transitions plot on parallel lines indicating that the reaction points proceed with similar kinetic factors in the basins studied which have similar present day thermal gradients.

geologic space than those which occur at 300°C. This explains why the I/S series occurs over kilometers in diagenetically altered sediments whereas reaction boundaries in metamorphic rocks are most often found over several metres distance.

9.3 Chemical effect on clay stability

9.3.1 Reaction rate: diagenetic conditions

Up until now, we have considered the chemistry of the clay mineral system to be constant and only time and temperature have varied. However, if the chemical activity of a critical component is changed in a given system, the rate of the reaction will be influenced. Consider the potential reaction smectite to illite as being:

$$\text{smectite} + K^+ = \text{illite} + \text{silica} \tag{3}$$

instead of:

$$\text{smectite} \rightarrow \text{illite} \tag{4}$$

In the first reaction it is obvious that if one changes the activity of potassium in the system, the reaction will proceed to the illite side. In order to see if the second reaction is pertinent to smectite-to-illite conversion the results of experiments reported by several authors (Howard and Roy, 1981; Whitney and Northrup, 1988; Roberson and Lahann, 1981) are considered, all using the same or very similar, natural smectite mineral starting materials but with a different potassium content. Assuming a first-order reaction

$$dx/dt = k(a/(a-x)) \quad \text{and} \quad kt = \ln a/(a-x) \tag{5}$$

where $a = 100\%$ smectite and $x =$ the proportion of illite produced by the reaction for experiments at 250°C, a measure of the rate of reaction progress (dx/dt) will give an equilibrium constant, k. Reaction times considered were up to 40 days. The deduced reaction constant for smectite $+ K^+$ can be seen to follow the overall concentration of potassium in the experimental charges (Figure 9.8). This shows that if the potassium concentration is increased by 5, the reaction will proceed 25 times faster at the same temperature. It is evident that firstly, the reaction is not a simple smectite to illite conversion and secondly, the overall composition of the system will influence the rate of reaction of clays.

When the concentration of potassium is decreased, the reaction will proceed more slowly. Thus one would expect to find that sediments with different

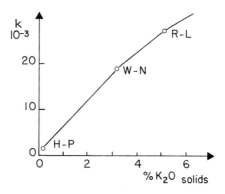

Figure 9.8 Plot of calculated reaction constant (k) for smectite–illite at 250°C as a function of the K_2O content of the solids for experiments using natural smectite starting materials. Data: H–P (Howard and Roy, 1985); W–N (Whitney and Northrop 1988); R–L (Roberson and Lahann 1981).

potassium contents or with different potassium activities in the fluids will show different stages of reaction progress in the smectite to illite conversion. This has in fact been noted several times (Velde and Nicot, 1985; Hoffman and Hower, 1979) for sandstones when compared to adjacent shale layers. It seems that the shales progress further in the conversion than do the sandstones on the average. This might well be explained by the difference in the potassium available in the shales which commonly contain abundant detrital muscovite, biotite, and K-feldspars, all potential sources of potassium for the conversion of smectite to illite.

A second example of difference in the rate of conversion of smectite to illite in natural systems is the comparison of bentonite beds with enclosing shales (Hoffman and Hower, 1979). The more dense, potassium-deficient bentonite beds form an illite-rich I/S phase through diffusion of potassium from the shale layers into the tuffaceous material. The reaction almost always progresses more slowly in the bentonites than in the adjacent shales which serve as a source of potassium.

Therefore, by both experiments and natural examples, it is evident that the reaction rate can be influenced by the chemical composition of the system in which it occurs.

9.3.2 Mineral chemical stability: weathering

The rate of change of a mineral reaction can be influenced by the chemistry of a system as it affects a reaction equilibrium as demonstrated in Section 9.3.1. The effect of chemical activity is even more striking when one looks at weathering profiles. In the lower levels of the profile, the clay phases are relatively rich in alkalis and alkaline earth elements. In the upper levels in mature profiles the clays are composed of the alkali-free phase kaolinite (Al–Si) and in the most mature stages only hydrated oxides of Fe and Al are present. The sequence can be attributed to a decreasing activity of Na, K, Mg, and Ca in the solutions as one moves up in the profile whereas the activity of Al, Fe (and to a lesser extent Si) remains constant throughout. The clays present in the lower parts of the profile are destabilized as a function of the chemistry of the solutions. The chemistry of the solutions is a function of the ratio of rock to water which interacts to approach a chemical equilibrium of the whole system. As rainwater penetrates into the soil profile, its residence time increases and the ratio of water to rock decreases whereupon the activity of elements in solution becomes more and more dominated by the minerals in the rock.

Such a situation can be illustrated by a phase diagram using the variables of composition and chemical activity. This diagram can be compared to a typical soil profile such as shown schematically in Figure 9.9. The description which follows is taken from Meunier (1980). At the top of the soil profile one finds the most advanced stages of weathering, and the clay minerals which are often the products of change in other clays. The example shown is a granite which contains feldspar, quartz, and, to simplify, muscovite. In zone 1, the rock is little or unaltered. Alteration takes place along cracks and fissures. Intensification of this process produces a rock with heterogeneous mineralogy, the clays being developed along channel ways where water flow is greatest. The first clays produced are illites, occurring at potassium feldspar – muscotive grain contacts. As one moves upward in the profile, certain zones show significant alteration; smectite-like vermiculites are found as is kaolinite (zone 2). Further up in the profile kaolinite is much more common and eventually in the upper reaches, the soil zone, mainly kaolinite and perhaps hydrated aluminum oxide are present.

The soil clay assemblages are shown in the potassium activity-composition (Al and Si) diagram. It is apparent that the initial illite is eventually lost to a certain extent in the upper reaches as is a portion of the smectitic material. The new and predominant clay is kaolinite. It is evident that the potassic (and calcic in multicomponent systems) clays become unstable in the upper regions of the profile where potassium activity in solution is lowered due to the ratio of water to dissolved solids which is the result of the proximity to the entering rainwater. Here the stability of the clays is a function of the chemistry of the system.

The development of a soil profile, the transformation of unstable rock minerals into a vertical sequence of clay-dominated mineral assemblages, is known to be a function of time and climate. Climate is a function of rainfall and temperature and since rainfall determines the bulk chemistry of the system one finds the same variables of time, temperature, and composition responsible for the stability of clays at the earth's surface.

The time of development of soil profiles is of the order of 10^5 years or less (see Birkeland, 1984, Chapter 8). The rate of transformation of a rock into a

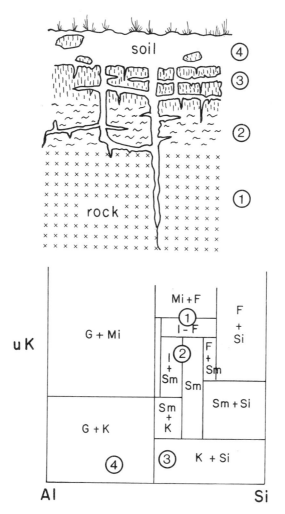

Figure 9.9 Diagram showing the relation of soil-mineral reactions and position in a soil profile formed on a two mica granite (after Meunier, 1980 and Velde, 1985). Circled numbers indicate same reaction stages in the alteration profile and on a chemical potential (uK^+) – composition (Al to Si) diagram. Clay mineral stabilities are indicated by horizontal lines on the activity diagram. G = gibbsite, Mi = mica, Sm = smectite, K = kaolinite, F = potassium feldspar, Si = silica in an amorphous or cryptocrystalline form.

weathering soil profile of clay assemblages is of course not only a function of the exterior parameters affecting the material but also the nature of the specimen. For example, a glacial deposit rich in phyllosilicates similar to clays, muscovite and chlorite, will give fewer new clay minerals over the same period of time than a granite. However, there is a significant amount of soil development in the time range of hundreds of thousands of years. Figure 9.10

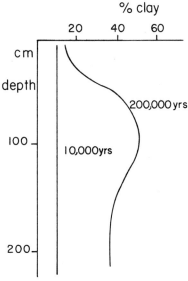

Figure 9.10 Representation of measured clay content for two different, dated soil profiles on glacial till in North America. This is used as an indication of clay forming reactions in soils as a function of time. The 10,000 year profile shows little clay content while 200,000 year profile shows a typical clay content maximum in the 100 cm depth region. This gives an order of magnitude for reaction times in the clay forming process at surface conditions. Weathering can form clay concentrations in greater or shorter periods of time, depending upon the weathering intensity (rainfall and temperature) and chemical instability of the starting material.

shows the development of clays in a glacial till in a temperate North American climate as a function of time. The typical curve of clay distribution on the right-hand side of the figure indicates that there is a significant interaction of the new chemical parameters and the till materials over a period of 100,000 years. This can be used as an order of magnitude for the stability of clay minerals at surface conditions in weathering. More severe conditions of temperature and rainfall, will hasten the reactions whereas low-temperature conditions, alpine or arctic conditions will slow reactions. More unstable materials, such as tuffs or basalts will alter more rapidly than phyllosilicate-containing tills.

In the examples given above, the importance of chemistry is demonstrated. Changing the concentration of a critical element can destabilize a clay and following the reaction kinetics, this clay will be changed into another mineral form.

9.4 Phases and internal chemical equilibrium

The initial example of thermal clay-mineral stability (Section 9.2) showed the importance of time and temperature in determining the range over which a phase transition reaction can take place. Section 9.3 showed the importance of the chemistry of the system as it influences the reaction rate and hence the range of conditions over which a clay mineral will persist. We will now attempt to investigate what happens to a phase once it has formed.

It is generally considered that the phase which forms in a metamorphic or clay-mineral-bearing system will have a chemical composition which is dictated by the system in which it formed. In hydrous systems this composition is controlled by the chemical activity of its different component elements in the solutions which affect the system. These activities will be determind either by the phases in the immediate system or by resources outside of the system considered. There are systems which maintain the same phase under different pressure–temperature conditions but the phase has a different composition according to the $P-T$ conditions applied to it. If $P-T$ conditions change, the old phases (minerals) present will then have to change their composition in order to be in chemical equilibrium with new physical conditions. This is demonstrated in Figure 9.11. Assuming a system of three chemical components, a, b, and c, we will consider the relations of a two-phase assemblage X and Y. At conditions T_1, P_1 the composition of the phases for a bulk composition in the centre of the system (the dot in the figure) are X_1 and Y_1. In changing the physical conditions to P_2, T_2 the compositional field of the phases changes and the coexisting phases X and Y produced by the bulk composition at the dot are X_2 and Y_2 which are very different from those found at P_1, T_1. The system maintains the same two phases in equilibrium for a range of physical conditions but the compositions of the coexisting phases at equilibrium must change. If the minerals (phases) do not react to adjust to the new conditions, their compositions wll be heterogeneous in the sample, ranging from X_1 to X_2 and Y_1 to Y_2.

An example of this situation in natural mineral assemblages can be found in the compositions of chlorites found in diagenetic sequences. Detrital chlorites

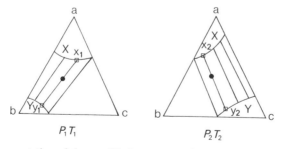

Figure 9.11 Representation of the possible importance of changing physical conditions to the chemical composition of coexisting phases. The specific composition of coexisting phases X and Y changes in going from conditions T_1, P_1 to T_2, P_2 at constant bulk composition, X_1 to X_2 and Y_1 to Y_2. The constant bulk rock composition is indicated by the large dot in the centre of the diagram.

in many sediments can be expected to have varied origins and compositions which depend upon the rocks which furnish the sediment. The compositions of detrital grains could be expected to be varied, as much as the differences in the source rocks. A second origin of chlorites in diagenetic sequences is that of mineral reaction of the sedimentary material as it is buried under increasingly higher temperatures. These processes have been observed to occur rather near the surface, as early as the 660 m depth level in 45 Ma samples taken from drill holes in the Texas Gulf Coast from the Peeler well of Figure 9.4 (Velde and Medhioub, 1988). The present-day temperatures are near 40°C. In these clay mineral assemblages I/S, kaolinite, and chlorite appear to form and coexist. The minerals illite, kaolinite, smectite, and chlorite can also be of detrital origin. X-ray-diffraction methods do not allow one to distinguish the detrital from authigenic minerals.

New chlorite can probably form from several reactions involving kaolinite and iron oxides or from a change in the bulk composition of the I/S mineral series as the reaction progresses towards an illitic end member. Various chlorites have been produced in this way in laboratory experiments (Velde, 1977). It is clear, then, that two kinds of chlorite may be found in sediments, that of detrital origin and that of authigenic origin which responds to new and changing thermal conditions.

The question can then be asked, if detrital chlorite is present when new authigenic chlorite is formed, how will the metastable detrital minerals react to the new chlorite-forming conditions? Will the old, detrital grains dissolve to form a new chlorite composition or will the 'phase' chlorite be respected as new chlorite is formed? Some information is available which helps to answer this question and which gives insight into the stability of this clay-mineral type.

Velde and Medhioub (1988) and Velde, et al. (1991) have determined the range of chlorite compositions in sediments of different ages which are undergoing diagenesis. It was found that the range in the ratio $Fe/Fe+Mg$ changes in sediments as a function of the age of the sedimentary rock more than as a function of the present-day burial temperature. The range in Fe content and average age of the sedimentary series are given in Figure 9.12. In all three series the temperature ranged from 50 to 110°C. Very old rocks (Belt Precambrian series in Western Montana, USA) give a value of 0.025, much greater than 1%, for sediments of 800 Ma and older. The process of homogenization, attainment of chemical equilibrium, is very slow indeed.

If diffusion, homogenization, is slow and it is reputed to be slower than dissolution and crystallization (Chapter 8 in Lasaga and Kirkpatrick, 1981) the process of chlorite formation, crystallization, must go on independently of the attainment of a homogeneous composition of the chlorites already present in the sediment. If this were not true, the chlorite compositions would be much more homogeneous than they are for they would be constantly undergoing dissolution and crystallization. The fact that one sees large ranges in chemical composition of the same phase indicates that this phase remains stable despite

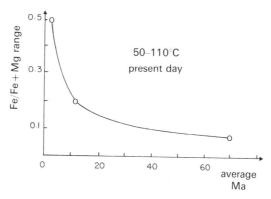

Figure 9.12 Range of Fe/(Fe + Mg) values in chlorites in samples from burial diagenesis series of equivalent depth (maximum burial temperature 50–110°C) as a function of the age of the series. It can be seen that as the average age of the sedimentary series increases, the composition of the chlorites becomes more homogeneous, and the range in Fe/(Fe + Mg) decreases with age of the sediment. Chemical equilibration in the chlorites is much slower than are the rates of crystallization. A phase forms before it comes into chemical equilibrium with the chemical system in which it formed.

the fact that the new crystals forming have a different composition from many others in the rock. This indicates that the crystallization of a phase and its persistence depends on the characteristics of the phase as a whole and not on its homogeneity nor upon its chemical composition compared to the thermodynamic values of elements in solution which would indicate full chemical equilibrium. A phase forms and remains present even though its composition is not that which is in chemical equilibrium. Thus, there is a sort of phase equilibrium in clay systems and a chemical equilibrium in clay systems. The latter is attained much more sowly than the former.

The stability (presence) of a clay can depend upon phase equilibria and not on internal chemical equilibria.

9.5 Conclusion

In the examples given, there are two different types of mechanism which have been investigated to show the limits of stabilities of clay minerals. In the first instance a mineral type became unstable with respect to another (the smectite-to-illite transition) through a change in thermal conditions. The gradual process of transfer of material from the old smectite to the new illite occurs through the production of an intermediate mineral (mixed-layered smectite/illite) whose nature changes during the reaction. The rate at which this reaction proceeds depends upon the temperature of the environment. If the temperature remains below 100°C, the reaction can take place over hundreds of millions of years.

The second variable is chemistry which is operative, of course, at all temperature conditions. Changing potassium content slows the formation of illite in the smectite-to-illite reaction and hastens the formation of kaolinite in soil clay-mineral formation, for example. Changing chemical conditions in weathering profiles can destabilize a clay-mineral species in periods of less than hundreds of thousands of years, greater than 10^3 faster than with the temperature instability.

Therefore we can say that the reactions of clay minerals in nature obey the forces of normal chemical parameters—temperature, time, and chemical composition of the sytem. The only difference between clays and other minerals is the temperature under which they form. The low temperatures of clay stability tends to slow reaction progress and give an appearance of metastability when in fact the phases are obeying thermodynamics as they should.

The last point is that once a phase is formed, or crystallized, it is not necessarily at chemical equilibrium with its neighbours. Phase equilibrium seems to precede chemical equilibrium.

References

Birkeland, B. W. (1984) *Soil and Geomorphology*, Blackwells, 540 pp.

Champion, D. and Lanson, B. (1991) I/S to illite transformation in diagenisis, *American Journal of Science*, **291**, 473–506.

Hoffman, J. and Hower, J. (1979) Clay mineral assemblages as low grade metamorphic geothermometers: application to the thrust fault disturbed belt of Montana, in Aspects of diagenesis, *Society of Economic Paleontology and Mineralogy Special Publication*, **26**, 55–81.

Howard, J. J. and Roy, D. M. (1981) Development of layer charge and kinetics of experimental smectite alteration. *Clay Mineralogy*, **33**, 81–8.

Inoue, A., Velde, B., Meunier, A., *et al.* (1988) Mechanism of illite formation during smectite to illite conversion in a hydrothermal system, *American Mineralogist*, **73**, 1325–34.

Jeans, C. V. (1989) Clay diagenesis in sandstones and shales, an introduction, *Clay Mineralogy*, **24**, 127–36.

Lasaga, A. C. and Kirkpatrick, R. J. (eds) (1981) *Kinetics of Geochemical Processes*, American Mineralogical Society, Reviews in Mineralogy, 8.

Meunier, A. (1980) Les méchanismes de l'altération des granites et le role des microsystèmes. *Memoir Societé Géologique France*, T LIX, 80 pp.

Reynolds, R. C. (1980) Interstratified clay minerals, in *Crystal Structures of Clay Minerals and their X-ray Identification*, (eds G. W. Brindley and G. Brown,), Mineralogical Society, London, pp. 249–303.

Roberson, H. E. and Lahann, R. W. (1981) Smectite to illite conversion rates: effects of solution chemistry. *Clays Clay Mineralogy*, **29**, 129–35.

Steinberg, M., Holtzapffel, T., and Rautureau, M. (1988) Overgrowth structures and local equilibrium around clay particles. *Clays Clay Mineralogy*, **35**, 189–95.

Velde, B. (1977) A proposed phase diagram for illite, expanding chlorite, corrensite and illite-montmorillonite mixed layer clays. *Clays Clay Mineralogy*, **25**, 319–23.

Velde, B. (1985) *Clay Minerals: A Physicochemical Explanation of their Occurrence*, Elsevier, Amsterdam, 421 pp.

Velde, B. and Espitalié, J. (1989) Comparison of kerogen maturation and illite/smectite composition in diagenesis. *Journal of Petrology and Geology*, **12**, 103–10.

Velde, B. and Medhioub, M. (1988) Approach to chemical equilibrium in diagenetic chlorites. *Contributions to Mineralogy and Petrology*, **98** 122–7.

Velde, B. and Nicot, E. (1985) Diagenetic clay mineral compositions as a function of pressure, temperature and chemical activity. *Journal of Sedimentary Petrology*, **55**, 541–7.

Velde, B., El Moutaouakkil, N., and Iijima, A. (1991) Compositional homogeneity in low temperature chlorites. *Contributions to Mineralogy and Petrology*, **107**, 21–26.

Weaver, C. W. (1989) *Clays, Muds and Shales*, Elsevier, Amsterdam, 819 pp.

Whitney, G. and Northrop, H. R. (1988) Experimental investigation of the smectite to illite reaction: dual reaction mechanisms and oxygen isotope systematics, *American Mineralogist*, **73**, 77–90.

Index

Acoustic modes
  Debye model 138
  Kieffer model 143–4, 148
  Salje–Werneke model 158
Acoustic velocities 146
Acoustic waves, propagation through isotropic solid 139–40
Adjacency matrix 27, 27, 30
Adularia, modulated structures in 218
Akermanite 319
Albite
  breakdown to jadeite and quartz 158
  phase transitions in 186–7, 187, 197, 198, 202, 203
Alkali feldspars
  displacive phase transitions in 179, 192, 192, 193
  heat capacity of 198
  modulated structures in 217–18
  see also Potassium feldspar
Alloy systems and cation packing 6–7
Al_2SiO_5 polymorphs, phase equilibria of 158
Amarantite 54–5
Anapaite 50, 50, 52, 53
Anatase
  density-of-states diagram for 104
  incorporation of lithium into lattice 103–4
Anchor data 298
Andalusite, vibrational density of states of 159, 159
Andalusite–sillimanite phase equilibrium boundary 159
Angular overlap model 109, 110
  d-orbital energies calculated using 126
  energy variation with bond angle investigated by 121–2
  molecular-orbital diagram using 122
  octahedron-distortion energies calculated using 130
Anhydrite ($CaSO_4$), instability of 44–5
Anion lattices, radii of interstices 9

Anions 36, 41
  bond strength around 37–8
  bond-valence
    requirements 47
    variations around 42
  formal valence of 8
  Lewis base strength of 42
ANNNI model 17, 18
Anorthite 8
  Al/Si order/disorder transition in 187, 188, 210, 210
  displacive transitions in 204–5
Anti-bonding orbitals 90, 91
Apparent Gibbs free energy 275
Aqueous species 296–7, 307
  data set comparisons 320
Aragonite, phase equilibria of 159, 172, 173
Aragonite–calcite phase equilibrium 172, 173, 199
Argentian tetrahedrite–tennantites, thermodynamic properties of 251–9
Artinite 62–3
  bond valence structure **63**
As_2O_5, ferroelastic transition in 197
Atomic radii 34
Atomic size 9–10
Atomic vibrations 16
Atomistic simulations
  interatomic potential parameters used in, $MnTiO_3$ simulation **162**
  lattice dynamics using 161–8

Bafertisite 53, 58
Band gap 13
Band model see energy-band model
Band-structure calculations 36
Basic radii, Born–Mayer theory 10
Bentonite beds 343
Benzene, orbitals of 92
Bermanite 53, 59
Beryl, containing occluded (H_2O) groups

INDEX

65–6
Beryllia, molecular-orbital energies affected by interatomic Be–Be separation 120–1, *121*
α-Beryllia
  wurtzite-type structure 115–16, *116*
  coordination geometry for *118*
β-Beryllia
  coordination geometry for *118*
  structure of 118–20, *120*
Bilinear coupling 201–3
'Billiard-ball' model 88, 103
Binary compounds 12–15
  tetrahedral structures 12
Binary structural representation 67–8
Biopyriboles 17
Biotite 316, 343
Biquadratic coupling 203–5
$BiVO_4$, tetragonal–monoclinic transition in 182
Block orbitals 32
Bloedite *50*, 50, *52*, 53
Bonattite, structure of 57, *59*
Bond connectivity, stability of 59
Bond directionality, modelling of 161
Bond length
  cation–anion division 10
  mean bond length 35
  prediction of 36
  variation with electron configuration 114–15
Bond network families 4
Bond networks 4, 25
  in goedkenite 67, *68*
  topological features of 27
Bond overlap populations
  beryllium–oxygen (Be–O) linkages 120
  bond strength estimated using 106
  definition of 106
  lead–oxygen (Pb–O) linkages *107*, 107–8
Bond strength 38, 40, 41
Bond topology
  and electric energy density of states 33
  and energy of the structure 47
  energetic content of 75
  ionicity and covalency 36–7
  oxysalt minerals 60

Bond-breaking sequences 11
Bond-valence 38
  bond-length relationship 45
  bond-strength relationship 38
  curves, universal *40*, 40, **41**
  distribution, hydrogen-bearing groups *60*, 61
  mean 41–2
  as a molecular orbital model 45
  relationships 37–40
    conceptual basis of 41–4
  theory 8–9, 37–45, 47, 69, 75
  transformer 64
  (H_2O) as 66–7, 69
Bonding 16
  and structure diagrams 12–15
  orbitals *90*, 91
Bonds
  directed, formation of 12
  s–p, metallization of 12
Borazine molecule, orbitals of 94, *95*
Born model of solids 161
Boron hydride (BH_3)
  localization of delocalized orbitals of *93*
  molecular orbitals generated for 91–2, *92*
  molecular-orbital theory applied to 91–3
Botryogen 54–5
  structural unit of 69, **70**
Bowen's discontinuous reaction series 45–6, **46**
Brackebuschite, calculation of structural unit basicity **71**, 71
Brackebuschite group 54–5
Bragg–Williams model 193–4
  compared with Landau approach 194–7, *195*
  order parameter as function of temperature *180*, 194
Brandtite, calculation of structural unit basicity **71**, 71
Brassite 54–5
Brillouin zone
  first 96
  mean value point of 130
  phonon frequencies determined for points within 163

353

INDEX

replacement by sphere 140, 143
Brillouin-zone centre 96
  Landau theory applied to ordering at 222–4
  phase transitions associated with 185, 233
    free energy 201
    strain 186
Buckingham potential 161
Butlerite *54–5*, 58

C_3 operation 89, 95
Ca_2SiO_4 44
Ca intersitial atoms 64–5
cadmium halide sheet structure 130
caesium chloride (CsCl) structure *107*
$CaGeO_3$ polymorphs, phase equilibria of 158
Calcite
  order/disorder transition of CO_3 groups in *182*, 182–3, 196, 197
  phase equilibria of 159, 172, *173*
Calcite–aragonite phase equilibrium 172, *173*, 199
Calorimeters, adiabatic 278
'Calorimetric' Debye model *see* Debye model
Calorimetry 278–9
$CaSO_4 \cdot 2H_2$ (gypsum) 45
$CaSO_4$ (anhydrite) 44–5
CATCH Tables 270
$CaTiO_3$ 168
Cation–anion bond, electrostatic bond strength of 8
Cation–cation repulsions 115, 120
Cation-centred tetrahedra *115*, 117
Cations 36
  bond-valence requirements 47
  Lewis acid strength of **42**, 42, *43*, 68, 69
  mean bond-valence of 42
  octahedral
    divalent 49
    trivalent 49
  packing 6–7
  substitution 18, 20
  valence of 41
  see also Interstitial cations
Chain condensation 53

Chain structures 4
  differing linkages 53, 57
  Se 11
  vertex-sharing 58
Chalcanthite group minerals *54–5*
Charge transfer 12
Chemical bonding 11, 27, 29–30
  in inorganic structures 37
  valence-matching principle 44
Chemical changes, irreversible 275
Chemical potentials, tetrahedrite–tennantite components 249, 254
Childrenite group *54–5*, 58
Chlorite
  authigenic 348
  compositions in diagenetic sequences 347–8
  detrital 347, 348
  formation process 348–9
Chlorotionite *54–5*
Chromium dioxide (CrO_2), distortion modes in 128
Chrysotile 315
Clay reactions, kinetics of 333–42
Clay stability
  change in reaction rate 342–3
  and phase equilibria 348–9
  time-temperature control of 330–42
Clay-mineral assemblages
  and burial depth 329
  as function of age and burial history 333–42
Clays
  destabilized 344
  dvelopment of in glacial till 345–6
  origins of 330
Climate 344
Clinochlore 319
Cluster condensation 53
Complexes 4
Composition, coupling with order parameter 188–93
Compositional variability 297
Connectivity 51
  local 34
Consistency, formal, partial or full 277
Consistency requirements, thermodynamic data bases

276–7, 289–91
Coordination geometry 117–18
Coordination numbers 34, 35
Coordination polyhedra
  clusters of 47
  polymerization of with higher bond-valences 46–7
Cordierite
  hexagonal–orthorhombic transition in 184
  modulated structures in 240
  phase equilibria of 172, *173*
Correlation lengths 177, 178
Couer d'Alene deposits 245, *261*
Coulomb integral 28
Coulombic forces 161
Coulombic repulsions, destabilizing effects of 115
Covalency 36–7
Covalent bond fraction 13
Covalent bonding 161
Covariance 290–1
Covariance matrix 290, 291
Crystal chemistry, topological aspects of 34–7
Crystal field stabilization energy (CFSE)
  compared with MOSE 113–14, *114*
  molecular-orbital equivalent of 113
  slope applied to plot *114*, 114
Crystal field theory 14–15, 109–15
  compared with molecular-orbital theory 112–14
Crystal
  lattices, eutactic 6
  radii 10
  stability and thermodynamics 15–21
  structure prediction strategy, difficulties of 1–2
Crystal structures
  geometrical constraints on 3–10
  Goldschmidt's analysis 2
  stability, general principles behind 2–3
  theories of 2–3
  see also Modulated crystal structures
Crystallization/alteration sequences 26
Crystals 30–2
Cyclic molecules, molecular orbitals generated for *90*, 90

Cyclobutadiene 29–30

D–s (orbital) mixing, effect of 127
Darken equation, extended form of 249, 254
Data analysis 282
  goals 282–3
  methods of 283–8
Data
  quality 282
  sufficiency 282
Debye frequency 140
Debye (lattice vibration) model 133, 139–43
  'calorimetric' model 149, *151*, *154*, 155–6
    thermodynamic functions calculated *152*, 152–3, *153*, *156*, *157*
  deviations from 142–3
  'elastic' model 149, *151*, *152*, 152, 153, *154*, 155–6
    thermodynamic functions calculated *152*, 152–3, *153*, *156*, *157*
  low quartz 149, *151*, 152–3
    thermodynamic functions calculated *152*, 152–3, *153*
  rutile *154*, 155–6
    thermodynamic functions calculated *156*, *157*, 157
Debye oscillators 160
Debye solid, thermal energy of 141
Debye temperature
  calorimetric 142, 149
  elastic 141, 142, 149
Debye–Hückel law 296, 307
Delocalized energy bands *102*, 103
  localization of 103
Delocalized interaction 88
Delocalized orbitals 91–3
  localization of 92, *93*
Density of states
  electric energy 32
  electronic energy 36–7
  vibrational 136–9
    approximations for 139–46
    see also Vibrational density of states
Density-of-states diagram 29
  titania *104*

INDEX

Descloizite, structure of 59, 59
Diagonalization 32, 33
Diatomic molecular structure 11
Differential scanning calorimetry 278–9
Diffraction theory, modulated structures described by 216–17, 234
Dimensionality, of structural unit 63
Diopside 8, 317
  bond-strength 37, **38**
  bond-valence 38, **39**
Discrepancies, in data sets 315–19, 320, 321
Disorder states, quenched 279
Dispersion relation 136, *137*
  in Kieffer approximation *144*
Displacive transformation 17
Dissolution and precipitation 339
Double-sheet GaSe structure 11
Drop calorimetry 279
  enthalpy measured by 198
Dulong–Petit limit 142

Edge-sharing tetrahedra, instability of 8, 34, 115–23
Edges 27
Einstein oscillators 145, 148, 160
'Elastic' Debye model *see* Debye model
Electric energy density of states 32
Electron configuration, bond-length affected by 114–15
Electron density considerations 92, 117
Electronegativity 12
  and mean bond-valence 42
Electronegativity difference, effect of 123
Electroneutrality 34, 49, 70
  law of 41
Electronic energy density of states 36–7
Electrostatic valence rule 35–6, 116
  extended rule 8
Electrum–chalcopyrite–pyrite–sphalerite, equilibration with tetrahedrite–tennantites 258
Element zoning, crystallochemical control of, tetrahedrite–tennantites 259–63
Elements, electronegativity scale of (Pauling) 15–16
Energetic terms, important 34

Energy of activation 341
Energy band of orbitals
  band-folding process 99, 101
  compared with molecular-orbital theory 98
  definition of 96
  effect of doubling of unit cell 99, 99
  effect of electronegativity difference 99, *100*
  heteroatomic chain 98
  homoatomic chain 97
  metallic oxides 103–5
  silicates 100–3
Energy bands of solids 94–9
Energy-band model, compared with localized model 103, 128
Energy-difference curves
  octahedral structure *127*
  rutile structure *129*
Enthalpy
  Bragg–Williams model compared with Landau approach 196
  second-order transition 180
Enthalpy data 274, 306
  problems with 280, 281
Entropy
  Bragg–Williams model compared with Landau approach *195*, 196
  comparison of calculations by various vibrational models
  low quartz 149–53, *153*
  rutile 156–7, *157*
  harmonic expression for **135**
  low quartz 149–53, *153*
  $MnTiO_3$ 165–6, **167**
  molar vibrational, tetrahedrite–tennantites 247, 254
  rutile 156–7, *157*
  second-order transition 180
  uncertainties in 279
Entropy of mixing 18
Epithermal veins, tetrahedrites crystallized in 245, 256
EQ3/6 data base 304–5, 306
  cf. MINTEQ data base 315–16, 320
  cf. SUPCRT data base 319, 320
Equations of state 272
  for carbon dioxide 296

356

for fluids 292–3
for fluids and aqueous species 291–7
for water 293, *294–5*
Error
  analysis 289
  distribution, specific 278–81
  propagation 283
  propagation analysis 298
  and internal consistency 289–91
Error propagation theory 290
Errors of measurement 277
Eutaxy 5
EXAFS (extended X-ray absorption fine structure) spectroscopy, tetrahedrites studied using 257
Excess functions, thermodynamic 20
Exchange-potential calculations, tetrahedrite–tennantites 260–1, *261*
Extended electrostatic valence rule 8

Face-sharing, stability affected by 8, 34, 115
Fahlores *see* Tetrahedrite–tennantites
Fairfieldite group *54–5*
Fe-oxides 316
Feldspars 298, 317, 344
  see also Alkali, Plagioclase, Potassium feldspars
Ferrichromite spinels 257
Ferrichromites, Fe–Mg exchange with olivines 257
Ferricyanide complexes, bond lengths compared with ferrocyanide complexes 114
Ferrierite 4, 5
$FeTiO_3$, high-pressure transformations of 167
Fibbroferrite *54–5*
Finite cluster structures 49–51, **78**
First-order transition 180
Fleischerite group 49
Fluids 291–2
Formal consistency 277, 305
Fornacite group *54–5*
Forsterite
  bond-strength 37, **38**
  bond-valence 38, **39**

vibrational density-of-states approximations for 160
Framework silicates 146
  phase transitions in 200
  strain fields in 177–8
Framework structures, 4, 57–8, **82–4**
Free energy 18
  crystals with two phase transitions 201–2, 203–4
  determination of 132, 135
  Landau expansion applied 172, 179, 220
    composition considered 189, 191
    gradient interactions 207, 231, 232
    kinetics considered *207*
    with order-parameter coupling 201, 202, 203, 204, 205, 206
    pressure effects 181, 183
    strain energy included 185, 186
    symmetry constraints 184
  $MnTiO_3$ 166–7, *168*
  see also Vibrational Gibbs energy
Free energy
  of formation 281
  of transition 16
Full consistency 277, 282
Fundamental building blocks (FBB) 30, 47, 53, 58, 59
  in infinite chain structures 51

Garnet, structure of 6–7
GASES 320
GE0CALC **268**, **271**, 304
Geochemical programs 307
Gibbs free energy 15, *19*
  determination of 132, 135
  see also Vibrational Gibbs energy
Ginzburg interval 177, *178*
Ginzburg–Landau equation 207–8
  solution with fluctuating order parameter 210
  solution using Landau expansion 208
Goedkenite, bond network in 67, *68*
Gradient coupling 206–7, 209
Gradient invariants 227, 228, 229–33
Gradient ploy 233
Granite 344, *345*
Graph theory 26–7
Graphs, and subgraphs 30, *31*

INDEX

Group representation tables 221
    β-quartz **234**
    potassium feldspar **239**
    space group $P6_2 22$ **234**
    space group Pmm2 **223, 225**
    space group 2/m **239**
Group theory 219
    phase transformations studied using 221, 222–6
Growth kinetics 339
    Guildite *54–5*

H_2O
    as bond-valence transformer 66–7
    as ligand 63–4
    structural role of 25,
H_2O groups
    bonded to interstitial cations 63–5
    hydrogen-bonded 65
    occluded 65–6
    transformer effect of 67
    and unconnected polyhedra 48–9
Haar equation 293
Haematite 317
Half-reversals 269–70
    in MAP and REG 286–7
Hamiltonian matrix, diagonalization of 32
Harmonic approximation 133–4
Harvie–Møller–Weare models 306
Heat capacity 16, 281, 319
    Debye solid 142
    determination using lattice dynamics 132
    harmonic expression for **135**
    low quartz 149–53, *152*
    $MnTiO_3$ 165–6, **167**
    molar, Kieffer model used 145–6
    rutile *156*, 156–7
    second-order transition 180
Heats of formation 16
Heitler–London wave function 105
Henry's law 278
Heteroatomic chain
    atom locations of 96–8, *98*
    s and p orbitals of *101*
Heteropolar compounds, ionic contribution to banding 13
HF-solution calorimetry 280

High quartz, structure of *234*
High temperature metamorphic reactions 333
Hohmannite *54–5*
Homoatomic chain
    atom locations of 96, *97*
    s and p orbitals of 100–1, *101*
Homogenization process 348
Homogenous solution 18–19
Homopolar crystals 13
HP-DS data base 304, 306, 321
    cf. UBC data base 308–12, 319
    weaknesses 321
Hückel
    approximation 29
    model 90
    molecular orbitals 30
    theory 30
Humite group 17
Hydration 45
Hydrogen acceptor anion *60*, 60
Hydrogen bond 60
Hydrogen bonding 48, 53, 63
Hydrogen coordination 59, *60*
Hydrogen donor anion *60*, 60, 63
Hydrogen-bearing groups *60*, 60–1
Hydrogen-bonded networks 63, 69
H_2 molecule, bonding orbitals of 106
H_3^+ molecule, bonding orbitals of 90–1, *91*, 92, 106
Hydromagnesite 315
Hydronium jarosite 60
Hydrothermal fluid, exchange of metals with tetrahedrite–tennantites 260–2
Hydrous systems, phase composition 347
Hydroxyacid molecules, molecular orbital calculation on 9

I/S clays, under burial diagenesis 334–7
I/S grains *340*
    stability of 339
I/S minerals 333, 335, 338, 339, 348
I/S-illite transformation 341
Identity, uncertainties of 277–8
Illite 338–9, 344, 348
Ilmenite ($FeTiO_3$), high-pressure transformations of 167

358

Image analysis and grain abundance 338–9
Incommensurate (IC) phase transformations 17
　theory of 226–8
Incommensurate (IC) structures 216
　Landau expansion for 206, 232
　quartz 233–8
　see also Modulated crystal structures
Inelastic neutron scattering, phonon dispersion relations determined by 138, 146, *147*, 148, 154, 160
Inequality constraint 275
Infinite chain structures 51, *54–5*, **78–9**
Infinite sheet structures 53, *56*, 57, *58*, **80–1**
Infinite systems and method of moments 33–4
Infrared spectroscopy, vibrational density-of-states determined using 160
Interatomic distances 9–10, 41, 45
　bond overlap populations affecting 106–8
　in a coordination polyhedron 35
Interatomic potentials, lattice dynamics described using 160, 161, **162**, 168
Internal consistency 321
　and error propagation analysis 289–91
Interstitial cations 49, 50, 68
　bond-valence controls on 70–2
　change in character of 73–4
　(H_2O) groups bonded to 63–5
　interaction with structural unit 67
　low-valence 47, 48
Ionic crystals 7
Ionic model 88, 103
　compared with orbital model 120, 123
Ionic pair potential model, $MnTiO_3$ high-pressure polymorphs simulated using **162**, 162
Ionic radii 9, 35
　estimations 9–10
　perturbation of 10
Ionicity 12, 13, 36–7
　critical 13
Irreducible representations 221, 227, 240
Isomers

geometric 53
graphical 53, 57

Jadeite–augite system, phase transitions in 193, *194*, 199–200
Jahn–Teller effect 123–30
　distortion of octahedral complexes predicted by 125
　second-order effect 107, 127
Jahnsite group *54–5*
JANAF Tables 270
Jurbanite, finite cluster structure 49, *50*

K-feldspar *see* Potassium feldspar
Kaolinite 317, 344, 348, 350
$K_2Cd_2(SO_4)$ cubic–tetragonal transition in 187
Khademite 49
Kieffer (lattice vibration) model 133, 143–6
　andalusite *159*, 159
　application to phase equilibria 158, *159*
　low quartz 148–9, *149*–52, *151*
　thermodynamic functions calculated 149–52, *152*, *153*
　rutile *154*, 155
　thermodynamic functions calculated *156*, 156, *157*, 157
Kinetics 20–1
　phase transitions 207–11
　　order parameter fluctuations/inhomogeneities affecting 209–11
　　remaining homogeneous 208
Known minerals 25
Krausite *54–5*
Krohnkite chain 53
Krohnkite group *54–5*

$LaAlO_3$, displacive transition in 177, *178*
Landau free energy expansion 179–84
　compared with Bragg–Williams model 194–7, *195*
　effect of composition 188–93
　effect of pressure 181–4
　effect of strain 185–8
　gradient terms in 206, 227–8, 231, 232

INDEX

incommensurate structures 206
linear-coupling terms in 185
quadratic-coupling terms in 185
symmetry constraints in 184
Landau regime 177, *178*
Landau theory (of phase transformations) 17, 20, 219–26 298
  application to β-/α-symmetry point transformation 234
  inadequacy at low temperatures 178
  principle element of 172
  special point transformations 219–22
  symmetry point structures 222–6
Lattice dynamics
  application of 132–3
  atomistic simulations used 161–8
  determination of distribution function 137–8
Lattice translation symmetry 228
Lattice vibration frequency distribution
  approximations of 133, 139–46
  Debye approximation 133, 139–43
  definition of 134
  Kieffer approximation 133, 143–6
  low quartz 138, *150–1*
  one-dimensional linear chain 136–7
  rutile *154*
  Salje–Viswanathan approximation 158–9, *159*, 160
  thermodynamic functions calculated from 134–5
    three-dimensional crystal, approximations for 139–46
Lattice vibrations 16
  and polymorphic phase transitions 16–17
Laueite group 53, *56*, 58
Layer structures 4
Lead oxide (PbO), structure of *107*, 107
Leonite *50*, 50, *52*, 53
Leucite, cubic–tetragonal transition in 205
Ligands, tied-off 61–2
Likelihood function 290, 291
Linarite group 53, *54–5*, 59
Linear Combination of Atomic Orbitals (LCAO) 28
Linear programming 283
Linear regression 284

Liroconite *54–5*
Localized bonds 92, 103
Log K data
  comparison between 299, 302–20
  MINTEQ-EQ3/6 **300–2**
  SUPCRT-MINTEQ **313–14**
  UBC-MINTEQ **315**
Low quartz
  comparison of vibrational models 146–53
  entropies of *153*
  heat capacities of *152*
  structure of *235*
  vibrational density of states for *147*
Low-temperature data sets 312, **313–14**, *315*, *316*, *317*, *318*

Maddox, J., quoted 1
Madelung energies 117, 120
Magnesium sulphate minerals, paragenesis of 72–5
Magnetite 317
Maier–Kelly function 319
Mandarinoite, structure of 65, *66*
MAP *see* Mathematical programming
Margarite breakdown reaction 312
Margules parameters 20
Mathematical programming 283, 284–8, 305
Matrix formulation 288–9
Melilite 8
Merlinoite, layer structure 4, *5*
Merwinite 319
  group 53, *58*
Metallic materials 11
Metallic oxides, localized/delocalized bonds in 103–5
Metamorphism and clay mineralogy 329–50
Metavauxite group 53, *56*, 58
Metavoltine *50*, 50
Meteoric water and the weathering profile 344
Methane, tetrahedral structure of 118, 120
Method of moments 32–4
$MgGeO_3$ polymorphs, phase equilibria of 158
Mg_2GeO_4 polymorphs, phase equilibria

INDEX

of 158
Mg$_2$SiO$_4$ polymorphs, phase equilibria of 158
Microcline 317
Microstructural identity 278
Mineral data 273
Mineral paragenesis, structural control on 45–6, 72–5
Mineral solution phases 322
Mineral stability 3, 4
  application of vibrational models to 158–60
  electronic factors 10–15
Mineral structure, molecular-orbital viewpoint 36
Minerals
  interstitial (H$_2$O) in 69
  other 25, 26
  rock-forming 25, 26
MINTEQ data base 304–5, 306–7
  cf. EQ3/6 data base 315–16, 320
  cf. SUPCRT data base 316–17, 320
  cf. UBC data base 317–18
Minyulite 53, *56*
Miscibility gaps, argentian tetrahedrites *256*, 256–7, 259
MnTiO$_3$
  high-pressure polymorphs 161–2
    entropies 166, **167**
    free energies 166–7, *168*
    heat capacities **167**
    observed and simulated structures compared **163**, 163–5, **164**
    simulation of 162–3
    thermodynamic properties calculated 165–8, **167**, 168
    vibrational density of states *166*
  high-pressure transformation of 161
  LiNbO$_3$ structure 161–2, **163**, 164, *166*, 166–7, **167**, *168*
  perovskite structure 162, **164**, 164–5, *166*, 166–7, **167**, *168*
Modulated crystal structures 216
  examples of 233–40
  factors creating 217–18, 231–2
  potassium feldspar 238–40
  symmetry of 228–9

Molar volume data 273
Molecular building blocks 30
Molecular orbital
  highest occupied (HOMO) *108*, 108, 109
  lowest unoccupied (LUMO) *108*, 108
Molecular-orbital calculation 9
Molecular-orbital diagrams
  OBe$_4$ unit *122*
  PbO$_8$ unit *108*
Molecular-orbital energies 29, 31
  stabilization energy (MOSE) 113
    compared with CFSE 113–14, *114*
    slope applied to plot *114*, 114
Molecular-orbital theory 27, 89–94
  compared with crystal field theory 112–14
  compared with electrostatic model 106–7
  one-electron model 89–91
  solid-state equivalent of 96
  topological aspects of 28–34
  two-electron model 91–3
Molecular-orbital wave function 28
Molecule-crystal difference 32
Molecules
  giant, energy levels *31*
  in molecular-orbital theory 28–30
Monticellite 319
Morinite *52*
  interstitial cations and hydrogens bonds *50*, 50
MRK–equation 296
Mulliken–Hund model 105
Mullite, incommensurate structures in 218
Murnaghan equation 16
Muscovite 343, 344

Na$_4$SiO$_4$ 44
NaNO$_2$, incommensurate structure of 240
Natural assemblages 322
Nepheline, low-temperature structural state of 218
Neutron diffraction 59
Newberyite *56*, 57
  structure of 61–2
Non-bonded radii 10

361

INDEX

Non-stoichiometric solid solutions, onset of modulation in 218
Nyholm–Gillespie model 121

Octahedral complexes 111–12
  CFSE compared with MOSE *114*
Octahedral structure
  d-orbital energies affected by distortion 126–7
  distortion calculated using angular overlap model *130*
  distortion predicted by Jahn–Teller effect 125
Octet compounds 13, 14
Olivine-spinel transition, high pressure 7
Olivines, Fe–Mg exchange with ferrichromites 257
Olmsteadite 53, *58*
Omphacite, cation-ordering transition in 199, 200, 208, *209*
One-angle radii *see* Non-bonded radii
Optic continuum 145, 148
Optic modes
  Debye model 138
  Kieffer model 144–5
  Salje–Werneke model 158–9
Optimization methods 267–8
Optimization theory, use of MAP and REG 284–7
Orbital energy, total 32
Orbital model 88
  compared with ionic model 120, 123
  see also Molecular orbital
Order parameter 174–9, 220
  Bragg–William-model solution 194
  characteristic regimes of 177
  components of 184
  coupling with composition 188–93
  coupling with spontaneous strain 185–8
  definition of 175
  kinetics when fluctuating 209–11
  kinetics when homogeneous 208
  phase transitions involving one 197–200
  saturation regime of 177, *178*
  series expansions in 179, 181
Order–disorder/displacive transition 17
Order–phonon microstructures 240

Order–parameter coupling 200–7
  bilinear coupling 201–3
  biquadratic coupling 203–5
  gradient coupling 206–7, 209
  linear-quadratic coupling 205–6
Order–disorder transitions 17, 298
  Bragg–Williams and Landau treatments compared 193–7
  kinetics of 207–11
  Landau treatment 179–84
  order parameter used 174–9
  coupling with strain and composition 185–93
  order-parameter coupling 200–7
Ordering and mixing 298
Orthoclase, modulated structures in 218
Orthogonal transverse waves in potassium feldspar *239*, 239
Oxide-melt-solution calorimetry 280
Oxyanions 42–4
  bond valence structure of (SO_4) *43*, 43–4
  Lewis basicities of **44**, 44–5
  polymerization of 45–6
Oxysalt structures, (OH) and (H_2O) in 59–66

Parabutlerite *54–5*, 58
Parameter optimization 290
PARAPOCS computer code 162, 163
Pargasite 319
Partial consistency 277, 306
Paths, stongly bonded 34
Pauli
  principle 104
  repulsion forces 13
  repulsions 121
Pauling's rules 2, 8–9, 34, 88
  first rule 8, 9, 34, 35
  second rule 8, 34, 35–6, 37–8
  third rule 8, 34, 36, 115–23
  fourth rule 8, 34, 36
  and bond topology 35–6
  electrostatic valence rule 116
Periodic packing, equivalent spheres 4, **5**
Perturbation theory 109
  interaction energy between orbitals evaluated by 109–10
Peruvian (tetrahedrite–tennantite

dominated) deposits *261*, 262
Phase changes, kinetics of 341–2
Phase equilibria
  in HP-DS cf. UBC 308, *309*–11
  in 'static' space 331–3
  use of vibrational models 158, 159
Phase stabilities, prediction of 132
Phase transformations 16–17
Phase transitions 279, 297, 298
  effect of composition on transition temperature 189–93
  one order parameter involved 197–200
  pressure-induced 183
    $MnTiO_3$ 161–2
  thermodynamics of 172–211
  two order parameters used 200–7
Phase-equilibrium data 269, 275, 283, 284
  low temperature 322
  MAP and REG compared 284–8
  uncertainties in 280–1
Phases and internal chemical equilibrium 346–9
Phillips–Van Vechten diagram 13
Phlogopite 305, 319
Phonon density of states, determination of 138, 146–7, 154–5, 163, 165
Phonon dispersion curves 137–8
Phonon frequencies
  determination by solution of equations of motion 163
  spectroscopic measurement of 158
Phosphate minerals, classification of 46
Phosphate paragenesis 26
Phosphorroesslerite 49
Phosphorus, bond-valence of 41
π–σ mixing 129
π-bonding 92
  d-orbital energies affected by 126
π-donor ligands **126**, 127
Pitzer equation 296–7, 307
Plagioclase feldspars
  Al/Si ordering transition in *192*, 192
  incommensurate structures in 218
Point group-symmetry elements 222
  group representation table for 222, **223**
Polarizability, modelling of 161
POLY-II **268**
Polyacetylene, energy band for *97*

Polyhedra
  coordinated 8
  unconnected 48–9, **77**
Polymerization
  control by (OH) and (H_2O) groups 61
  valence-sum rule 47
Polymetallic base-metal sulphide deposits, tetrahedrite–tennantites in 243, 245, 259
Polysomatic series, structural relations 17–18
Polysomes 17
  structural stability of 18
Polytypism 17
Potassium concentration and smectite-illite reaction 342–3, 350
Potassium feldspar 317–18, 343
  displacement waves in
    geometry of *239*
    symmetry of **239**
  modulated structure in 238–40
  tweed structure in *240*
Primary data 274
  uncertainty in 277–8
Properties, estimation of 281
Pseudolaueite group 53, **56**, 58
Pseudopotential radii 3, 10, 13–14
PTAX **268**
Pyrite 316
Pyrophyllite 315
Pyroxene transformations 16–17
Pyroxenes, vertex-sharing chain in *100*, 100
Pyroxenoid transformations 16–17
Pyrrhotite 319

Quantum-mechanical methods, bonding properties studied by 161
Quantum-mechanical models 88
Quartz 298, 344
  incommensurate structure of 206, 218, 233–8, *238*
    dominant structure 236–7, *238*
    phase relationship of dominant to servient structures 237, *238*
    servient structure *237*, 237
    symmetry groups for **235**, **236**, *236*
  $P3_221$ form 233, 237, *238*

INDEX

P6$_2$ form 233, *237*, 237, *238*
  see also High quartz; Low quartz

Radius ratios 8, 9, 35
Rajpura–Dariba polymetallic deposits 250
Raman spectroscopy, vibrational density-of-states determined using 160
Reaction rates, changes in 342–3
Reactions (of minerals), types of 172
Red lead oxide *107*, 107
Redlich–Kwong equation 292, 296
REG *see* Regression analysis
Regression analysis 283, 284–8
Resonance integral 29
Reversals 269, 287
  weighting for 287
Rhomboclase 53, *58*, 60
Roemerite *50*, 50, 53
Roesslerite 49
Rotation, angle of, SrTiO$_3$ 175, *176*
Rozenite group 49–50, *50*, *52*, 53
Rundle–Pimentel model 121
Rutile
  comparison of vibrational models 154–6
  density-of-states diagram for *104*
  entropies of *157*
  heat capacities of *157*
rutile structure 123, *124*, 146
  band structure calculations *129*
  edge-sharing chains in *124*, 146
  transition-metal oxides and fluorides **124**
  Type 1 distortion 123, **124**, *129*
  Type 2 distortion 123, **124**, *129*

Salje–Viswanathan (lattice vibration) model 158
  andalusite *159*, 159
Sanidine 317
Sansomite *54–5*
Schertelite *50*, 50, *52*, 53
Second-order transition 180
Secular determinant 29, 31
  special points method solution 32
Sedimentary basins, differing burial histories 336–7

Segelerite group *54–5*
Sepiolite 315, 319
Shales 343
Shell models
  ionic polarizbility modelled by 161
  vibrational density of states calculated from *147*, 147, 154, 155
Si–Mg repulsion 7
Si–O–Si bonding potential curves 105
Sideronatrite *54–5*
Silicate framework, polymerization of in silicate compounds 11, **12**
Silicate materials, as polymerizations of coordination polyhedra 34
Silicates
  localized/delocalized bonds in 99–103
  structure of 105
Sillimanite 8
Silver reserves, estimation of 263
Silver-poor tetrahedrite–tennantites, thermodynamic properties of 246–51
Silver-rich tetrahedrite–tennantites, thermodynamic properties of 251–9
Single layer As structure 11
SiO$_2$ polymorphs, phase equilibria of 158
Six-atom ring, orbitals of 93–4, *95*
Smectite 334, 348
Smectite-to-illite conversion 333–4, 341
Smectite-to-illite reaction, initiation and termination of 335–6
Sodium nitrite (NaNO$_2$), incommensurate structure of 240
Soil profiles 344–6
Soil-mineral reactions 344, *345*
SOLGASMIX **268**
Solid solutions 18–21, 278
  ideal solution 18–19
  multicomponent 19–21
Solution models (activity models) 297–8, 322
Solutions, excess behaviour of 20
Space group, meaning of term 94
Space group P6$_2$22, group representation table for **234**
Space group Pmm2

364

INDEX

group representation tables for **223, 225**
ordering patterns for *223, 225*
Space-filling postulate, Laves 4
Special point transformations, Landau
    theory used 219–22
Specific heat
    Debye T^3 law for 142
    low quartz 148
    variations from Debye model 143
Sphalerites, Fe–Zn exchange with
    tetrahedrites 249–50, 254
Sphere packing 4–6
Spinel structures, sorting of *14*, 14–15
Spontaneous strain
    coupling with order parameter 185–8
    definition of 185
$SrTiO_3$, cubic–tetragonal transition in
    175, *176*
Stability
    compositional effects on 17–20
    relative 269, 274
Stacking sequences 18
Standard-state thermodynamic properties
    (SSP) 306
    and solution models 298
Stereoisomerism 53
Stewartite group *56*, 58
stoichiometry, and crystal Structure 17
Strain-coupling terms (in Landau
    expansion) 185, 187
Stretching modes, in Kieffer
    approximation 145
Stringhamite, structure of 64–5
Strontianite 319
Structural building process 53
Structural disorder 278
Structural hierarchies 45–59
    a general hypothesis 46–7
    structural specifics 47–8
Structural mapping 12–15
Structural sorting diagrams 3
Structural stability and bond topology 37
Structural unit fragments, identification
    of 53
Structural units
    change in character of *74*, 74
    as complex oxyanion 67
    condensed 59

depolymerization of 73, 74
dimensionality of 63
increased polymerization 53
Lewis basicity of 67, 69
mode of polymerization 47–59
(OH) and (H_2O) as components of 61–3
Structure, influence on mineral
    paragenesis 45–6, 72–5
Structure diagrams, use of
    pseudopotential radii *14*, 14
Structures
    close-packed 4–5
        cubic close-packed (ccp) 5, 6
        hexagonal close-packed (hcp) 5, 6
    as graphs 26–7
    polymerization of 34
Strunzite group 53, 58
Struvite, hydrogen bonding 49
Sulphosalts 244
    see also Tetrahedrite–tennantites
SUPCRT data base **271**, 304, 305
    cf. EQ3/6 320
    cf. MINTEQ data base 316–17, 320
    cf. UBC data base 319
Superconductors 105, 125
Symmetry point structures
    gradient interactions between pair of 232
    Landau theory for 222–6, 227
    modulation of 229
    symmetry properties of 229–30, *230*
Synthesis data 269

t_{2g} configurations, effect of distortion
    energy on electronic
    configuration **126**, 127
t_{2g}–e_g mixing 128–9
t_{2g} orbitals, change in energies on
    distortion 125–8
Talc 315, 316
Talmessite group *54–5*
Tancoite *54–5*
Taylor expansion, vibrational Gibbs
    energy of tetrahedrite–
    tennantites described using
    247, 248, **249**
Temperature and clay-mineral change 330
Tetrahedral chain, modelling of backbone
    100

INDEX

Tetrahedral complexes 112, *113*
Tetrahedrite–tennantite solid solution 20
Tetrahedrite–tennantites
  argentian
    incompatibility between Ag and As 258
    incompatibility between Ag and Zn 255
    miscibility gaps for *256*, 256–7, 259
    reciprocal ordering reactions of **253**
    thermodynamic properties of 251–9
  characteristics of 244–5
  composition-curve calculations 262, *263*
  deposition of 245–6
  element zoning of 241, 258–63
  equilibration with electrum–chalcopyrite–pyrite–sphalerite 258
  exchange of metals with hydrothermal fluids 260–2
  exchange-potential calculations 260–1, *261*
  Fe–Zn exchange with sphalerites 249–50, 254
  formulae for 244, 246
  occurrence of 245
  reaction between components 246
  silver–copper (Ag–Cu) exchange in 258, 259
  silver-poor (Ag-poor), thermodynamic properties of 246–51
  solid solutions 20
  thermodynamic properties of 246–59
Thenardite 44
THERIAK 268
Thermal energy
  Debye solid 141
  harmonic expression for **135**
Thermal gradient, in sedimentary basins 334, 335
THERMO-CALC 268
Thermobarometry 322
THERMOCALC 268, **271**
Thermochemical data 273–4
Thermochemistry, tetrahedrite–tennantites 246–63
Thermodynamic data
  measurements 274–81
  need for 268–9
  the problems 269–70
  types and sources of 272–4
Thermodynamic data bases 270–2
  comparison of 299–320
  data retrieval and data analysis 281–9
  Type K 270, 299, 304
  Type S 272, 299, 304
Thermodynamic parameters 306
  constraining values of 269
Thermodynamic properties
  determination using atomistic simulations, $MnTiO_3$ high-pressure polymorphs 165–8
  determination using Debye model
    low quartz *152*, 152–3, *153*
    rutile *156*, 157, *157*
  determination using Kieffer model
    low quartz 149–52, *152*, *153*
    rutile *156*, 156, *157*, 157
  determination using lattice dynamics 132, **135**
  determination using VDOS models
    low quartz 149, *152*, *153*
    rutile *156*, 156, *157*, 157
  optimized, quality of 289–91
  tetrahedrite–tennantites 246–59
Thermodynamics
  and crystal stability 15–21
  third law of 178
Tight-binding method 88, 96, 105
Time-temperature space 330–3
Titania
  density-of-states diagram for *104*
  incorporation of lithium into lattice 103–4, 123
  see also Anatase; Rutile
Titanite, structure of 58, *59*
Topology and eutaxy 3–7
Tornebohmite 47, *48*
Transition metals 279
Transition-metal complexes
  comparison between molecular-orbital and crystal-field approaches 112–14
  octahedral complexes 111–12
  CFSE compared with MOSE *114*
  tetrahedral complexes 112, *113*

366

Transition-metal fluorides
  coordination geometry for **124**
  distortion in 128
Transition-metal oxides
  coordination geometry for **124**
  distortion in 128
Translation group 94
  characters of 95–6
  symmetry elements of 94
Translational symmetry 32
Transmission electron microscopy (TEM) 339
  shape of clay particles 338, *340*
Tremolite 315, 317, 319
Tricritical transitions, 180, 197, 204
Tsumcorite 53, 59
Two-centre two-electron bonds 100, 103
Type 1 distortion 123, **124**, *129*
Type 2 distortion 123, **124**, *129*

UBC data base 304, 305
  cf. HP-DS data base 308–12, 319
  cf. MINTEQ data base 317–18
  cf. SUPCRT data base 319
  main goal 321–2
Uklonskovite *54–5*, 58
Uncertainties 289
Uncertainties of identity 277–8
UNIX operating system 303

Valence electrons 11, **12**
  and chemical bonds 45
  excess 11
Valence-matching principle 67
  applications of 44–5, 69
  applied to binary representation of structure 70–1
Valence-shell–electron-pair repulsion (VSEPR) model 121
Valence-sum rule 41, 47
Van der Waals equation 292, 296
Van der Waals interaction 65
Van der Waals solid Xe structure 11
Vanadium carbonyls
  bond lengths in 115
  distortion in 127–8
van't Hoff equation 302, 306
Vauquelinite group *54–5*

VDOS (lattice vibration) model
  rutile *154*, 155
  thermodynamic functions calculated *156*, 156, *157*, 157
VDOS-1 (lattice vibration) model
  low quartz 148, 149, *150*
  thermodynamic functions calculated 149, *152*, *153*
VDOS-2 (lattice vibration) model
  low quartz 148, 149, *150*
  thermodynamic functions calculated 149, *152*, *153*
VERTEX **268**
Vertex-sharing tetrahedra 99–100, *100*
Vertices 27
Vibrational density of states 136–9
  andalusite *159*, 159
  approximations used 139–46
  forsterite 160
  low quartz *147*, 147–52, *150–1*
  $MnTiO_3$ polymorphs 165, *166*
  rutile 155–7
Vibrational Gibbs energy
  Ag-poor tetrahedrite–tennantites 247, 248
  argentian tetrahedrite–tennantites 252
Vibrational models
  application to mineral stability 158–60
  comparision of 146–57
  low quartz 146–53
  rutile 154–7
  Debye model 133, 139–43
  low quartz 149, *151*, *152*, 152–3, *153*
  rutile *154*, 155–6, *156*, *157*, 157
  Kieffer model 133, 143–6
  low quartz 148–9, 149–52, *151*, *152*, *153*
  rutile *154*, 155, *156*, 156, *157*, 157
  Salje–Viswanathan model 158
  andalusite *159*, 159
  VDOS model, rutile *154*, 155, *156*, 156, *157*, 157
  VDOS-1 model, low quartz 148, 149, *150*, *152*
  VDOS-2 model, low quartz 148, 149, *150*, *152*
Virial equations 293
Volume data 279–80

Wairakite 316
Water of hydration 75
    botryogen **70**
Wave equation 89
Weak bonds 61, 62, 67
Weathering profiles, showing effect of chemical activity 343–6
Weighted least-squares regression 284, 287
Weighting, in MAP 287
Weighting factors 291
Whitmoreite group 53
Wurtzite structure 11, 115–16, *116*
    coordination geometry for *118*
    relative energies for 118, *119*

X-ray diffraction (XRD) 2, 333, 339, 348
    showing structure types 338
X-ray fluorescence chemical analysis (XRF), composition of individual grains 338

Yavapaite *58*
Yftisite *54–5*

Zeolite frameworks 4, *5*
Zero-point energy 134